Computer-Based Microscopic Description
of the Structure and Properties of Materials

MATERIALS RESEARCH SOCIETY SYMPOSIA PROCEEDINGS

ISSN 0272 - 9172

Volume 1—Laser and Electron-Beam Solid Interactions and Materials Processing, J. F. Gibbons, L. D. Hess, T. W. Sigmon, 1981

Volume 2—Defects in Semiconductors, J. Narayan, T. Y. Tan, 1981

Volume 3—Nuclear and Electron Resonance Spectroscopies Applied to Materials Science, E. N. Kaufmann, G. K. Shenoy, 1981

Volume 4—Laser and Electron-Beam Interactions with Solids, B. R. Appleton, G. K. Celler, 1982

Volume 5—Grain Boundaries in Semiconductors, H. J. Leamy, G. E. Pike, C. H. Seager, 1982

Volume 6—Scientific Basis for Nuclear Waste Management, S. V. Topp, 1982

Volume 7—Metastable Materials Formation by Ion Implantation, S. T. Picraux, W. J. Choyke, 1982

Volume 8—Rapidly Solidified Amorphous and Crystalline Alloys, B. H. Kear, B. C. Giessen, M. Cohen, 1982

Volume 9—Materials Processing in the Reduced Gravity Environment of Space, G. E. Rindone, 1982

Volume 10—Thin Films and Interfaces, P. S. Ho, K.-N. Tu, 1982

Volume 11—Scientific Basis for Nuclear Waste Management V, W. Lutze, 1982

Volume 12—In Situ Composites IV, F. D. Lemkey, H. E. Cline, M. McLean, 1982

Volume 13—Laser Solid Interactions and Transient Thermal Processing of Materials, J. Narayan, W. L. Brown, R. A. Lemons, 1983

Volume 14—Defects in Semiconductors II, S. Mahajan, J. W. Corbett, 1983

Volume 15—Scientific Basis for Nuclear Waste Management VI, D. G. Brookins, 1983

Volume 16—Nuclear Radiation Detector Materials, E. E. Haller, H. W. Kraner, W. A. Higinbotham, 1983

Volume 17—Laser Diagnostics and Photochemical Processing for Semiconductor Devices, R. M. Osgood, S. R. J. Brueck, H. R. Schlossberg, 1983

Volume 18—Interfaces and Contacts, R. Ludeke, K. Rose, 1983

Volume 19—Alloy Phase Diagrams, L. H. Bennett, T. B. Massalski, B. C. Giessen, 1983

Volume 20—Intercalated Graphite, M. S. Dresselhaus, G. Dresselhaus, J. E. Fischer, M. J. Moran, 1983

Volume 21—Phase Transformations in Solids, T. Tsakalakos, 1984

Volume 22—High Pressure in Science and Technology, C. Homan, R. K. MacCrone, E. Whalley, 1984

Volume 23—Energy Beam-Solid Interactions and Transient Thermal Processing, J. C. C. Fan, N. M. Johnson, 1984

Volume 24—Defect Properties and Processing of High-Technology Nonmetallic Materials, J. H. Crawford, Jr., Y. Chen, W. A. Sibley, 1984

MATERIALS RESEARCH SOCIETY SYMPOSIA PROCEEDINGS

Volume 25—Thin Films and Interfaces II, J. E. E. Baglin, D. R. Campbell, W. K. Chu, 1984

Volume 26—Scientific Basis for Nuclear Waste Management VII, G. L. McVay, 1984

Volume 27—Ion Implantation and Ion Beam Processing of Materials, G. K. Hubler, O. W. Holland, C. R. Clayton, C. W. White, 1984

Volume 28—Rapidly Solidified Metastable Materials, B. H. Kear, B. C. Giessen, 1984

Volume 29—Laser-Controlled Chemical Processing of Surfaces, A. W. Johnson, D. J. Ehrlich, H. R. Schlossberg, 1984.

Volume 30—Plasma Processsing and Synthesis of Materials, J. Szekely, D. Apelian, 1984

Volume 31—Electron Microscopy of Materials, W. Krakow, D. Smith, L. W. Hobbs, 1984

Volume 32—Better Ceramics Through Chemistry, C. J. Brinker, D. E. Clark, D. R. Ulrich, 1984

Volume 33—Comparison of Thin Film Transistor and SOI Technologies, H. W. Lam, M. J. Thompson, 1984

Volume 34—Physical Metallurgy of Cast Iron, H. Fredriksson, M. Hillerts, 1985

Volume 35—Energy Beam-Solid Interactions and Transient Thermal Processing/1984, D. K. Biegelsen, G. Rozgonyi, C. Shank, 1985

Volume 36—Impurity Diffusion and Gettering in Silicon, R. B. Fair, C. W. Pearce, J. Washburn, 1985

Volume 37—Layered Structures, Epitaxy and Interfaces, J. M. Gibson, L. R. Dawson, 1985

Volume 38—Plasma Synthesis and Etching of Electronic Materials, R. P. H. Chang, B. Abeles, 1985

Volume 39—High-Temperature Ordered Intermetallic Alloys, C. C. Koch, C. T. Liu, N. S. Stoloff, 1985

Volume 40—Electronic Packaging Materials Science, E. A. Giess, K.-N. Tu, D. R. Uhlmann, 1985

Volume 41—Advanced Photon and Particle Techniques for the Characterization of Defects in Solids, J. B. Roberto, R. W. Carpenter, M. C. Wittels, 1985

Volume 42—Very High Strength Cement-Based Materials, J. F. Young, 1985.

Volume 43—Coal Combustion and Conversion Wastes: Characterization, Utilization, and Disposal, G. J. McCarthy, R. J. Lauf, 1985

Volume 44—Scientific Basis for Nuclear Waste Management VIII, C. M. Jantzen, J. A. Stone, R. C. Ewing, 1985

Volume 45—Ion Beam Processes in Advanced Electronic Materials and Device Technology, F. H. Eisen, T. W. Sigmon, B. R. Appleton, 1985

Volume 46—Microscopic Identification of Electronic Defects in Semiconductors, N. M. Johnson, S. G. Bishop, G. D. Watkins, 1985

MATERIALS RESEARCH SOCIETY SYMPOSIA PROCEEDINGS

Volume 47—Thin Films: The Relationship of Structure to Properties, C. R. Aita, K. S. SreeHarsha, 1985

Volume 48—Applied Material Characterization, W. Katz, P. Williams, 1985

Volume 49—Materials Issues in Applications of Amorphous Silicon Technology, D. Adler, A. Madan, M. J. Thompson, 1985

Volume 50—Scientific Basis for Nuclear Waste Management IX, L. O. Werme, 1985

Volume 51—Beam-Solid Interactions and Phase Transformations, H. Kurz, G. L. Olson, J. M. Poate, 1986

Volume 52—Rapid Thermal Processing, T. O. Sedgwick, T. E. Siedel, B. Y. Tsaur, 1986

Volume 53—Semiconductor-on-Insulator and Thin Film Transistor Technology, A. Chiang. M. W. Geis, L. Pfeiffer, 1986

Volume 54—Interfaces and Phenomena, R. H. Nemanich, P. S. Ho, S. S. Lau, 1986

Volume 55—Biomedical Materials, M. F. Nichols, J. M. Williams, W. Zingg, 1986

Volume 56—Layered Structures and Epitaxy, M. Gibson, G. C. Osbourn, R. M. Tromp, 1986

Volume 57—Phase Transitions in Condensed Systems—Experiments and Theory, G. S. Cargill III, F. Spaepen, K. N. Tu, 1986

Volume 58—Rapidly Solidified Alloys and Their Mechanical and Magnetic Properties, B. C. Giessen, D. E. Polk, A. I. Taub, 1986

Volume 59—Oxygen, Carbon, Hydrogen, and Nitrogen in Crystalline Silicon, J. W. Corbett, J. C. Mikkelsen, Jr., S. J. Pearton, S. J. Pennycook, 1986

Volume 60—Defect Properties and Processing of High-Technology Nonmetallic Materials, Y. Chen, W. D. Kingery, R. J. Stokes, 1986

Volume 61—Defects in Glasses, Frank L. Galeener, David L. Griscom, Marvin J. Weber, 1986

Volume 62—Materials Problem Solving with the Transmission Electron Microscope, L. W. Hobbs, K. H. Westmacott, D. B. Williams, 1986

Volume 63—Computer-Based Microscopic Description of the Structure and Properties of Materials, J. Broughton, W. Krakow, S. T. Pantelides, 1986

Volume 64—Cement-Based Composites: Strain Rate Effects on Fracture, S. Mindess, S. P. Shah, 1986

Volume 65—Fly Ash and Coal Conversion By-Products: Characterization, Utilization and Disposal III, G. J. McCarthy , D. M. Roy, 1986

Volume 66—Frontiers in Materials Education, G. L. Liedl, L. W. Hobbs, 1986

MATERIALS RESEARCH SOCIETY SYMPOSIA PROCEEDINGS   VOLUME 63

Computer-Based Microscopic Description of the Structure and Properties of Materials

Symposium held December 2-4, 1985, Boston, Massachusetts, U.S.A.

EDITORS:

J. Broughton
Department of Materials Science, State University of New York at Stony Brook, Stony Brook, New York, U.S.A.

W. Krakow
IBM Thomas J. Watson Research Center, Yorktown Heights, New York, U.S.A.

S. T. Pantelides
IBM Thomas J. Watson Research Center, Yorktown Heights, New York, U.S.A.

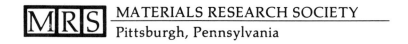

MATERIALS RESEARCH SOCIETY
Pittsburgh, Pennsylvania

This work was also supported in part by the Air Force Office of Scientific Research, Air Force Systems Command, USAF, under Grant Number AFOSR-85-0355.

Published by:

Materials Research Society
9800 McKnight Road, Suite 327
Pittsburgh, Pennsylvania 15237
telephone (412) 367-3003

Library of Congress Cataloging in Publication Data

Computer-based microscopic description of the structure and properties of materials.

(Materials Research Society symposia proceedings, ISSN 0272-9172 ; v. 63)
Bibliography: p.
Includes index.
1. Materials—Mathematical models—Congresses. 2. Microstructure—
Mathematical models—Congresses. 3. Materials—Data Processing—
Congresses. 4. Microstructure—Data Processing—Congresses.
I. Broughton, J. II. Krakow, William. III. Pantelides, Sokrates T. IV. Materials
Research Society
TA418.5.C66 1986   620.1'1'0285   86-12446
ISBN 0-931837-28-6

Manufactured in the United States of America

Manufactured by Publishers Choice Book Mfg. Co.
Mars, Pennsylvania 16046

Contents

PREFACE   xi

*STRUCTURAL, ELECTRONIC, AND MAGNETIC PROPERTIES OF
SURFACES, INTERFACES, AND SUPERLATTICES
    J. Freeman, C.L. Fu and T. Oguchi   1

DEFECT STRUCTURE AND DYNAMICS IN SILICON
    S.T. Pantelides, R. Car, P.J. Kelly and
    A. Oshiyama   7

ELECTRONIC PROPERTIES OF LIQUID SILICON
    J.Q. Broughton and P.B. Allen   13

THEORETICAL STUDY OF SEMICONDUCTOR INTERFACES
    G. Van De Walle and M. Martin   21

*THE STUDY OF SURFACE PHONONS BY ELECTRON ENERGY LOSS
SPECTROSCOPY: THEORY OF THE EXCITATION CROSS SECTION
    D.L. Mills   27

COMPUTER SIMULATION OF SPUTTERING FROM TWO COMPONENT
LIQUID METAL TARGETS
    M.H. Shapiro, D.Y. Lo and T.A. Tombrello   31

VIBRATIONAL PROPERTIES OF THE π-BONDED CHAIN MODEL OF
THE Si(111)2x1 SURFACE
    O.L. Alerhand, D.C. Allan and E.J. Mele   37

*COMPUTER SIMULATION OF ELECTRON MICROSCOPE IMAGES
FROM ATOMIC STRUCTURE MODELS
    W. Krakow   43

DEFECT BEHAVIOUR IN OXIDES
    A.N. Cormack, C.R.A. Catlow, G.V. Lewis
    and C.M. Freeman   55

*SIMULATION OF EQUILIBRIUM IN ALLOYS USING THE EMBEDDED
ATOM METHOD
    M. Foiles   61

THE SOLID-LIQUID INTERFACE: THEORY AND COMPUTER
SIMULATIONS
    B.B. Laird and A.D.J. Haymet   67

*FROM HAMILTONIANS TO PHASE DIAGRAMS
    J. Hafner   73

*DYNAMICS OF COMPRESSED AND STRETCHED LIQUID SiO_2, AND
THE GLASS TRANSITION
    C.A. Angell, P.A. Cheeseman and C.C. Phifer   85

*Invited Talk

CRYSTAL, LIQUID AND GLASS IN 2 DIMENSIONS.
ANALYSIS OF THE GLASS TRANSITION
    F. Lancon and P. Chaudhari   95

DYNAMIC PROPERTIES OF COMPOSITE MATERIALS USING A
T-MATRIX TO DESCRIBE MICROSTRUCTURE
    V.V. Varadan and V.K. Varadan   101

*PSEUDOPOTENTIAL CALCULATIONS OF STRUCTURAL PROPERTIES
OF SOLIDS
    M.L. Cohen   107

HUNTING MAGNETIC PHASES WITH TOTAL-ENERGY SPIN-
POLARIZED BAND CALCULATIONS
    P.M. Marcus, V.L. Moruzzi and K. Schwarz   117

*FRACTURE AND FLOW VIA NONEQUILIBRIUM MOLECULAR
DYNAMICS
    W.G. Hoover, G. De Lorenzi, B. Moran, J.A.
    Moriarty & A.J.C. Ladd   125

*INTERATOMIC FORCES AND STRUCTURE OF GRAIN
BOUNDARIES
    V. Vitek and J.Th.M. De Hosson   137

MONTE CARLO SIMULATION OF GROWTH OF CRYSTALLINE AND
AMORPHOUS SILICON
    W. Dodson and P.A. Taylor   151

COMPUTED STRUCTURES OF [001] SYMMETRICAL TILT
BOUNDARIES
    J.T. Wetzel, A.A. Levi and D.A. Smith   157

SURFACE RECONSTRUCTION OF Si(100)
    T.A. Weber   163

NEW SILICON (111) SURFACE (7x7) RECONSTRUCTION BENZENE-
LIKE RING MODEL
    Y.G. Hao and Laura M. Roth   167

CLASSICAL TWO AND THREE-BODY INTERATOMIC POTENTIALS
FOR SILICON SIMULATIONS
    R. Biswas and D.R. Hamann   173

*SPECIAL PURPOSE PROCESSORS FOR COMPUTING MATERIALS
PROPERTIES
    A.F. Bakker   181

*MATERIALS BY DESIGN--A HIERARCHICAL APPROACH TO THE
DESIGN OF NEW MATERIALS
    J. Eberhardt, P.J. Hay, and J.A.
    Carpenter, Jr.   191

*Invited Talk

PROGRESS IN SPIN GLASSES AND RANDOM FIELDS--RESEARCH
WITH A SPECIAL PURPOSE COMPUTER
    A.T. Ogielski   207

THE STONY BROOK MULTIPROCESSING MATERIALS SIMULATOR
    H.R. Carleton   213

A HIGHLY PARALLEL COMPUTER FOR MOLECULAR DYNAMICS
SIMULATIONS
    D.J. Auerbach, A.F. Bakker, T.C. Chen,
    A.A. Munshi and W.J. Paul   219

MICROSTRUCTURAL DYNAMIC STUDY OF GRAIN GROWTH
    M.P. Anderson, G.S. Grest and D.J. Srolovitz   225

MODELLING OF THIN FILM GRAIN STRUCTURES AND GRAIN
GROWTH
    H.J. Frost and C.V. Thompson   233

COOPERATIVE PREMELTING EFFECTS ON A (110) FCC SURFACE:
A MOLECULAR DYNAMICS STUDY
    V. Rosato, V. Pontikis and G. Ciccotti   241

HARTREE-FOCK CLUSTER COMPUTATIONS FOR IONIC CRYSTALS
    J.M. Vail and R. Pandey   247

THEORETICAL INVESTIGATION OF THE JAHN-TELLER VIBRONIC
$^1E \otimes \epsilon$ STATE IN THE NEUTRAL VACANCY OF SILICON
    J.C. Malvido, P.V. Madhavan and J.L. Whitten   253

*CALCULATIONS OF RESISTIVITY AND SUPERCONDUCTING T_C
IN TRANSITION METALS
    P.B. Allen, T.P. Beaulac, F.S. Khan, W.H.
    Butler, F.J. Pinski and J.C. Swihart   259

*MICROSCOPIC PHENOMENA OF MACROSCOPIC CONSEQUENCES:
INTERFACES, GLASSES, AND SMALL AGGREGATES
    U. Landman, R.N. Barnett, C.L. Cleveland,
    W.D. Luedtke, M.W. Ribarsky, D. Scharf and
    J. Jortner   273

AUTHOR INDEX   285

SUBJECT INDEX   287

*Invited Talk

Preface

This volume contains the papers that were presented at the 1985 MRS Fall Meeting in Symposium R, entitled "Computer-based Microscopic Description of the Structure and Properties of Materials." The purpose of the symposium was to bring together the various communities that use computers to predict, describe, or simulate the atomic and electronic structure and properties of materials. It was the first such symposium at an MRS meeting. It brought together those whose main focus is the electronic structure and those whose main focus is the dynamics of atomic motion.

Electronic structure approaches, based on the quantum mechanical wave equation (Hartree-Fock, density functional theory, pseudopotentials, band-structure theory, molecular-orbital theory, etc.) have been quite successful in obtaining very accurate descriptions of the electronic properties of given atomic arrangements, but have been limited in searching for stable atomic configurations or doing dynamics at finite temperatures. On the other hand, "molecular dynamics" approaches have traditionally absorbed all electronic structure information into phenomenological interatomic potentials, which makes dynamical calculations at finite temperatures feasible and efficient. These two communities have been evolving separately, but, in the last few years, there have been tendencies to borrow from each other or merge the two approaches. The symposium was meant to foster this interchange of ideas, and to bring these two main communities together with others, including those who design and use special-purpose computers, those who simulate experimental data, and other related fields.

The symposium was sponsored by the NSF (Steven R. Williams), the ARO (John C. Hurt), NASA (Joseph R. Stephens), Oak Ridge National Laboratory (through Energy Conversion and Utilization Technologies; Joseph A. Carpenter, Jr.) and AFOSR (John E. Lintner). We are thankful for their generous support.

Last of all, we would like to thank Joan Pidot, Kitty Boccio, Barbara Cardella, Lorraine Miro and Janet Hutsko for their professional and supportive secretarial services.

Jeremy Broughton
William Krakow
Sokrates Pantelides

1

STRUCTURAL, ELECTRONIC, AND MAGNETIC PROPERTIES OF
SURFACES, INTERFACES, AND SUPERLATTICES*

ARTHUR J. FREEMAN, C.L. FU AND T. OGUCHI
Department of Physics and Astronomy, Northwestern University, Evanston,
IL 60201

ABSTRACT

Advances in all-electron local density functional theory approaches
to complex materials structure and properties made possible by the
implementation of new computational/theoretical algorithms on
supercomputers are exemplified in our full potential linearized augmented
plane wave (FLAPW) method. In this total energy self-consistent approach,
high numerical stability and precision (to 10   in the total
energy) have been demonstrated in a study of the relaxation and
reconstruction of transition metal surfaces. Here we demonstrate the
predictive power of this method for describing the structural, magnetic
and electronic properties of several systems (surfaces, overlayers,
sandwiches, and superlattices).

I. INTRODUCTION

Recent progress in the fabrication and property modifications of
artificial materials, such as thin films, sandwiches, and modulated
structures, have generated a great deal of interest. The excitement
surrounding this development lies in the possible discovery and synthesis
of new materials with desired properties to specification, and permitting
new phenomena to be investigated and novel devices to be made.
Concurrently, advances in computational/theoretical approaches have
developed capabilities of yielding precise results and making successful
predictions possible. In coordination with experimental studies, these
theoretical efforts serve to advance the appearance of new and quite
unexpected phenomena and to promote their understanding.
In this paper, we study theoretically the structural and magnetic
properties of (1) 3d transition metal surfaces and (2) 3d transition
metals as overlayers or sandwiches and superlattices with noble metals.
For the surfaces and sandwiches, we solve the local spin density
functional equations[1] self-consistently using the full-potential
linearized augmented plane wave[2] (FLAPW) method; and the linear
muffin-tin orbital[3] (LMTO) method is employed for CrAu(001)
superlattice with various thicknesses.

II. TRANSITION METAL SURFACES

We start with the discussion of surface magnetism of 3d transition
metals. Since the surface atoms have lower coordination and reduced
symmetry, it is now widely recognized that there is an enhancement of the
magnetism near the surface layer as a result of localized surface
states[4]. In Table I we summarize the calculated results of the magnetic
moment for surface atoms and, for the sake of comparison, their bulk
counterparts.
For ferromagnetic Fe and Ni (100) and (110) surfaces, the magnetic
moments are enhanced by about 10 ~ 30% from their bulk values. It is
not surprising that the (100) surface has the largest enhancement simply

because the surface atom has the lowest coordination numbers.

A more dramatic case is Cr(100), which undergoes a surface ferromagnetic phase transition with a largely enhanced moment of $2.49\mu_B$. Recently, this has been confirmed by angle-resolved photoemission experiments[5]. By contrast to the results obtained for Cr(001), using a spin-polarized total energy approach and a study of the multilayer relaxation of V(100), we predict a stable paramagnetic state for this surface. Further, for the paramagnetic V(001), we obtain a 9% contraction of the topmost interlayer spacing accompanied by an expansion of the second interlayer spacing by 1%, which is in excellent agreement with LEED[6]. Analysis of the electronic structure of V(100) shows that the surface states which are characteristic of the occurrence of surface magnetism in bcc Cr and Fe is located about 0.3 eV above E_F. A comparison between Cr and V also serves to illustrate the role played by these surface states in the enhancement of the surface magnetism in the case of Cr(001).

Table I. Calculated magnetic moments (in μ_B) for 3d transition metals

		V	Cr	Fe	Ni
	(100)	0	2.49	2.98	0.68
surface					
	(110)			2.63	0.63
bulk		0	0.59	2.15	0.56

III. METALLIC OVERLAYERS AND BIMETALLIC SANDWICHES

Having illustrated the effect of surface states on the magnetic and electronic properties of transition metal surfaces, our next goal was to explore the possibility of inducing 2D magnetism in a controlled way. It turns out that the transition metal and noble metal interfaces are of special interest[7,8] because: (1) a close to perfect epitaxial growth due to a nearly perfect match of the 2D lattice translational vectors for (001) Cr (or Fe) and Au (or Ag) (the (100) plane of the transition metal layers are rotated by 45 with respect to that of the noble metal, and the stacking has atoms in the four-fold hollow site of adjacent atomic planes); and (2) the dissimilarity of the electronic structures between transition metals and noble metals which ensures minimum hybridization between localized surface (interface) states near E_F and the underlying non-magnetic noble metal d-band.

In order to understand the electronic and magnetic structures at the interface, we first examined the crystal structure. This was carried out using a total energy approach for both Au/Cr/Au(001) sandwiches and AuCr coherent modulated structures. We find that the interaction between noble and transition metals is local in nature; this manifests itself in an

Au-Cr interlayer spacing which is an average of the bulk fcc Au and bcc Cr spacings; the bulk Cr and Au structures are essentially unperturbed away from the interface.

Selected results for the magnetic moments calculated within each atomic sphere at the Au/Cr interface are presented in Table II. As an aid in understanding the underlying physics and to emphasize its 2D nature, we focus first on results for the experimentally unattainable free monolayer Cr(001) film. A large magnetic moment of $4.12\mu_B$ is obtained which is close to the atomic limit and substantially larger than that of bulk antiferromagnetic Cr metal ($0.59\mu_b$). Surprisingly, when a monolayer of Cr(001) is deposited onto Au(001), the magnetic moment of the Cr overlayer decreases by only a small amount from the free monolayer value to $3.70\mu_B$. This extremely large moment – the largest value reported for a transition metal other than for Mn – is 50% greater than that of the surface layer in Cr(001) predicted theoretical ($2.49\mu_B$) or derived experimentally ($2.4\mu_B$). Also surprising is the finding that this substantially enhanced moment is only moderately reduced (to $3.10\mu_B$) when the Cr overlayer is itself covered by a Au layer. Apparently, the hybridization between Au and Cr is far less than expected. A very similar result was obtained, $2.95\mu_B$ per Cr atom, when the (1x1) Cr/Au coherent modulated structure was studied with the linear muffin-tin orbital approach (Table II) to be discussed later. Since Cr is a notorious getter, the retention of this enhanced 2D magnetization in either the single sandwich or superlattice structures might make its observation much easier.

Similar investigations have also been performed for monolayers of Fe and V adsorbed on Ag(001). For Fe-Ag there is a close matching of the lattice constants but a 5% mismatch for V-Ag or -Au. As in the case of Cr/Au(001), large magnetic moments, $2.96\mu_B$ (Fe) and $2.0\mu_B$ (V), are found for the adsorbate monolayers. The magnetic moment of the Fe overlayer on Ag(001) is remarkably close to the theoretical magnetic moment of the surface layer of an Fe(001) film ($2.98\mu_B$) – again indicating a lack of interaction with the substrate. The result for V/Ag(001) is much more surprising since, like the bulk, the surface layer of V(001) is not magnetic. We thus have the remarkable prediction that an overlayer of V on Ag(001) is magnetically ordered with a sizable magnetic moment ($1.98\mu_B$) which is almost as large as the moment of Fe in bulk Fe. (If confirmed, this will be the first solid material for which elemental vanadium demonstrates magnetic ordering.) The origin of magnetic ordering is not negative pressure since, in fact, the matching to the Ag(001) substrate results from a reduced lattice constant for V by 5%. For both the Fe and V overlayers we find – despite some hybridization between the d bands of the adsorbate and substrate – that the adsorbate-localized surface state bands retain their quasi-2D behavior for these systems. This behavior is further demonstrated for a monolayer of Fe sandwiched by Ag(001). The magnetic moment of Fe is only slightly decreased from that of the overlayer of Fe/Ag(001) to $2.80\mu_B$. [Similarly, a moment of $2.92\mu_B$ on the Fe site is found for an Au/Fe/Au(001) sandwich.] However, when a monolayer of V(001) is sandwiched by Ag(001), the vanadium layer becomes paramagnetic.

The sensitivity of the magnetic ordering of V to its metallic environment is further illustrated by calculations with two layers of V on Ag(001): In this case the interface layer atoms remain essentially in the paramagnetic state and the surface layer has a moment of $1.15\mu_B$ which is substantially reduced from that of the single-overlayer value ($1.98\mu_B$). We thus have the result that a Ag substrate is more amenable for the magnetism of a V monolayer than is another V layer.

IV. COHERENT MODULATED STRUCTURES (SUPERLATTICES)

In order to examine the magnetic coupling between Cr layers in the multilayer structure, thickness effects have been examined by varying the Cr and Au layer numbers in the CrAu(001) coherent modulated structure.

We start with the simplest case – a CrAu [1*1] superlattice (hereafter we specify a system by a notation [n*m], where n and m denote the number of Cr and Au layers). A non-spin polarized calculation shows a very high density of states at E_F associated with Cr d states, and a Stoner instability to a ferromagnetically ordered state since the calculated Stoner factor $N(E_F)I_{xc} = 1.34$. In a spin polarized calculation, a sizable magnetic moment ($2.94\mu_B$) is actually found on the Cr atoms while a small moment ($0.10\mu_B$) is ferromagnetically induced on the Au atoms. This value of the magnetic moment is very close to that obtained ($3.1\mu_B$) for a monolayer of Cr sandwiched by Au (cf. Table II).

Table II. Theoretical layer by layer magnetic moments (in μ_B) for specified cases (with estimated uncertainties of $\pm 0.03\mu_B$). S and (S-n) indicate surface and subsurface layers. The last column shows the spin-polarized energy (in eV). CMS, coherent modulated structure.

	Cr	Nearest Au	E(para.) $- E$(spin-pol.)
Cr monolayer	4.12	. . .	1.69
1 Cr/Au(001) overlayer	3.70	0.14	0.78
2 Cr/Au(001)	2.90 (S),	-0.08	0.60
	-2.30 (S-1)		
Au/Cr/Au(001) sandwich	3.10	0.14 (S),	0.38
		0.13 (S-2)	
(1×1) Au/Cr CMS	2.95	0.10	0.25
	Fe	Nearest Cu, Ag, Au	
Fe monolayer	3.20	. . .	1.34
1 Fe/Cu(001)	2.85	0.04	0.70
1 Fe/Ag(001)	2.96	0	1.14
2 Fe/Ag(001)	2.94 (S),	0.05	1.15
	2.63 (S-1)		
Ag/Fe/Ag	2.80	0	0.88
Au/Fe/Au	2.92	0.08	0.97
	V	Nearest Ag, Au	
1 V/Au(001)	1.75	0.04	0.10
1 V/Ag(001)	1.98	0.06	0.14
2 V/Ag(001)	1.15 (S),	0	0.08
	< 0.05 (S-1)		
Ag/V/Ag	0	0	. . .
Cr/Fe/Au(001)	3.10 (Cr),	-0.04 (Au)	0.68
	-1.96 (Fe)		
Fe/Cr/Ag(001)	2.30 (Fe),	-0.09 (Ag)	0.52
	-2.40 (Cr)		

For superlattices with thicker constituent layers, a reduction of the magnetic moment at the interface may be expected due to greater hybridization. The calculated magnetic moments for some CrAu(001)

Table III. Magnetic moments (in μ_B) within atomic spheres for CrAu (001) superlattices with various thicknesses. Each layer is specified by atomic type, Cr or Au, ordering assumed (F=ferro and AF=antiferromagnetic) and interplanar distance from the interface layer (i).

superlattice type	layer	magnetic moments
[1*1]F	Cr_i	2.94
	Au_i	0.10
[1*1]AF	Cr_i	3.00
	Au_i	0.00
	Cr_i	-3.00
	Au_i	0.00
[1*3]	Cr_i	3.01
	Au_i	0.07
	Au_{i-1}	0.04
[3*3]	Cr_{i-1}	-1.07
	Cr_i	1.89
	Au_i	0.09
	Au_{i-1}	0.01
[5*5]	Cr_{i-2}	0.68
	Cr_{i-1}	-0.79
	Cr_i	1.65
	Au_i	0.08
	Au_{i-1}	0.00
	Au_{i-2}	0.01

superlattices with various thicknesses is listed in Table III. Interestingly, in the case of CrAu [1*3], the magnetic moment on Cr is not reduced but even slightly enhanced by an additional Au layer. The interface Au d band becomes wider than that of the Au monolayer but not enough to overlap with the Cr d band region; the Cr d band is essentially

6

the same as that of CrAu [1*1]. We may conclude that the Cr monolayer
sandwiched by Au has a sizable magnetic moment $(3\mu_B)$ independent of
Au thickness.

For sandwiches with thicker Cr, the magnetic moment of the interface
Cr is substantially reduced from that of the monolayer Cr with Au but
still much larger $(1.89\mu_B$ and $1.65\mu_B$ in CrAu [3*3] and [5*5],
respectively) than that of bulk Cr $(0.59\mu_B)$. The large reduction of
the interface moment is due to broadening of the interface Cr d band
caused by hybridization with inner Cr d states. We can expect such a
large moment on the interface Cr atoms $(1.5\mu_B)$ even for thicker
sandwiches than five layers of Cr, because the innermost Cr atoms in CrAu
[5*5] already have a magnetic moment very close to the bulk value.

Finally, we discuss the effect of magnetic interactions between the
layers. As shown in Table III, interplanar magnetic interactions between
the Cr layers always result in antiferromagnetic couplings, while those
between the interface Cr and Au layers result in ferromagnetic ones. In
order to study the interplanar magnetic interaction in more detail, we
compare the total energies for CrAu [1*1] superlattice where the
ferromagnetic Cr(001) monolayers are coupled ferromagnetically ([1*1]F)
or antiferromagnetically (in which case the unit cell is doubled,
[1*1]AF) along the [001] direction. We found that the antiferromagnetic
coupling is preferred by 11 mRy/formula unit. This difference will, of
course, vary with Au thickness and reach a point - with perhaps 3-5 Au
layers separating the Cr monolayers - where the interaction becomes so
weak that a reasonable external magnetic field would align the monolayers
ferromagnetically. The magnetic moment of the interface Cr in [1*1]AF is
slightly larger than that in [1*1]F (see Table III). This enhancement of
the moment and also the similar enhancement in [1*3] mentioned above can
be understood by the fact that the antiferromagnetic coupling between the
ferromagnetic Cr(001) monolayers works destructively on the magnitude of
the moments under the ferromagnetic constraint along the [001] direction.

References

1. P. Hohenberg and W. Kohn, Phys. Rev. 136 , B864 (1964); W. Kohn and
   L.J. Sham, Phys. Rev. 140 , A1133 (1965).
2. E.Wimmer, H. Krakauer, M. Weinert, and A.J. Freeman, Phys. Rev. B
   24 , 864 (1981).
3. O.K. Andersen, Phys. Rev. B 12 , 3060 (1975).
4. A.J. Freeman, C.L. fu, S. Ohnishi, and M. Weinert, in Polarized
   Electrons in Surface Physics , edited by R. Feder (to be published
   by World Scientific Publishing Co.) and the references therein.
5. L.E. Klebanoff, S.W. Robey, G. Liu, and D.A. Shirley, Phys. Rev.
   B 30 , 1048 (1984).
6. V. Jensen, J.N. Andersen, H.B. Nielsen, and D.L. Adams, Surf. Sci.
   116 , 661 (1984).
7. M.B. Brodsky et al., Solid State Commun 43 , 675 (1982).
8. C.L. Fu, T. Oguchi, and A.J. Freeman, Phys. Rev. Lett. 54 , 2700
   (1985).

*Supported by the Office of Naval Research and the National Science
Foundation.

Defect Structure and Dynamics in Silicon

S. T. PANTELIDES, R. CAR,* P.J. KELLY** and A. OSHIYAMA***
IBM T. J. Watson Research Center, Yorktown Heights, NY 10598

ABSTRACT

This paper gives a brief account of recent calculations of equilibrium configurations, formation energies, and migration energies of intrinsic lattice defects (vacancies, self-interstitials) and complexes of dopant impurities (phosphorus, aluminum) with these defects. The results have been used to provide a comprehensive interpretation of low- and high-temperature diffusion data.

The electronic properties of solids have been successfully studied for years by using an effective single-particle potential to describe the many-electron system. Norm-conserving pseudopotentials,[1] density functional theory and the local density approximation for exchange and correlation[2] have been especially successful in providing parameter-free descriptions of metals and semiconductors.

For perfect crystals, the Bloch theorem provides a convenient framework and makes the "size" of the calculation be determined by the size of the unit cell. Over the years, electronic structure calculations provided details of the energy bands, charge density, and, more recently of quantities that depend on the total energy as a function of lattice spacing (cohesive energy, bulk modulus, etc.).[3]

In terms of difficulty, the next frontier was surfaces, where periodicity is broken in only one out of three dimensions. Most calculations are done by imposing an artificial periodicity in this third dimension by repeating a sequence of crystalline slabs and vacuum. One usually has to deal with fairly large unit cells, but the approach has been quite practical and successful. As in the case of bulk materials, it has been possible to get energy-level dispersions (surface bands) and also properties that depend on the total energy, e.g., surface reconstruction.[4]

Point defects are an even harder problem since periodicity is now broken in all three dimensions. One approach is to impose artificial periodic boundary conditions. Thus one ends up studying an infinite crystal which contains a periodic array of point defects. By choosing a fairly large unit cell (supercell) one might hope to isolate these point defects from one another. The method proved inadequate for energy levels since practical unit cells yield large dispersions for bound states.[5] More recently, however, the method was shown to be more successful for total energies.[6]

Since 1978, Green's function techniques were shown to be quite powerful in solving the single-particle problem for a point defect in an otherwise perfect infinite crystal.[7] These techniques exploit the fact that the underlying crystal was initially infinite and periodic and the fact that the point defect introduces a disturbance that is naturally localized. Thus, one first solves the perfect crystal problem using standard methods for bulk crystals, and then determines the changes induced by the point defect. The "size" of the problem is then naturally determined by the extent of the disturbance. Calculations can be done self-consistently at the same level of sophistication as bulk and surface calculations.

For several years, most Green's function calculations for point defects were limited to energy levels, charge densities, and wavefunctions. Recently, it has been possible to carry out total-energy calculations, which allowed the determination of lattice relaxation, ground-state atomic configurations, energies of formation, migration barriers, etc. We have carried out such calculations for intrinsic lattice defects (vacancies, self-interstitials) and dopant impurities, including complexes of impurities and intrinsic lattice defects. These results have been published in two papers.[8] Here, we give a brief summary of our key findings.

Self-interstitials in Si: It was found that the self-interstitial has roughly the same total energy at several sites (tetrahedral, hexagonal, bond center, split) even though lattice relaxation is substatntially different at each of these sites (virtually no relaxation for the tetrahedral site, very moderate for the hexagonal site, quite large for the bond-centered and the split configurations). Small differences in these total energies, however, are quite significant and give rise to fascinating properties. For example, the ground state configuration of the self-interstitial was found to depend strongly on the charge state. (Fig. 1) In fact we found that the self-interstitial is a negative-U center: When the Fermi level is in the lower part of the energy gap, the self-interstitial is stable as doubly positively charged at the tetrahedral interstitial site. When the Fermi level is in the upper part of the gap, the self-interstitial is stable as neutral at either the hexagonal or the bond-center site (within the uncertainty of the calculation the two sites have the same energy). Positively charged self-interstitials are not the ground state for any Fermi-level position.

Self-interstitials were found to have another remarkable property. In p-type material, where the equilibrium state is doubly positively charged, capture of one or two electrons makes the center unstable (the electron or electrons are captured in a state of p-like symmetry, whereby the partial filling leads to a Jahn-Teller instability), leading to motion toward another site, such as the hexagonal, bond-centered site, etc. Subsequent loss of the extra electron(s) at that site makes the interstitial move to one of the original stable sites, namely to a tetrahedral site. Thus, in the presence of excess electrons and holes, a self-interstitial can migrate athermally by successive capture of electrons and holes (Bourgoin-Corbett mechanism[9]). Such an athermal migration is possible along several paths.[8 &s,.10] Some of these paths are simple (in the sense that one atom follows a given path), others involve exchanges with lattice atoms. This results provides detailed theoretical understanding of the experimental observation that Si self-interstitials are capable of migrating quite efficiently at very low temperatures ($\approx$4 K) in the presence of ionizing radiation.[11]

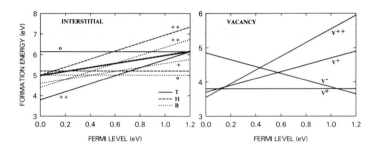

Fig. 1 Formation energies for various charge states of the self-interstitial (at sites T=tetrahedral, H=hexagonal, B=bond center) and the vacancy in Si as functions of the Fermi level position in the energy gap. The zero of energy is at the top of the valence bands. From Ref. 8.

At high temeperatures, thermally created interstitials can mediate self-diffusion, i.e., the long-range migration of host atoms occupying normal atomic sites. The self-diffusion activation energy has been measured to be about 5 eV,[1&2]. but differing analyses of experimental data have led to conflicting conclusions regarding the role of self-interstitials. Theoretically, the activation energy is simply the sum of formation and migration energies of the defect that mediates the process. In order to reconcile low- and high-temperature data, it has been argued that the structure if self-interstitials changes rather drastically as a function of temperature, giving rise to changes in formation and migration energies. Early theoretical estimates either found activation energies that are too small (of order 2 eV and thus in conflict with experiment) or too large (of order 8 eV), ruling out any role for the interstitial.[12] Our results (Fig. 1), however, show activation energies between 5 and 6 eV, leading us to conclude that self-interstitials indeed participate in self-diffusion. More significantly, our results provide a consistent interpretation of both low- and high-temperature data: At low temperatures, in experiments where self-interstitials are introduces by irradiation, the formation energy is irrelevant and migration energies are small so that charge-state alternation allows athermal migration as described above. At high temperatures, where self-interstitials must be created thermally, the formation energy is large, about 5 eV, leading to the observed large self-diffusion activation energy.

Vacancies: The vacancy in Si has been known to be a negative-U center.[13] Our calculations confirmed this result. Furthermore, we obtained the first parameter-free result for the formation energy of the vacancy (Fig. 1). The neutral vacancy has a formation energy of 3.8 eV, whereas the formation energies of the vacancies depend on the Fermi level. Combining our formation energies with experimental migration energies measured at low temperatures (typically 0.3-0.4 eV; see Ref. 11) we we find activation energies for vacancy-mediated self-diffusion to be between 4 and 5 eV. Thus, vacancies also contribute to self-diffusion. As in the case of the self-interstitials, our results established for the first time that low- and high-temperature data can be accounted for in a simple way, without having to invoke strong temperature dependence for formation and migration energies (earlier estimates of the vacancy formation energy gave about 2.5 eV, requiring a high-temperature migration energy of 1.5 to 2 eV in order to agree with self-diffusion data).

Dopant Impurities: We found that substitutional phosphorus and aluminum form stable pairs with vacancies with typical binding energies of order 2 eV. Combining this with the experimental result that vacancy-impurity pairs migrate with a migration energy of roughly 1 eV, we predicted activation energies for impurity diffusion approximately 1 eV smaller than the self-diffusion activation energy, in agreement with experiment.[12] We further predicted that these impurities can bind a self-interstitial and migrate as a pair. However, these pairs are unstable against ejection of the impurity in the interstitial channels by the self-interstitial. We found that this mode of impurity diffusion (known as kick-out, but believed to be active only for heavy impurities such as gold) also contributes with an activation energy in the range of observed values.[12]

This work was supported in part by the Office of Naval Research under Contract No. N00014-84-C-0396.

REFERENCES

* Present address: International School for Advanced Studies, Trieste, Italy.
** Present address: Philips Research Laboratory, Eindhoven, Netherlands.
*** Present address: NEC Coproration, Tokyo, Japan.
[1] D. R. Hamann, M. Schluter and C. Chiang, Phys. Rev. Lett. 43, 1494 (1979).
[2] P. Hohenberg and W. Kohn, Phys. Rev. 136, B864 (1964); L. Sham and W. Kohn, Phys. Rev. 140, A1133 (1965); for recent reviews, see The Inhomogeneous Electron Gas, edited by N. H. March and S. Lundqvist (Plenum, New York, 1984).
[3] See article by M. L. Cohen, this volume.
[4] E.g., see article by K. C. Pandey, this volume.
[5] S. G. Louie, M. Schluter, J. R. Chelikowsky, and M. L. Cohen, Phys. Rev. B 13, 1654 (1976).
[6] Y. Bar-Yam and J. D. Joannopoulos, Phys. Rev. Lett. 52, 1129 (1984).
[7] J. Bernholc, N. O. Lipari, and S. T. Pantelides, Phys. Rev. Lett. 41, 895 (1978); G. A. Baraff anf M. Schluter, Phys. Rev. Lett. 41, 892 (1978).
[8] R. Car, P. J. Kelly, A. Oshiyama, and S. T. Pantelides, Phys. Rev. Lett. 52, 1814 (1984); ibid. 54, 360 (1985).
[9] J. Bourgoin and J. W. Corbett, Phys. Lett. 38A, 135 (1972).
[10] Y. Bar-Yam and J. D. Joannopoulos, Phys. Rev. B 30, 2216 (1984).

[11] G. D. Watkins, in <u>Radiation Damage in Semiconductors,</u> (Dunod, Paris, 1964), p. 67; see also G. D. Watkins, in Lattice Defects in Semiconductors - 1974, edited by F. A. Huntley, Institute of Physics Conference Series No. 23, 1975, p. 1.

[12] For a recent review of early theories, experimental data, and on-going debates see W. Frank, U. Goesele, H. Mehrer, and A. Seeger, in <u>Diffusion in Solids II,</u> edited by A. S. Nowick and G. Murch, (Academic, New York, 1984), p. 64. See also the review article by S. M. Hu, in <u>Atomic Diffusion in Semiconductors,</u> edited by D. Shaw, (Plenum, New York, 1973), p. 217.

[13] G. A. Baraff, E. O. Kane, and M. Schluter, Phys. Rev. B <u>21,</u> 5662 (1980); G. D. Watkins and J. R. Troxell, Phys. Rev. Lett. <u>44,</u> 593 (1980).

ELECTRONIC PROPERTIES OF LIQUID SILICON

J.Q. Broughton* and P.B. Allen[+]
*Department of Materials Science
[+]Department of Physics
State University of New York at Stony Brook
Stony Brook, New York 11794

ABSTRACT

The electronic properties of liquid silicon were computed by coupling molecular dynamics and tight binding methods. By employing the Stillinger-Weber potential, atomic configurations of liquid Si at 1740°C were generated by molecular dynamics. Tight binding parameters chosen to fit fcc,bcc, simple cubic and diamond cubic band structures of silicon, were then used to obtain the electronic properties of the system. All states within 10eV of the Fermi level are found to be delocalized, the density of states spectrum similar (but much broadened) to that of diamond cubic silicon and the optical conductivity is found to be almost featureless with no Drude behavior.

INTRODUCTION

Silicon exhibits at least four common dense phases; the diamond cubic, amorphous, β-Sn and liquid phases. The former are semiconducting and the latter are metallic. Whereas much experimental and theoretical work has been devoted to understanding the properties of the three low temperature phases, very little is known of the liquid. And yet, molten Si is operationally an important phase. Single crystal silicon is grown by Bridgeman or Czochralski methods from the liquid phase and novel dopant structures are generated by laser melting the surface of silicon. The metallic nature of the melt is used to determine the thickness of the laser melted region (high conductivity) and an alternating magnetic field is used to improve the quality of Czochralski growth. Silicon melts at 1410°C which makes it experimentally difficult to handle. This is clearly a case in which theory can be helpful. To our knowledge, the only other calculation of the electronic structure of molten Si is due to Gaspard et al.[1] who assumed the geometric structure to be dominated by simple cubic and β-Sn forms.

Below we describe first the molecular dynamics (MD) method used to generate liquid Si atomic configurations and then the tight-binding (TB) method for calculating one-electron wavefunctions. Finally we present our electronic property results.

MOLECULAR DYNAMICS SIMULATIONS

Structurally it is known by neutron/X-ray scattering experiments that the bond length in liquid silicon (2.5Å) is appreciably longer than in the diamond cubic structure (2.35Å) [2]. Further, the number of nearest neighbors increases from 4 to 6.5 upon melting resulting in a density increase. Many models have been used to explain these results but all assume the existence of two species of Si atoms in the melt or of two types of coordination [3]. These weaknesses are resolved by using the Stillinger-Weber (SW) potential which was parametrized to give approximately the correct melting point for Si and to fit the liquid and crystal structure factors [4].

The total SW potential energy of the system is given as the sum over all pairs of atoms of a Lennard-Jones-like term which smoothly goes to zero at a range near the second neighbor distance plus a sum over all triplets of a three-body term of the form:

$$\phi_3(r_{ij}, r_{ik}, \theta_{jik}) = \lambda \exp[\gamma(r_{ij}-a)^{-1} + \gamma(r_{ik}-a)^{-1}] \times [\cos\theta_{ijk} + 1/3]^2 \qquad (1)$$

This term vanishes if either r_{ij} or r_{ik} is greater than a. The angular term is zero at the ideal tetrahedral angle and positive otherwise.

We modelled a 216 atom periodically connected system by molecular dynamics simulation at the liquid Si density of 2.53 gm/cm^3. Newton's laws of motion were integrated by the Beeman algorithm and the force law between particles was assumed to be of SW form. The system was taken to $\sim 3000^{\circ}$C for 40,000 time steps ($\Delta t = 3.8 \times 10^{-16}$ sec) before being slowly cooled to 1740°C, the temperature guessed by SW as the melting point of their model. After equilibration at this temperature; the system was run for 36,000 Δt and twelve system configurations stored at invervals of 3000 Δt. These configurations were then used in the tight-binding calculations. The MD diffusion coefficient at this temperature is 0.98×10^{-4} cm^2/sec which implies a mean square displacement per particle over the 36,000 Δt of 38.9 Å^2 or a mean distance travelled of ~ 6Å. In other words, the length of the simulation spans significantly different parts of phase space.

TIGHT-BINDING METHOD

We use non-orthogonal Slater-Koster [5] parameterization

$$\psi_i = \sum_m C_i(m)|m> \qquad (2)$$

$$\sum_n (H_{mn} - E_i S_{mn}) C_i(n) = 0 \qquad (3)$$

where the basis functions are s and p atomic-like functions

$$|m> = |\underset{\sim}{R} \alpha > \qquad (4)$$

Here R represents the atomic site and α is the orbital quantum number. In two center approximation the matrix element H_{mn} is expressed as geometric factors times irreducible matrix elements $H_{ss\sigma}(r)$, $H_{sp\sigma}(r)$, $H_{pp\sigma}(r)$, $H_{pp\pi}(r)$ which are scalar functions of the interatomic separation $r=|\underset{\sim}{R}-\underset{\sim}{R}'|$. Similar scalar functions $S_{ss\sigma}(r)$ etc. determine the overlap matrix elements S_{mn}. Following Slater and Koster we treat these scalar functions as empirical parameters. Mattheiss and Patel [6] obtained a close fit to valence and conduction bands of Chelikowski and Cohen [7] for diamond cubic Si, fitting the scalar parameters at the three discrete distances r_i corresponding to the first three neighbor distances (2.3Å < r < 4.6 Å). We have found values at an additional 6 intermediate r_i's by fitting McMahan's [8] LMTO calculations for hypothetical metallic sc, bcc, and fcc Si at the same density as diamond cubic, and using first and second neighbors only. The final fitted values $H_{ss\sigma}(r_i)$ etc. scatter somewhat around smoothly decreasing curves. We have drawn in smooth compromise curves and found polynomial representations. This yields a model for electronic structure of arbitrary geometrical arrangements of Si atoms, which we then apply to liquid Si.

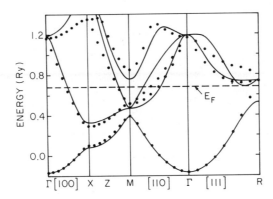

Fig. 1. Electron energy bands of silicon in a simple cubic lattice with
a = 2.715Å which corresponds to the normal density of diamond
structure Si. The "data" points are LMTO calculations (ref. 8)
and the solid line is derived from "universal" tight-binding
parameters. A much better fit can be obtained by a small variation of these parameters.

Figure 1 gives an example of the band structure of simple cubic
silicon obtained from these polynomials. The band structures for the other
crystalline states are equally good. The agreement with the LMTO results is
poor high up in the conduction band because we chose not to fit the band
structure here. At higher energies the l=2 component of ψ_i increases and
our sp basis becomes increasingly artificial.

ELECTRONIC PROPERTIES

We evaluated the eigenvalues and vectors of the (864x864) matrix on the
CRAY-XMP at Boeing, Seattle for each of the 12 MD-generated configurations.
Instantaneous properties for each configuration were obtained and then
ensemble-averaged together. Densities of states, participation ratios and
optical conductivities were evaluated. Bearing in mind that in our tight-
binding method, the basis functions are not rigorously defined, it is
necessary to find approximate expressions for the last two quantities.

The participation ratio (P) of state i is usually defined as:

$$P_i = [\int |\psi_i(r)|^2 \, dr]^2 / \Omega \int |\psi_i(r)|^4 \, dr \qquad (5)$$

and equals zero for a totally localized state and unity for a fully
delocalized state. If the basis orbitals had been orthogonal then the
probability (p) that an electron in the ith state is on a given atom would
have been:

$$p(i|R) = \sum_\alpha |C_i(R\alpha)|^2 \qquad (6)$$

Making this approximation, we write the participation ratio as:

$$P_i = [\sum_R p(i|\underset{\sim}{R})]^2/N \sum_R [p(i|\underset{\sim}{R})]^2 \qquad (7)$$

The optical conductivity (σ) is given by the standard Kramers-Heisenberg formula for each configuration:

$$\sigma_{xx}(\omega) = \frac{he^2}{\Omega m^2} \sum_{ij} \frac{f_i - f_j}{\varepsilon_j - \varepsilon_i} |<i|p_x|j>|^2 \delta(\hbar\omega - \varepsilon_j + \varepsilon_i) \qquad (8)$$

where i and j label eigenstates, f are Fermi functions and ε are eigenvalues. In order to evaluate the dipole matrix elements we make the on-site approximation, namely

$$<\underset{\sim}{R}\alpha|p_x|\underset{\sim}{R}'\beta> = \delta_{\underset{\sim}{R}\underset{\sim}{R}'} \; p_0[\delta_{\alpha s}\delta_{\beta p} + \delta_{\alpha p}\delta_{\beta s}] \qquad (9)$$

and p_0 is a constant to be determined.

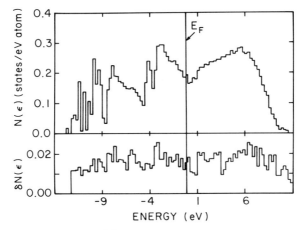

Fig. 2. Density of states $N(\varepsilon)$ of liquid Si. The vertical line shows the Fermi energy at -0.13eV. The calculation averaged 12 configurations of 216 atoms with periodic boundary conditions. The rms fluctuation in the different configurations is $\sim$ 10% as shown in the lower panel which means that the numerical error (standard deviation) in $N(\varepsilon)$ is $\sim$ 3%.

The results of our calculations are given in figures 2,3 and 4. Figure 2 shows the density of states (DOS). The system is clearly metallic with a minimum at the Fermi level. Comparison with the density of states of diamond cubic Si shows striking similarities in the energy range of the principal features. The simple cubic DOS of Gaspard et al [1] is very different. The only data with which to compare our DOS are the X-ray emission results of Hague et al [9] which pick out the 3p partial DOS in the VB. The DOS maximum just below E_f is of pure p-character. The crystalline and amorphous phases show sp mixing at about 4.5 eV below E_f

whereas the liquid does not. We intend to project out the s and p components of our calculated DOS to compare with experiment. The inset at the bottom of figure 2 shows the root mean square (RMS) fluctuation in the DOS which indicates that the DOS calculated for each configuration is essentially self averaging.

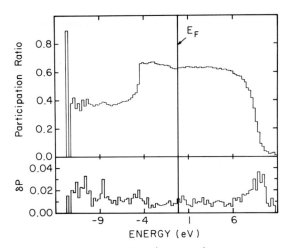

Fig. 3. Average participation ratio (eqs. 5-7) versus energy for liquid Si. The rms fluctuation over the 12 configurations is ∿ 5% as shown in the lower panels.

Figure 3 shows the participation ratio as a function of energy. Each state of the system is seen to be delocalized. The rise near -4eV is probably due to a change in wavefunction character from s to p and does not represent a significant change in localization. Our next step will be to test, by considering a 512 particle system, that the delocalized nature of these states is independent of system size.

Finally, in figure 4, the real part of the optical conductivity (in relative units) is shown. It represents an average over the diagonal elements of the conductivity tensor (Off-diagonal elements are more than a factor of ten smaller but with the same RMS fluctuation.) The conductivity spectrum is featureless; the most important observation being a complete lack of Drude behavior. The Drude model, representing the classical behavior of an electron in an applied field, predicts a conductivity which varies as:

$$\sigma_{Drude}(\omega) = \sigma_{DC}/(1 + \omega^2\tau^2) \tag{10}$$

where τ is the relaxation time. Such behavior is common in liquid metals. Hg, Al, Sn, alkali metals and even Ge all exhibit this behavior, whereas Te does not. Our results indicate one of two possibilities. Either scattering is so strong that σ remains at its DC value independent of ω or liquid silicon retains some of its semiconducting characteristics in which only interband transitions are to be expected. Such has been the argument advanced to explain the similar behavior of liquid Te [10].

18

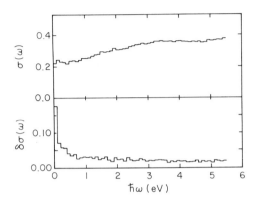

Fig. 4. ac conductivity versus applied frequency ℏω for liquid Si.
The rms fluctuation shown in the lower panel is typically
∿ 10% but increases for small ω.

In liquid Te, it is thought that spiral chains of atoms in solid Te
persist to some extent and that the semiconducting nature of solid Te is not
completely lost in the liquid. By analogy this would imply that the structure
of liquid silicon implied by the SW potential, despite its average six
nearest neighbor behavior, may have local semiconducting structures (perhaps
like those of amorphous silicon). Such behavior returns us to the early
models of liquid silicon which were based on an admixture of different types
of coordination. This suggests an analysis of the distribution of near
neighbors in the SW liquid.

There is one final possibility in explaining the lack of Drude tail;
namely that the on-site approximation is too crude. It is not completely
obvious that eq. 8 contains a Drude contribution under any circumstances.
Therefore we took a metallic periodically connected 216 atom simple cubic
Si system with atoms displaced Gaussianly from ideal lattice sites to
represent a hot solid, and did   obtain Drude behavior at low
frequencies. We are currently examining the on-site approximation by
checking that it gives the correct $\sigma(\omega)$ for diamond-cubic Si.

ACKNOWLEDGEMENTS

This work was supported in part by DOE grant No. DE-FG02-85ER45218 (JQB)
and NSF grant No. DMR8420308 (PBA) and by the NSF office of advanced
scientific computing. We thank A. Silverstein for time on the SUNY-Chemistry
VAX-11/780, and P.V. Madhavan for help with numerical analysis.

REFERENCES

1. J.P. Gaspard, Ph. Lambin, C. Mouttet and J.P. Vigneron, Phil. Mag. 50,
    103, (1984).

2. J.P. Gabathuler and S. Steeb, Z. Naturf. 34, 1614,(1979).

3. See, for example, reference 1 in ref. 1.

4. F.H. Stillinger and T.A. Weber, Phys. Rev. B31, 5262, (1985).

5. J.C. Slater and G.F. Koster, Phys. Rev. 94, 1498, (1954).

6. L.F. Mattheiss and J.R. Patel, Phys. Rev. B23, 5384, (1981).

7. J.R. Chelikowsky and M.L. Cohen, Phys. Rev. B10, 5095, (1974).

8. A.K. McMahon, Phys. Rev. B30, 5835, (1984); and private communication.

9. C.F. Hague, C. Sénémaud and H. Ostrowiecki, J. Phys. F., 10, L267,(1980).

10. J.N. Hodgson in "Liquid Metals. Chemistry and Physics". Ed. S.Z. Beer. Publ. Marcel Dekker, New York, 331, (1972).

THEORETICAL STUDY OF SEMICONDUCTOR INTERFACES.

CHRIS G. VAN DE WALLE[a,b] AND RICHARD M. MARTIN[b]
[a] Stanford Electronics Laboratories, McCullough 422, Stanford, CA 94305
[b] Xerox Palo Alto Research Center, 3333 Coyote Hill Road, Palo Alto, CA 94304

ABSTRACT

We perform self-consistent density functional calculations on semiconductor heterojunctions, using *ab initio* nonlocal pseudopotentials, and derive valence band discontinuities for many different lattice matched interfaces. Spin-orbit effects are included *a posteriori*. A comparison is made with previous calculations, which used empirical pseudopotentials, and with other heterojunction theories. We find reasonable agreement with reported experimental values, and derive some important conclusions about the nature of the lineups.

INTRODUCTION

Knowing the band discontinuities at heterojunction interfaces is essential for predicting the properties of a variety of novel semiconductor structures, such as quantum well lasers and high-speed modulation-doped field-effect transistors. Measured experimental values for band lineups are not well established yet,[1] even though considerable progress has been made in growth and analysis techniques.[2] A number of theories[3-5] have been developed to attempt to predict these important properties; however, they yield very different values, and are based upon certain untested assumptions. In a theoretical approach, the central problem is the derivation of the potential shift which occurs at the junction of two materials with different characteristics (e.g. different band gaps). In the neighborhood of the interface, the electron distribution will clearly differ from the bulk. The only way to obtain a full picture of this effect is to perform a calculation in which the electrons are allowed to adjust to the specific environment around the junction. This can be accomplished by performing self-consistent density functional calculations, as was done by Pickett *et al.*[6] These methods have been applied to a wide variety of solid state problems, and provide a fundamental theoretical framework to address the problem without making any uncontrolled assumptions. The earlier work followed this *ab initio* approach, except for one aspect: it made use of empirical pseudopotentials. In the present study, we will use the more recent *ab initio* pseudopotentials.

We have previously performed calculations of this type on pseudomorphic Si/Ge interfaces, where we studied the effects of the induced strain upon the heterojunction discontinuities.[7] Here, we will concentrate upon lattice matched systems. In the next section, we will describe our methods, which have previously been applied to a wide variety of problems in many different types of solids, and illustrate them with a typical calculation. We will then give an overview of the broad range of lattice matched systems that we studied. Finally, we will compare the results with other theories and with experiment, discuss some of the problems, and draw some important and general conclusions.

METHODS

The derivation of interface properties requires that one first determine the positions of the atoms of minimum energy, i.e. the structure of the interface. For the lattice matched systems that we consider in the present paper, we assume that the atoms occupy the ideal positions of the bulk lattice structure up to the interface. We have carried out full calculations of total energy, forces, and stresses for representative cases and have found that the expected deviations from this structure will be small (less than 0.05 Å). We have also checked that such small displacements of the atoms near the interface will have a negligible effect on the band lineups.

A major problem that has to be faced in calculating the electronic structure of an interface is the loss of translational symmetry, which is essential for using a reciprocal space formulation of the problem. The actual calculations are therefore performed on a superlattice, consisting of slabs of the respective semiconductors in a particular orientation. A typical (110) interface between two semiconductors, GaAs and AlAs, is sketched in Fig. 1.a. We also indicate a supercell appropriate for calculating the properties of this interface; it contains 12 atoms and 2 identical interfaces. Of course, what we emphasize here are the results for an isolated interface. These can be derived from our calculations to the extent that the interfaces in the periodic structure are well

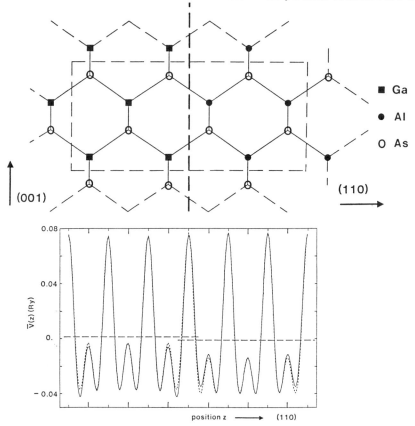

FIG. 1.a: Schematic representation of a GaAs/AlAs (110) interface. The supercell used in the calculations is indicated in dotted lines; it contains 12 atoms and 2 interfaces.

FIG. 1.b: Variation of the averaged l=1 component of the total potential $\bar{V}(z)$ (as defined in Eq. (1)) across the (110) interface. The dashed lines represent the corresponding potentials for the bulk materials. These coincide with $\bar{V}(z)$ in the regions far from the interfaces. However, the average levels of the two bulk potentials (broken horizontal lines) are shifted with respect to each other.

separated. We establish *a posteriori* that this is the case, by examining charge densities and potentials in the intermediate regions, and showing them to be bulklike.

The self-consistent calculations are performed within the framework of local density functional theory,[8] applied in the momentum space formalism,[9] and using nonlocal norm-conserving pseudopotentials.[10] We use the most recent, so called *ab initio* pseudopotentials; this term indicates that these potentials are generated using only theoretical calculations on atoms, without introducing any type of fitting to experimental band structures or other properties. This implies that all elements are treated in the same way, which is particularly important when we want to include different materials in the same calculation, as for an interface. This is not true for the empirical pseudopotentials which have been used in previous interface calculations.[6] Self-consistent solutions are obtained for the charge density and the total potential, which is the sum of ionic, Hartree and exchange-correlation potentials. The latter is calculated using the Ceperley-Alder form.[11] In each cycle of iteration a potential is used for generating the charge density, from which a new potential can be calculated. This is then used for constructing the input for the next cycle. The first cycle requires a trial potential, for which we choose the ionic potential screened by the dielectric function of a free electron gas. Convergence of the self-consistent iterations is obtained with the help of. the Broyden scheme.[12]

Plane waves with kinetic energy up to 6 Rydberg were included in the expansion of the wavefunctions (corresponding to more than 500 plane waves in some cases). 4 special points were used for sampling k-space. In the final self-consistent solution, a redistribution of electrons occurs which changes the electric dipole in the interface region. The resulting self-consistent potential across the supercell is plotted in Fig. 1.b, for the example of GaAs/AlAs. Because the *ab initio* pseudopotentials are non-local, the total potential consists of different parts corresponding to different angular momenta l. We only show the $l=1$ part of the potential here, which is the most important one in determining the lineup of the p-like valence bands. In the plot, the variation of the space coordinate $\bar{r}$ is limited to the component perpendicular to the interface, and values of the potentials are averaged over the remaining two coordinates, i.e. in the plane parallel to the interface:

$$\bar{V}(z) = 1/(Na^2) \int V(\bar{r}) \, dx \, dy. \qquad (1)$$

In the regions far from the interface, the crystal should recover properties of the bulk. Therefore we also plot (broken lines) the potentials determined separately from calculations on bulk GaAs and AlAs. One sees that already one layer away from the interface the potential assumes the form of the bulk potential. Similar results hold for the charge density. This confirms, *a posteriori*, that the two interfaces in our supercell are sufficiently far apart to be decoupled, at least as far as charge densities and potentials are concerned. The average levels of the potentials which correspond to the bulk regions are also indicated in Fig. 1.b. We denote these average levels by $\bar{V}_{GaAs}$ and $\bar{V}_{AlAs}$, and define the shift $\Delta\bar{V} = \bar{V}_{GaAs} - \bar{V}_{AlAs}$.

To get information about band discontinuities, we still have to perform the band calculations for the bulk materials. These were carried out with a 12 Ry cutoff; tests have shown that the choice of this cutoff is not critical for deriving the valence band lineups. We find that the valence band maximum in GaAs is 9.60 eV above the average potential $\bar{V}_{GaAs}$. In AlAs, the valence band occurs at 9.29 eV above $\bar{V}_{AlAs}$. From Fig. 1.b, we find $\Delta\bar{V}=0.032$ eV. This leads to a discontinuity in the valence band of $\Delta E_v=0.35$ eV (upward step in going from AlAs to GaAs). We estimate the inaccuracy of our calculations (due to finite separation between interfaces in the supercell, energy cutoff, number of special points,...) to be on the order of 0.05 eV. We did not include spin-orbit splitting in our density functional calculations. These effects can be included *a posteriori*, by using experimental values for spin-orbit splittings.[13] The results which we will give in the next section have been adjusted to include these effects.

RESULTS

We have studied a variety of lattice matched (110) interfaces, the results for which are given in Table I. For interfaces between a group IV element and a III-V compound, the (110) orientation is the only one which avoids charge accumulation without the need for mixing at the interface.[14] The column "empirical pseudopotentials" in the table contains values derived by performing self-consistent density functional calculations very similar to ours, but with empirical pseudopotentials.[6] We also list values obtained by means of the heterojunction theories of Harrison[4] and Tersoff.[5] Also shown in Table I are experimental data from various sources. At the present time, it is far from clear how reliable these values are. A striking example is that of the GaAs/AlAs interface, for which "Dingle's 85/15 Rule" had become widely accepted: $\Delta E_v = 0.15 \Delta E_g$, where ΔE_g is the difference in direct band gaps (most experiments are done on GaAs/$Ga_{1-x}Al_xAs$ interfaces, in a region of alloy compositions $x < 0.44$ where the band gap of the alloy is direct). Since last year, however, this result has been challenged and several groups have come up with results which now seem to indicate that more than 35% of the discontinuity is in the valence band.[1] Care should be taken in extrapolating this result to the GaAs/AlAs interface ($x=1$), since ΔE_g is not linear as a function of alloy composition x, for $x > 0.44$. Therefore we only compare our theoretical result to the values derived from experiments on the "pure" ($x=1$) interface.[15,16]

TABLE I.

Heterojunction	Present results	Empirical pseudopot.[6]	ΔE_v (eV) Harrison theory[4]	Tersoff theory[5]	Experiment
AlAs/Ge	1.05		0.45	0.87	0.95 [a]
GaAs/Ge	0.63	0.35	0.41	0.52	0.56 [b]
AlAs/GaAs	0.37	0.25	0.04	0.35	0.45 [c] - 0.56 [d]
GaP/Si	0.61		0.50	0.45	0.80 [e]
InAs/GaSb	0.38		0.52	0.43	0.51 [f]
ZnSe/GaAs	1.59	2.0±0.3	1.05		1.10 [g]
ZnSe/Ge	2.17	2.0±0.3	1.46		1.52 [g]

[a] M. K. Kelly, D. W. Niles, E. Colavita, G. Margaritondo, and M. Henzler (unpublished).
[b] J. R. Waldrop, E. A. Kraut, S. P. Kowalczyk and R. W. Grant, Surface Sci. 132, 513 (1983).
[c] Reference 15.
[d] Reference 16.
[e] P. Perfetti, F. Patella, F. Sette, C. Quaresima, C. Capasso, A. Savoia, and G. Margaritondo, Phys. Rev. B 30, 4533 (1984).
[f] Reference 19.
[g] S. P. Kowalczyk, E. A. Kraut, J. R. Waldrop, and R. W. Grant, J. Vac. Sci. Technol. 21(2), 482 (1982).

We find a valence band discontinuity for a GaAs/AlAs (110) interface, after including the spin-orbit splitting effects, of 0.37 eV. This value differs significantly from the one calculated by Pickett et al.,[6] who found ΔE_v=0.25 eV for a (110) interface. The difference is due to our use of *ab initio* pseudopotentials, compared to their empirical pseudopotentials (fitted to reproduce experimental bandstructures). If we use those same pseudopotentials in our calculations, we reproduce their result (within the numerical accuracy of 0.05 eV). This indicates that the essential difference is in the choice of the pseudopotential - the *ab initio* pseudopotential providing a better justified starting point. The value we obtain in this way is certainly closer to the value reported from recent experiments.[15,16]

We have also studied other interface orientations for this system. In particular, for the (100) interface we find a valence band discontinuity of 0.37 eV, the same as the value for the (110) interface. For (111), we find ΔE_v=0.39 eV. This indicates that ΔE_v does not depend on interface orientation, a result that was also found experimentally.[17] Let us note that this is not necessarily valid for pseudomorphic strained-layer systems, in which different strains associated with different interfaces can have sizeable effects on the lineups, as discussed in Ref. 7. It also has been shown that rearrangements of atoms at polar interfaces can change the offsets.[18] Within such limitations, we believe that the result that the offset is orientation independent can be considered an important general result for suitably chosen lattice matched interfaces.

It is interesting to examine our results to establish the extent to which theory supports the proposition that the band offsets for any pair of semiconductors can be expressed as a difference of numbers intrinsic to each material. This has been observed from experiment,[2] and is an implicit assumption in theories such as Refs. 3-5. It is clear that our full interface calculations do not assume linearity, i.e. we do *not* postulate that our heterojunction lineups be given by the difference of two numbers which would each be characteristic for a particular semiconductor, independent of which heterojunction it is used in. A *posteriori*, however, we can check how close our results are to linearity, by examining transitivity, i.e. whether the following equation is satisfied:

$$\Delta E_v(A,B) + \Delta E_v(B,C) = \Delta E_v(A,C) \qquad (2)$$

where $\Delta E_v(A,B) = E_v(B) - E_v(A)$. As can be verified by inspection of the values in Table I, the transitivity rule is satisfied to within the numerical accuracy of the calculations. This result also applies to a large class of experimental values for well-prepared heterojunctions, indicating that the deviations from linearity are expected to be small.[2] Together with the orientation independence, we believe that this is indicative of the intrinsic nature of the band offsets for large classes of lattice matched systems.

A complete picture of the band lineups should also include the conduction bands. It is a well known problem of density functional calculations that absolute positions of conduction bands are not given accurately; band gaps are typically underestimated by 50%.[19] In our calculations we have derived the valence band lineup at the interface. The occupied states in the valence bands give rise to a charge density and set up dipoles, which are ground state properties of the system, and are therefore expected to be reliably given by a density functional calculation. The necessary corrections to the absolute positions of the bands are the subject of intense investigations at this time.[19] For the present purposes the most important point is that, so long as these corrections do not change the dipole, they always amount to additive shifts in the bulk band structures, which can be made a *posteriori*. For now, we include the conduction bands in the picture by using experimental values for band gaps. The conduction band lineup is found by subtracting the valence band discontinuity from the experimental band gap difference. For the GaAs/AlAs interface we find ΔE_c=0.34 eV (higher in AlAs). A very interesting case is that of InAs/GaSb, in which experimentally a "broken-gap" lineup" was detected,[20] meaning that the conduction band in InAs is lower in energy than the valence band in GaSb. We find that $\Delta E_c(\text{InAs})-\Delta E_c(\text{GaSb})=0.03$ eV, i.e. we find that the bands almost overlap, and only a small change in ΔE_v is needed to agree with experiment.

In summary, we have presented self-consistent *ab initio* calculations of lattice matched semiconductor interfaces. This approach is the only one which takes the full atomic and electronic structure of the interface into account for deriving heterojunction band discontinuities. In addition to the many specific results, we consider the most important general conclusions: the transitivity property and the orientation independence of the lineups. The insight into the mechanism of the lineups that we gained from these calculations is very helpful in the construction of a new and simpler approach to derive band lineups, which we are currently developing. Preliminary results from this method encourage us to believe that predictions of band lineups can be made using only simpler calculations on the bulk materials. Deviations from this ansatz are to be expected, however; only the complete self-consistent calculations can incorporate all the details of the effects around an interface.

ACKNOWLEDGMENT

This work was partially supported by ONR Contract No. N00014-82-C0244.

REFERENCES

1. G. Duggan, J. Vac. Sci. Technol. B3, 1224 (1985).
2. A. D. Katnani and G. Margaritondo, J. Appl. Phys. 54, 2522 (1983); Phys. Rev. B 28, 1944 (1983).
3. W. R. Frensley and H. Kroemer, J. Vac. Sci. Technol. 13, 810 (1976).
4. W. A. Harrison, Electronic Structure and the Properties of Solids (Freeman, San Francisco, 1930), p. 253.
5. J. Tersoff, Phys. Rev. B 30, 4874 (1984).
6. W. E. Pickett, S. G. Louie, and M. L. Cohen, Phys. Rev. B 17, 815 (1978); J. Ihm and M. L. Cohen, Phys. Rev. B 20, 729 (1979); W. E. Pickett and M. L. Cohen, Phys. Rev. B 18, 939 (1978).
7. C. G. Van de Walle and R. M. Martin, J. Vac. Sci. Technol. B3, 1256 (1985).
8. P. Hohenberg and W. Kohn, Phys. Rev. 136, B864 (1964); W. Kohn and L. J. Sham, Phys. Rev. 140, A1133 (1965).
9. O. H. Nielsen and R. M. Martin, Phys. Rev. B 32, 3780 (1985).
10. G. B. Bachelet, D. R. Hamann and M. Schluter, Phys. Rev. B 26, 4199 (1982).
11. D. M. Ceperley and B. J. Alder, Phys. Rev. Lett. 45, 566 (1980); J. Perdew and A. Zunger, Phys. Rev. B 23, 5048 (1981).
12. P. Bendt and A. Zunger, Phys. Rev. B 26, 3114 (1982).
13. Landolt-Börnstein, Numerical Data and Functional Relationships in Science and Technology (Springer, New York, 1982), Group III, Vol. 17.a-b.
14. R. M. Martin, J. Vac. Sci. Technol. 17, 978 (1980).
15. W. I. Wang and F. Stern, J. Vac. Sci. Technol. B3, 1280 (1985).
16. D. Arnold, A. Ketterson, T. Henderson, J. Klem, and H. Morkoç, J. Appl. Phys. 57, 2880 (1985).
17. W. I. Wang, T. S. Kuan, E. E. Mendez, and L. Esaki, Phys. Rev. B 31, 6890 (1985).
18. K. Kunc and R. M. Martin, Phys. Rev. B 24, 3445 (1981).
19. C. S. Wang and W. E. Pickett, Phys. Rev. Lett. 51, 597 (1983); L. J. Sham and M. Schlüter, Phys. Rev. Lett. 51, 1888 (1983); M. S. Hybertsen and S. G. Louie, Phys. Rev. B 30, 5777 (1984); Phys. Rev. Lett. 55, 1418 (1985).
20. J. Sakaki, L. L. Chang, R. Ludeke, C. - A. Chang, G. A. Sai-Halasz, and L. Esaki, Appl. Phys. Lett. 31, 211 (1977); L. L. Chang and L. Esaki, Surf. Science 98, 70 (1980).

THE STUDY OF SURFACE PHONONS BY ELECTRON ENERGY LOSS
SPECTROSCOPY: THEORY OF THE EXCITATION CROSS SECTION

D. L. Mills
Department of Physics, University of California, Irvine,
Irvine, California 92717

ABSTRACT

    In the past few years, experimental developments in
electron energy loss spectroscopy have allowed study of the
dispersion relations of surface phonons, on clean and adsorbate covered surfaces. Also, theoretical analyses of the
angle and energy variation of the excitation cross sections
have been developed. These have guided the choice of
scattering geometry, and also have assisted directly in the
interpretation of the data. We review the recent work, with
emphasis on the theoretical side.

Introduction

    This paper presents a summary of recent studies of sur-
face phonons by the electron energy loss method. An interplay
between a quantitative theory of the surface phonon excitation
cross section and the experimental effort has proved important.
Since these developments have been reviewed very recently
[1,2], here we present only a brief discussion with reference
to the relevant papers.

General Discussion

    For some years, electron energy loss spectroscopy has
proved a versatile tool for probing the vibrations of atoms in
the surface of crystals, and of atoms and molecules adsorbed
on the surface. However, until recently, attention of the
experimentalist has been directed exclusively to the study of
small angle deflections from the specular direction, where
scattering from the long-ranged electric fields generated by
the oscillating surface dipoles leads to large vibrational
excitation cross sections, particularly at low impact energies
(1eV-10eV) where the coulomb matrix element is large, and the
crystal is highly reflecting to the incident electrons. The
theory of dipole scattering was developed some years ago, and
rather general expressions for the excitation cross sections
follow from macroscopic considerations [3]. Kinematical
considerations show that in such experiments, surface excita-
tions with wave vector $Q_\parallel \cong 10^6$ cm^{-1} are excited, so only the
near vicinity of the origin of the Brillouin zone is explored
by the technique, used in this manner. It necessarily follows
that the dispersion relation of surface phonons can't be
explored by this traditional form of electron energy loss
spectroscopy.

    A short time ago, Ibach and his collaborators [4]
resolved the energy spectrum of electrons scattered far off

the specular, where the momentum transfer $Q_\parallel$ is the order of $10^8 cm^{-1}$. They were then able to explore the surface phonon dispersion relations throughout the two-dimensional Brillouin zone, for clean Ni(100) [4,5] and for the Ni(100) surface covered with ordered adsorbate layers of oxygen [6], sulfur [7], and carbon [8].

These experiments employed electron beams with much higher energy than employed in the dipole scattering studies. The high energies, in the 50eV-300eV range typical of low energy electron diffraction surface studies, allow quantitative calculation of the angle and energy variation of the surface phonon excitation cross sections, without the formidable complications of the image potential, which influences the calculated cross sections dramatically at low energies [9], and whose form is poorly understood in the near vicinity of the surface. A formalism to calculate the large angle cross section was developed, and an early set of calculations outlined, in general terms, the trends to be expected, and the selection rules which operate for non-dipole active modes [10].

Recently, a new series of calculations have brought the theory into quantitative contact with experimental data. There have been two series of calculations.

The first was directed toward the clean Ni(100) surface [5]. The theory shows that on the (100) surface of an fcc crystal, there are two surface phonons at the $\bar{X}$ point of the two dimensional Brillouin zone [11]. One, the S_4 mode, has displacement normal to the surface in the outermost atomic layer. This is the mode which in the long wavelength limit is the Rayleigh surface wave. Then there is a second surface phonon, the S_6 mode, for which the displacements are parallel to the surface in the outermost layer. In the original studies of surface phonons on the clean Ni(100) surface [4], the S_6 mode was not observed. This is not surprising, since a simple kinematical estimate suggests that the S_6 excitation cross section should be smaller than for S_4 by roughly two orders of magnitude. The full multiple scattering analysis [5] predicts that there are three narrow "energy windows" between 50eV and 300eV within which the S_4 and S_6 cross sections are comparable. Subsequent experiments observed this mode, at precisely the predicted energies. The theory also provides a quantitative account of the energy variation of the S_4 excitation cross section, over a wide energy range.

The theory also has been applied to the c(2x2) sulfur overlayer on Ni(100). This system has been explored in detail by electron energy loss spectroscopy [7]. A lattice dynamical analysis based on a simple nearest neighbor model places the surface overlayer 1.45 A above the surface [7], a distance in disagreement with the earlier height determination of 1.35 A provided by low energy electron diffraction and EXAFS. A detailed study of excitation cross sections of the surface phonons localized in the sulfur overlayer shows that the systematics are incompatible with the 1.45 A height but accounted for perfectly by the 1.35 A position [12]. Evidently there are small contributions to the force constants of non-

central character, and when these are incorporated into the
lattice dynamics, a good account of the measured dispersion
curves is obtained with sulfur at the 1.35 A vertical height.

REFERENCES

1. D. L. Mills and S. Y. Tong, Phil. Trans. R. Soc. London,
   to be published.

2. D. L. Mills, S. Y. Tong and J. E. Black, Chapter 5 of
   Surface Phonons, edited by W. Kress and F. W. de Wette
   (Springer-Verlag, Heidelberg, to be published).

3. For a review, see Chapter 3 of Electron Energy Loss
   Spectroscopy and Surface Vibrations, H. Ibach and D. L.
   Mills (Academic Press, San Francisco, 1982).

4. S. Lehwald, J. M. Szeftel, H. Ibach, T. S. Rahman, and D.
   L. Mills, Phys. Rev. Letters 50, 518 (1983).

5. M. L. Xu, B. M. Hall, S. Y. Tong, M. Rocca, H. Ibach and
   J. E. Black, Phys. Rev. Letters 54, 1171 (1985).

6. J. M. Szeftel, S. Lehwald, H. Ibach, T. S. Rahman, J. E.
   Black, and D. L. Mills, Phys. Rev. Letters 51, 268 (1983).
   See also Talat S. Rahman, J. E. Black, J. M. Szeftel, S.
   Lehwald, and H. Ibach, Phys. Rev. B30, 589 (1984).

7. S. Lehwald, M. Rocca, H. Ibach and Talat S. Rahman, Phys.
   Rev. B31, 3477 (1985). For an erratum, see Phys. Rev.
   B32, 1354 (1985).

8. Talat S. Rahman and H. Ibach, Phys. Rev. Letters 54, 1933
   (1985).

9. Burl M. Hall, S. Y. Tong and D. L. Mills, Phys. Rev.
   Letters 50, 1277 (1983).

10. S. Y. Tong, C. H. Li and D. L. Mills, Phys. Rev. Letters
    44, 407 (1980); C. H. Li, S. Y. Tong and D. L. Mills,
    Phys. Rev. B21, 3057 (1980); S. Y. Tong, C. H. Li and D.
    L. Mills, Phys. Rev. B24, 806 (1981).

11. For a discussion of these modes within the framework of a
    simple model, see Chapter 5 of reference 3.

12. M. L. Xu and S. Y. Tong, to be published.

COMPUTER SIMULATION OF SPUTTERING FROM
TWO COMPONENT LIQUID METAL TARGETS

M.H. SHAPIRO*, D.Y. LO** AND T.A. TOMBRELLO**
*Physics Dept., California State University, Fullerton, CA 92634, U.S.A. and
Division of Physics, Mathematics and Astronomy, Caltech, Pasadena, CA 91125,
U.S.A.
**Division of Physics, Mathematics and Astronomy, Caltech, Pasadena, CA
91125, U.S.A.

ABSTRACT

The sputtering of In and Ga atoms from a "liquid" target composed of gallium covered by a surface monolayer of indium by incident 5 keV Ar^+ ions was simulated using the multiple interaction molecular dynamics technique. Yields, energy distributions, and angular distributions of sputtered atoms were obtained at a temperature above the melting point for the eutectic alloy. Similar information was obtained for a pure gallium and a pure indium target. Our results for layer yield ratios and angular distributions are in good qualitative agreement with Dumke's experimental data for the Ar^+, In-Ga system. Absolute yields, however, were found to be sensitive to the detailed nature of the two-body potentials used to describe the atom-atom interactions.

INTRODUCTION

Although most experimental sputtering data have been obtained from solid targets, important experimental information about the fundamental nature of the sputtering process has been obtained by sputtering from liquid metal targets. For example, Dumke [1] has shown by bombarding liquid In-Ga eutectic targets with 15 and 25 keV Ar^+ ions that (1) Gibbsian surface segregation resulted in the formation of a surface monolayer of indium that was maintained during bombardment, and (2) that the large majority of sputtered atoms had their origin in this surface layer. In two previous papers [2,3] we have investigated the generation of thin liquid Cu targets for use in sputtering simulation, and we have used them to simulate sputtering by normally incident 5 keV Ar^+ ions. Generally, those results were in good qualitative agreement with Dumke's experimental data. The present study is an attempt to more closely simulate the conditions of Dumke's experiment in order to determine which features of his results were general and which were dependent on the specific system which he studied.

SIMULATION CODE

Our simulation results have been obtained with the multiple interaction code SPUT1 which has been used previously to simulate sputtering from liquid and crystalline copper [2,3,4,5]. The code was run on the CYBER 730/760 system at the State University Data Center in Los Angeles. In SPUT1 the system of particles is assumed to interact classically under the influence of pair potentials. The Newtonian equations of motion are integrated with a simple predictor-corrector algorithm. The time step is adjusted during the integration to optimize the computational speed while maintaining a pre-determined level of accuracy.

The pure gallium liquid target used in these simulations was generated with semi-periodic boundary conditions using a modified version of the SPUT1 code [2]. The Ga atoms initially were located on an fcc lattice with the lattice constant chosen to produce the appropriate density for liquid Ga. Initial velocity components were assigned randomly to each atom from a Gaussian random number distribution. The target was then allowed to evolve under the influence of pair potentials. Heat was added by uniformly scaling the velocities until a liquid state was reached. Velocity scaling also was used to adjust the temperature of the target once melting had occurred.

Periodicity was demanded in two directions, while in the direction normal to the sputtering surface only the surface tension provided by the interatomic forces constrained the target atoms. The target was allowed to evolve until quasi-equilibrium was obtained as indicated by a zero temperature gradient, with reasonable temperature fluctuations. To minimize the computer time needed to simulate sputtering from the liquid surface, the liquid target configuration (position and velocity components for each atom) were saved in a file which served as input to the SPUT1 program.

The pure indium target was prepared in the same manner as the pure gallium target, while the In-Ga eutectic target was made by changing the mass of atoms within one atomic layer of the front surface from that of gallium to that of indium after the target was equilibrated. While the latter procedure was somewhat artificial, it allowed us to investigate the effects of monolayer indium coverage within the constraints of available computer resources.

POTENTIALS

In our previous studies [2,3] we had found that it was not possible to obtain reliable two-body potentials for liquid Cu by inverting structure factor data. However, semi-empirical potentials (a Moliere core joined to a Morse well by a cubic spline) produced reasonable results in the copper case. The same approach was used for this study. The two-body atom-atom potentials were obtained by joining standard Moliere potentials [6] to Morse wells with cubic splines. Since no parameters for the Morse potentials for the three needed cases (Ga-Ga, In-In, and In-Ga) were readily available in the literature, an extrapolation procedure was used to obtain values for the various parameters.

We found that when parameters for atoms in the same column of the periodic table were plotted against the atomic number (see Table 4.1 of ref. [6]), the variation in the parameter values was small. Plots were made of the values of the three parameters, and the results were found to lie on smooth curves that were roughly parallel. Using the parameters for aluminum to establish the location of similar curves for the Al-Ga-In case, we were able to obtain Morse potential parameters for gallium and indium. The In-Ga parameters then were obtained by interpolation. Standard Moliere potentials were used to simulate the ion-atom interactions. The potential parameters are given in Table I.

Table I
Potential Parameters

Ion-atom

$$V_{ij} = (A/r)[0.35e^{-0.3r/B} + 0.55e^{-1.2r/B} + 0.1e^{-6r/B}] \qquad r < r_a$$

$$V_{ij} = 0 \qquad r \geq r_a$$

$A = 8046.$ eV Å $B = 0.10223$ Å $r_a = 2.868$ Å   Ar-Ga

$A = 12701$ eV Å $B = 0.09335$ Å $r_a = 2.868$ Å   Ar-In

Atom-atom

$$V_{ij} = (A/r)[0.35e^{-0.3r/B} + 0.55e^{-1.2r/B} + 0.1e^{-6r/B}] \qquad r < r_a$$

$$V_{ij} = C_0 + C_1 r + C_2 r^2 + C_3 r^3 \qquad r_a \leq r < r_b$$

$$V_{ij} = D_e[e^{-2b(r-r_e)} - 2e^{-b(r-r_e)}] \qquad r_b \leq r < r_c$$

$$V_{ij} = 0 \qquad r \geq r_c$$

Ga-Ga:

$A = 13857.$ eV Å $B = 0.93947$ Å

Table I (cont.)

$C_0 = 60.59$ eV $C_1 = -56.83$ eV/Å $C_2 = 17.8921$ eV/Å^2

$C_3 = -1.900$ eV/Å^3 $D_e = 0.29$ eV $b = 1.11$ Å^{-1}

$r_e = 3.45$ Å $r_a = 2.20$ Å $r_b = 3.10$ Å

$r_c = 5.25$ Å (sputtering) $r_c = 12$ Å (target generation)

In-Ga:

A = 21874. eV Å   B = 0.08667 Å

$C_0 = 4.742$ eV $C_1 = -0.8610$ eV/Å $C_2 = -0.7460$ eV/Å^2

$C_3 = 0.1659$ eV/Å^3 $D_e = 0.30$ eV $b = 1.09$ Å^{-1}

$r_e = 3.57$ Å $r_a = 2.75$ Å $r_b = 3.25$ Å

$r_c = 5.25$ Å (sputtering) $r_c = 12$ Å (target generation)

In-In:

A = 34574. eV Å   B = 0.08065 Å

Spline 1:

$C_0 = 6.197$ eV $C_1 = -3.3233$ eV/Å $C_2 = -0.3681$ eV/Å^2

$C_3 = 0.01475$ eV/Å^3 $r_a = 2.75$ Å $r_b = 3.25$ Å

Spline 2:

$C_0 = 6063.32$ eV $C_1 = -11667.$ eV/Å $C_2 = 7506.9$ eV/Å^2

$C_3 = -1607.42$ eV/Å^3 $r_a = 1.03$ Å $r_b = 1.36$ Å

$D_e = 0.31$ eV $b = 1.07$ Å^{-1}

$r_e = 3.57$ Å

$r_c = 5.25$ Å (sputtering) $r_c = 12$ Å (target generation)

RESULTS

    Because of space limitations, in this paper we shall limit our discussion
primarily to the In-Ga case. Figure 1 shows the angular distributions of
sputtered In and Ga atoms from the In-Ga target resulting from 800 impacts
with 5 keV Ar$^+$ ions. In this simulation the indium atoms are constrained to
arise from the first layer, while almost all sputtered gallium atoms arise from
the second layer. The indium angular distribution is considerably "over-cosine"

in good agreement with Dumke's experimental observation (at 15 keV bombarding energy) that the sputtered indium angular distribution was fitted well with a $\cos^2\theta$ curve. The gallium angular distribution is even more sharply forward peaked, which is in good agreement with the $\cos^4\theta$ distribution measured by Dumke.

Yield information is given in Table II. Because of the low-yield obtained with the pure indium target, layer yield information is given only for the pure gallium and indium-gallium eutectic cases.

Table II
Sputtering Yields

Target	# of atoms	Temp (K)	Layer	Yield	Yield Ratio	Experimental Value [1]
Ga	612	401.	All	2.06	1.00	
			1	1.81	0.88	
			2	0.22	0.11	
In-Ga	171 (In)		All	1.44	1.00	
	441 (Ga)	401.	1	1.26	0.88	0.85[1]
			2	0.15	0.10	0.15[2]
In	612	474.	All	0.60[3]		

[1]Includes all sputtered In atoms.
[2]Includes all sputtered Ga atoms.
[3]Obtained using spline 1.

DISCUSSION

Our liquid In-Ga eutectic simulation results support the argument that the large majority of atoms sputtered by low energy ion bombardment arise from the first layer of the target. The angular distributions of sputtered indium and gallium atoms show a behavior quite similar to the experimental results of Dumke [1]; however, the absolute yields given in Table II do not follow the same pattern observed by Dumke. In his experimental work, there was very little difference in total absolute yields for the pure liquid gallium and indium targets and the liquid indium-gallium eutectic target. In fact a slightly higher yield was observed for the pure indium case in the experimental work.

36

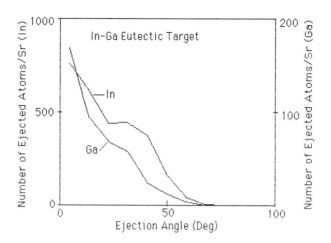

Figure 1
Angular Distributions of Sputtered Indium and Gallium Atoms

We observed very little difference in absolute yields for the two different splines used with the indium potentials, and we suspect that the yield problems is caused by improper depths for three potentials used. Further work is underway to investigate this possibility[4]. In addition, work also is underway to simulate the formation of liquid indium-gallium eutectic targets directly.

[4])Note added in proof:
Although changes in potential depth altered absolute yields slightly, a significant increase in yield for the In case was obtained by increasing the number of time steps allowed during program execution. Owing to the low surface binding energy for In (and Ga), some low energy sputtered atoms were not counted in the early computer runs.

References

1. M.F. Dumke, PhD thesis, Caltech, 1982; M.F. Dumke et al., Surface Sci., 124, 407 (1983).
2. D.Y. Lo, M.H. Shapiro, T.A. Tombrello, Nucl. Inst. and Meth. B, submitted.
3. M.H. Shapiro, D.Y. Lo, P.K. Haff, T.A. Tombrello, Nucl. Inst. and Meth. B, in press; Caltech preprint BB-31 (1985).
4. M.H. Shapiro et al., Radiation Effects 89, 234 (1985).
5. M.H. Shapiro et at., Nucl. Inst. and Meth. B12, 137 (1985).
6. I.M. Torrens, interatomic Potentials, (Academic Press, New York, 1972).

VIBRATIONAL PROPERTIES OF THE π-BONDED CHAIN MODEL OF THE
Si(111)2x1 SURFACE

O.L. ALERHAND, D.C. ALLAN and E.J. MELE
Department of Physics, University of Pennsylvania, Philadelphia,
PA 19104

ABSTRACT

A tight-binding theory is used to study vibrational excitations of the π-bonded chain model of the Si(111)2x1 surface.
Some aspects of the surface phonon spectrum are discussed. We
study the charge fluctuations driven by the vibrational excitations and the surface conductivity associated with the phonons.
We find that a longitudinal optical phonon on the surface chains
dominates the surface conductivity.

In this paper we present some recent results of our studies
of vibrational excitations of the π-bonded chain model of the
Si(111)2x1 surface. We have studied the vibrational spectrum of
the surface and the dipole activity associated with the intrinsic
surface phonons. We find a spectrum of localized surface vibrations which take the form of elementary structural excitations of
the surface chains. The most interesting of these is a longitudinal optical mode of the surface π-bonded chains which is found
to be anomalously strongly dipole active due to coupling to
virtual inter-surface state transitions. This feature provides a
microscopic model of the anomalously strong surface phonon frequently seen in high resolution inelastic electron scattering
spectra for Si(111)2x1, first reported by Ibach in 1971[1]. An
interesting consequence of this assignment is that the dynamic
dipole is oriented parallel to the surface along the chain direction, so that angle resolved EELS spectra can be used to test
this model.

In order to do realistic calculations of vibrational excitations on a variety of Si surfaces we use a model Hamiltonian that
is based on a microscopic quantum-mechanical description of the
valence electrons, but is at the same time computationally tractable for complex structures. This model is based on a Hamiltonian that treats the electronic degrees of freedom in a sp^3 tight-
binding theory with parameters that scale as d^{-2} with interatomic
bond lengths, plus a two-body elastic potential:

$$H = \Sigma_{ij} h_{ij} a_i^{\dagger} a_j + \Sigma_i U_1 x_i + U_2 x_i^2 \tag{1}$$

where $X_i = (d_i/d_o - 1)$, d_o is the equilibrium bond length, U_1 and
U_2 are used to fit d_o and the optical phonon at Γ for bulk Si
and h_{ij} are electronic tight-binding parameters. A detailed
description of the theoretical model and comparison with experiments and other calculations appears elsewhere[2], as well as
some results for the Si(100)2x1 surface[3]. We have extensively

tested this Hamiltonian obtaining generally good results[3,4].
Here we will only briefly describe how we apply it.

The first step in the calculation consists in finding the
equilibrium geometry of the surface[5]; this is done by relaxing
the atoms along the net forces on the ions derived from (1). We
then calculate the dynamical matrix through perturbation theory:

$$D_{\mu\upsilon}(q) = \frac{\partial^2 Uelast}{\partial x^\star_{\mu,q} \partial x_{\upsilon,q}} + 2 \Sigma_i f_i \langle i | \frac{\partial^2 Hel}{\partial x^\star_{\mu,q} \partial x_{\upsilon,q}} | i \rangle$$

$$+ 2 \Sigma_{ij} \frac{f_i(1-f_j)}{E_i - E_j} \langle i | \frac{\partial Hel}{\partial x^\star_{\mu,q}} | j \rangle \langle j | \frac{\partial Hel}{\partial x_{\upsilon,q}} | i \rangle \quad (2)$$

where f_n are Fermi factors. In Figure 1 we show the surface
vibrational modes and resonances along two symmetry directions in
the surface Brillouin zone. The π-bonded chain model of the
reconstructed Si(111)2x1 surface involves strong bonding along
the surface chains and weak coupling between chains. This manifests in the strong dispersion of the surface localized modes
along $\bar{\Gamma}$ to $\bar{J}$ (chain direction) and the weak dispersion along $\bar{J}$
to $\bar{K}$ (perpendicular direction).

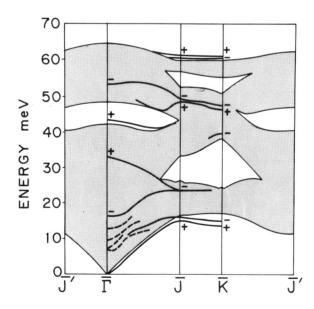

Fig. 1: Surface-localized vibrational modes and resonances for
the π-bonded chain model of the Si(111)2x1 surface along two
symmetry directions in the Brillouin zone.

At the zone center we obtain a spectrum of localized phonons which can be interpreted as elementary structural excitations of the surface chains[6]. This is not surprising considering the strong intrachain bonding and the radical difference of the reconstructed surface with respect to the bulk geometry. Of particular interest are an optical transverse and optical longitudinal modes on the surface chains ("rocking" and "dimer" modes); these occur at 33.3 and 53.5 meV respectively. The importance of these two modes is that they couple very strongly to the surface electron gap. A quantitative way of studying the coupling of vibrational modes to the electronic degrees of freedom is to calculate the linearized charge fluctuations driven by ionic displacements. We use this fluctuation to calculate the dynamic charges $\vec{e}_i^*$:

$$\vec{e}^* = i \int_0^\infty dt < [\vec{r}, h_i] >$$ (3)

where $h_i = (\partial H / \partial Q_i)$ and Q_i is an ionic degree of freedom. We can then project these dynamic charges onto the vibrational eigenmodes at the zone center to calculate the dynamic dipole of each mode and use this to calculate the real part of the conductivity due to phonons[6]. In Figure 2 we plot the surface conductivity; Figure 2a corresponds to the modes which are even under reflection through the mirror plane of the surface (polarized normal to the chain direction), and Figure 2b corresponds to the modes which are odd under the mirror reflection (polarized along the chain direction).

One strong peak associated with the longitudinal optical mode is obtained in the $\sigma_\parallel$ spectrum. Note that this peak completely dominates the spectrum for both polarizations (notice the different vertical scales in Figures 2a and 2b). The dimer mode, which is responsible for this feature, carries an unusually larger dynamic charge $e^* > 1e$. The "bare" energy of this mode has been shifted down from 53.5 meV by about 5 meV by introducing a correction due to dipole-dipole interaction and by tuning the electron gap to the observed experimental value[6]. Although Ibach originally observed the Si(111)2x1 surface phonon at ~ 55 meV, other more recent experiments[7] have reported this mode shifting down in energy by as much as 5 meV.

Although there has been considerable speculation as to the origin and polarization of this feature[8] our results indicate that such a large oscillator strength is possible only for polarization in the surface plane parallel to the chains, so that the excitations can linearly couple to the strong uniaxial surface electron polarizability[9]. An immediate consequence is that the loss feature should be sensitive to the azimuthal orientation of the scattering plane. Preliminary results for angle resolved EELS measurements support this assignment[10].

40

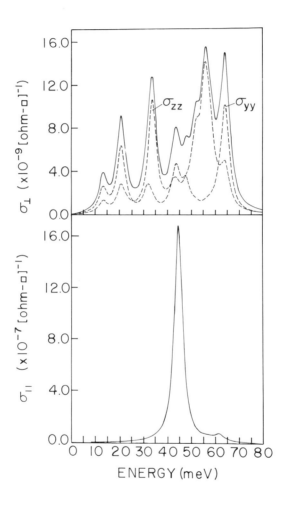

Fig. 2: Surface conductivity vs. energy. (a) polarization
perpendicular to the surface chains, (b) polarization parallel to
the surface chains.

The approach we take to study vibrational excitations of surfaces has proven to be computationally versatile and at the same time carries enough microscopic information to provide reliable results. As it is evident from this paper the study of vibrational properties of semiconductor surfaces can lead to new insights into their structure and provide a microscopic understanding of the coupling of electronic and structural degrees of freedom that occurs in these surfaces.

This work was supported by the Department of Energy under contact number DE-FG02-84ER45118. EJM also acknowledges support of a fellowship from the Alfred P. Sloan Foundation.

References

1.   H. Ibach, Phys. Rev. Lett. $\underline{27}$, 253 (1971).

2.   E.J. Mele, D.C. Allan, O.L. Alerhand, and D.P. DiVincenzo, J. Vac. Si. Tech. $\underline{B3}$, 1068 (1985).

3.   D.C. Allan and E.J. Mele, Phys. Rev. Lett. $\underline{53}$, 826 (1984) and Phys. Rev. $\underline{B31}$, 5565 (1985).

4.   D.J. Chadi, Phys. Rev. B$\underline{29}$, 85 (1984) and references therein.

5.   The calculations reported here were done for a 14-layer slab.

6.   O.L. Alerhand, D.C. Allan and E.J. Mele, Phys. Rev. Lett. $\underline{55}$, xxx (1985).

7.   G.J. Lapeyre (private communication).

8.   E. Evans and D.L. Mills, Phys. Rev. B$\underline{5}$, 4126 (1972).

9.   P. Chiradia, A. Cricenty, S. Selci, G. Chiarotti, Phys. Rev. Lett. $\underline{52}$, 1145 (1984); M.A. Olmstead and N.M. Amer, Phys. Rev. Lett. $\underline{52}$, 1148 (1984).

10. N.J. DiNardo, W.A. Thompson, H.A. Schell-Sorocin and J.E. Demuth (to be published).

COMPUTER SIMULATION OF ELECTRON MICROSCOPE IMAGES
FROM ATOMIC STRUCTURE MODELS

WILLIAM KRAKOW
IBM Thomas J. Watson Research Center, Yorktown Heights, NY 10598

ABSTRACT

It is generally the case that simple direct interpretation of high resolution electron microscope images is not possible due to the phase contrast imaging modes necessary to achieve atomic level spatial resolution. Therefore, an extensive number of computer programs have been developed to perform electron diffraction and image computations. Both single scattering or dynamical scattering processes can be simulated as well as any form of imaging mode currently available on most modern high performance transmission electron microscopes. Since one is interested in imperfections rather than perfect crystal structures, a large number of sampling points in real and reciprocal space are required. Often, large atom position arrays must be sampled requiring large mainframe computer memories and fast CPU's. High quality displays are also required for realistic image representations and even faster computational methods via television rate digital frame store devices. This paper will be centered about a number of materials areas requiring high resolution electron microscopy computer simulation from atomic structure models. These areas include: organometallic molecules, point defects, surface structure and reconstructions, amorphous thin films, quasi-crystals, semiconductor interfaces and grain boundary structure in metals.

INTRODUCTION AND THEORY

Under favorable imaging conditions, the conventional transmission microscope is capable of resolving point resolutions approaching 2Å and line resolutions which are considerably less. A wide range of materials and objects have been studied at or near these resolutions including organometallic molecules containing heavy atoms, amorphous materials, perfect crystalline structures, crystals containing small voids and point defects, and some extended planar defects. While the list is more extensive than indicated here, most of the studies have resorted to computer image modelling to interpret the phase contrast micrographs in all but the most simple and apparent cases. For the appropriate electron optical parameters as well as specimen parameters, it is possible to closely approximate the electron micrograph of any object. This statement is made with considerable caution since the computer simulations require a detailed knowledge of electron scattering formalism and the capability of computing the effect of the electron microscope on the transmitted and scattered waves emanating from the specimen.

Most diffraction computations fall into two distinct categories. The first of these is the weak phase object (WPO) scattering approximation, where single scattering is used and all atoms are considered to lie in a single object plane [1,2]. This is a good approximation for calculations of single-atom images and for small organometallic molecules. This type of approximation can also be extended to crystalline materials where the observation of surfaces is desired; here the atoms lie within one unit cell of the bounding surface. Also the contrast expected from the diffuse elastic scattering of displaced atoms in a crystalline lattice (defects) produced by the direct beam can be regarded as a single scattering event provided one discounts the contributions of the Bragg scattering into the subsequent diffuse scattering and the diffuse-diffuse scattering. Although the WPO approximation may give quantitative agreement with experiment, the diffraction intensities could be in considerable error. It is therefore necessary to use the second category of diffraction computation, multislice dynamical scattering, when accurate evaluations are required. A detailed description of this type of scattering is given elsewhere [3] and will only be mentioned briefly here. Basically, a specimen consists of a number of thin slices each of which is described as a WPO. Then, using Fresnel propagation, the scattering distribution emerging from a slice is convoluted with the scattering distribution of the next slice as shown in Fig. 1, *i.e.* multislice formalism. In other words a recursion relationship exists such that a wave in the n-1 layer propagates to the nth layer as defined by the convolution product:

$$\psi_n = [\psi_{n-1} \cdot P_{n-1}]^*Q_n \tag{1}$$

where ψ_n is the complex number diffracted wave array emerging from the nth layer after being convoluted with the WPO scattering of that layer Q_n. Here P_{n-1} and ψ_{n-1} are the propagator and scattered waves of the n-1 layer. As mentioned in Ref. 3, the multislice computational procedures used are suitable for generalized objects ranging from amorphous structures to perfect crystals. Therefore, the formalism employed here does not rely on periodic continuation used by others to restrict array sizes and hence computational time. Instead a truncated atomic model distribution is contained within the real space field of view of the image to be simulated. The multislice approach includes the curvature of the Ewald sphere and the diffuse elastic scattering where a large number of reciprocal space points are invoked for the complex number arrays of Eq. 1 (256×256).

After the diffraction computation, WPO or multislice, it is then possible to compute the electron microscope image for a variety of illumination conditions (bright-field or dark-field) depending whether the direct beam is included or excluded in the objective aperture. This subject has been discussed in Refs. 1-3 and involves taking the inverse Fourier transform of the scattering distribution ψ_n emerging from the crystal to obtain the image amplitude, $\psi_T(x_i, y_i)$ where:

$$\psi_T(x_i, y_i) = \frac{1}{\lambda}\mathscr{F}^{-1}\left[\psi_n(\overline{\eta})P'\left(\,|\,\overline{\eta}\,|\,\right)\right] \tag{2}$$

Here P' is the pupil function of the microscope which contains the objective aperture function and the phase shift function of the objective lens which includes defocus, spherical and axial astigmatism aberrations. The image intensity is the square modulus of the image amplitude. The relationship between the direct and scattered beams in the microscope relative to the optical axis of the objective lens are depicted in Fig. 2. In this case a dark-field imaging mode is illustrated since the direct beam is beyond the aperture edge.

For multislice dynamical diffraction, computational times based upon Eq. 1 average about six seconds per slice on an IBM 3081 processor. Additionally, the scattering distributions from any unique layer, *i.e.* a WPO, must be computed and can require up to several tens of minutes of CPU time. The scattering distribution is a summation of a direct beam term represented by a delta function and a weak phase object scattering distribution representing the atomic potential distribution of the object:

$$Q_n(\overline{\alpha}) = \delta(\overline{\alpha}) - iF_n(\overline{\alpha})\Delta Z \tag{3}$$

where ΔZ is the slice thickness and F_n is the scattering distribution of the layer, where

$$F_n(\overline{\alpha}) = \sum_{j}^{\text{all atoms}} f_j(\alpha) \exp\left[\frac{2\pi i}{\lambda}(\overline{\alpha} \cdot r_j)\right] \tag{4}$$

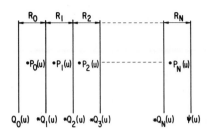

Fig. 1 Wave propagation schematic for multislice dynamical diffraction.

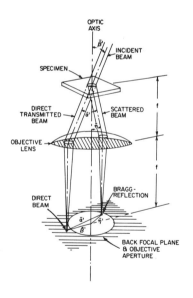

Fig. 2 Ray diagram of the microscope geometry for tilted beam imaging.

Here, f_j is the atomic scattering factor of the jth atom. The potential distribution is related to the scattering by the relationship

$$F(\bar{\alpha}) = \frac{2\pi m_0 e}{h^2} \mathscr{F}[V(\bar{r})] \tag{5}$$

It can be seen that solving Eq. 4 requires an evaluation of all atom positions for each scattering angle, $\bar{\alpha}$, of which there are 256×256 locations which is indeed cumbersome. Alternately, instead of computing the scattering distribution for each atom, a fast Fourier transform (FFT) similar to Eq. 5 can be obtained from a digitized image of the model. The model positions represent the projected potential distribution and need not be defined to as great a positional accuracy as used for the scattering amplitude computation. This is permitted since the computed image need only be accurate to a few tenths of an angstrom which is a fraction of the microscope resolution. The resultant gain in speed using Eq. 5 can be as great as a thousand-fold time savings which makes real time interactive computations possible for the WPO approximation. This fact allows the microscope parameters adjusted to match real conditions rapidly and images visualized on a digital frame store device. In the case of multislice results requiring hundreds of layers, overnight operation is generally required.

COMPUTATIONAL RESULTS AND DISCUSSION

Weak Phase Objects

For weak phase objects, one of the first areas where high resolution electron microscopy yielded structural information about atomic positions was that of the organometallic molecule 2,3,4,5-tetraacetoxymercurithiophene [5]. For real experimental micrographs, it was claimed that not only were the four Hg atoms on the acetate side groups visualized but the sulfur atom on the thiophene ring would be visible. However, WPO computer dark-field imaging experiments on a model consisting of the four Hg atoms as depicted in Fig. 3a and 3b demonstrated that "ghost" images are present [6]. This interference phenomena would preclude identifying the location of the sulfur atom and more importantly points out the difficulty in applying direct interpretation to high resolution microscope images. This situation is much more complex in the case of point defects in crystals which have accompanying strain fields. Figures 3c and 3d show the simulation results of a structurally relaxed [111] split crowdion self interstitial in tungsten [7]. Here the diffraction pattern of Fig. 3c is from the diffuse elastic scattering produced by the atoms surrounding the defect core deviating from their ideal lattice positions (182 atoms considered). As can be seen from the simulated dark-field image of Fig. 3d, the defect is extended by the strain field contribution and has a dimension of $\sim 5\text{Å} \times 10\text{Å}$ which is of much greater extent than the interstitial atom alone ($\sim 3\text{Å}$). Here, this example serves to demonstrate that identification of a wide range of point defects will depend upon the image contrast observed for a number of differing crystallographic views of the strain field [8].

Further image computations for WPO have been applied to imaging a (001) Au film surface [9]. For an unrelaxed surface model based upon a geometric stacking, an image simulation is displayed in Fig. 3e. This example shows the surface lattice periods of 2.87Å, characteristic of incomplete unit cells at the crystal's surface. Also the random granularity observed in most micrographs of thin crystals, which has been considered to be due to amorphous contamination, is in fact due to surface roughness of the crystal. This

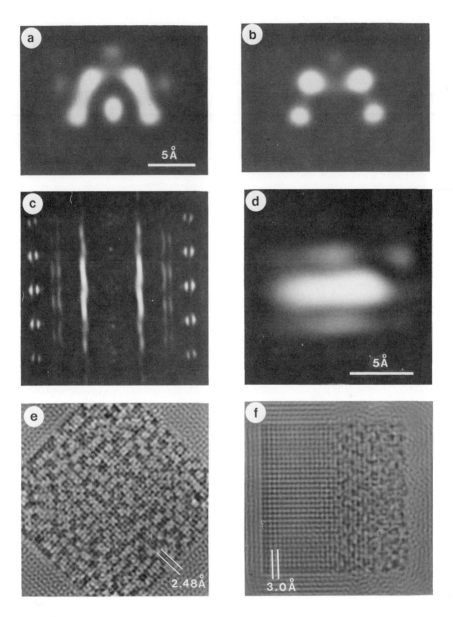

Fig. 3 (a) and (b) Computed dark-field images of the four Hg atoms located on the acetate side groups of the thiophene ring. (c) and (d) Diffuse scattering diffraction pattern and dark-field image of the [111] split crowdion interstitial in tungsten. (e) Bright-field image of (001) Au surface at 100kV using axial illumination. (f) Bright-field image of a faceted amorphous to crystalline transition boundary.

type of analysis indeed shows that surface topography may be more irregular than previously assumed for most surface studies.

Again for the WPO approximation, amorphous-to-crystalline transition boundaries have been studied by computer simulation [10]. Figure 3f is a computer generated example of the type of image expected for a 100kV microscope assuming a simple cubic structure on one side of the interface, a fully amorphous region on the other side and a faceted interfacial boundary. Here the boundary appears to be ill-defined to within perhaps 10 to 15Å of the interface because of the limitations of the microscope. Depending on the microscope aberrations and accelerating voltage this effect can be reduced to an accuracy of one monolayer. However, the question regarding the third dimension along the electron beam path requires that full dynamical scattering computations be performed.

Dynamical Scattering

For dynamical scattering a number of important materials areas have been explored where it was necessary to use relatively thick samples in actual experiments and hence in the subsequent computer analysis. An example of a model and bright-field image computation for columnar point defects in (110) Si is displayed in Figs. 4a and 4b respectively. Here the layer thickness of 3.8Å is the repeat slice distance displayed in the model and repeated approximately 53 times to produce the desired film thickness of ≈200Å. A comparison between computed micrographs and actual experiments confirmed that indeed these columnar defects existed in As^+ ion implanted samples [11,12,13]. There is little evidence of a long range strain in these specimens. Surprisingly, interstitials appear as reduced lattice contrast regions while vacancies appear as more intense whitened regions. Here, in projection, the structure repeats many times; and if the defect concentration is reduced by even 50% occupancy, i.e. returning to perfect crystal, defect visibility is almost totally diminished in the simulation.

It can be speculated that a non-repeating slice structure would indeed pose a formidable problem in attempting to identify atomic positions. An example of an image computation of this type is given in Fig. 4c [14] where an amorphous cylinder containing a relaxed screw dislocation model based upon dense random packing of atoms was employed [15]. The model consisted of ten unique layers 4Å thick where each layer required a large WPO diffraction computation. The important realization here is the projection of all the atoms does not permit identification of individual atoms. Only a faint indication of the dislocation line is present in the image running from the center to the right side of the object. This last case indicates perhaps the least probable chance of deducing atomic structure because of the true three-dimensional nature of the object.

There are classes of objects which pose somewhat less difficulty then amorphous structure such as interfaces where the materials are dissimilar in different specimen regions and hence repeatability of the slices must be evaluated. An example of this type is given in Fig. 4d for a Pd_2Si/Si interfacial structure image. Here the depth repeat distance between (110) Si and the properly oriented Pd_2Si structure is approximately three times the separation of {110} Si lattice planes. Further interesting aspects of this problem relate to varying the terminating plane of the Pd_2Si interface which can have two possible structures either Si rich or Si deficient. The interface plane position which also varies as a function of depth has also been included in the image simulations by introducing an additional translational shift factor in the scattering factor (see Eq. 3). The end result of these imaging experiments establish

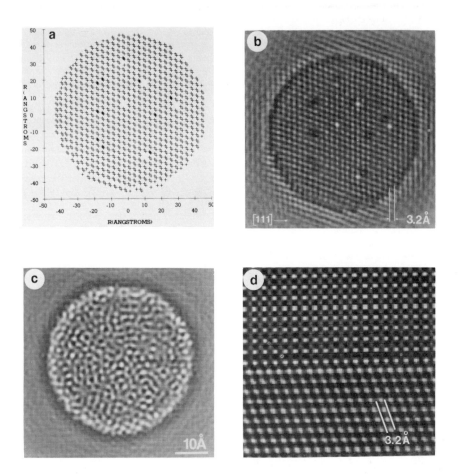

Fig. 4 (a) Model of various vacancy and di-interstitial chain configurations for <110> oriented Si. (b) Computed axial illumination image of (a). (c) Bright-field axial illumination image of an amorphous model. (d) Computed image of a Pd₂Si/Si interface.

that the interface seen edge-on is indeed rough and compositionally varies on an atomic scale [16].

Dynamical diffraction is also important for investigating the surface structure of thin films as well as inclined defects in the film. It is important to realize that surfaces may comprise only a small proportion of the total film thickness and surface roughness may persist on both upper and lower thin film surfaces. It has been found that strong diffraction intensities exist for partially filled surface layers and the degree of roughness can be ascertained from electron microscope images [17]. Two examples of computed surface images of a (111) Au film are shown in Fig. 5a and b where (a) is less rough than (b) and both are for the same microscope conditions. In both cases propagation through the upper surface layers, the bulk lattice and the bottom surface layers has been performed. Here the end result of the analysis is that surface roughness leads to a random or granular image not unlike that of a micrograph of an amorphous thin film. When the atomic site roughness is reduced well defined atomic features appear. If ideally flat surfaces are considered, it is possible to determine why contrast is black or white for image features of the type are displayed in (5a). Essentially one must evaluate the number of excess layers in the stacking sequence ABC..... for the entire film. If the relationship relating phases of different layers $\overline{A} + \overline{B} + \overline{C} = \overline{0}$ is considered one can deduce that a single excess layer gives black atoms, 2 excess layers give white atoms and no excess layers will null contrast, *i.e.* multiples of three layers give bulk lattice diffraction and no surface contrast. The same type of analysis can also be extended to include stacking errors in thin films such as a double positioning boundary, which is a special class of twin structure [17]. Here the planar defect occurs on a {111} plane which is inclined to the film (111) surface at $70.5°$. Because of the stacking error these defects exhibit stronger lattice fringe contrast than surfaces. Two examples of double positioning boundaries are given in Fig. 5c and d where the former case has a planar interface alone, while the latter case has a short (111) twinning plane segment. These examples represent the only cases yet evaluated of inclined defects by multislice dynamical diffraction computations.

Real Time Computations

As mentioned in the previous section, the scattering distribution of the object is the Fourier transform of its projected potential. Using a graphics display to represent atomic potentials and a real time image processing capability to compute images several results of different microstructures are displayed in Fig. 6 along with their model structures. Figures 6a and b are for an Al/Si interface structure where both the Al and Si have a (110) axis along the beam direction and {111} planes parallel to the interface [18]. Two different interfacial configurations were deduced in a matter of a few hours using the digital image processing capability. A second example of interfacial imaging in Fig. 6c and d is for a [110] tilt boundary in Au. Here the atom positions are the centroid positions of white dots in an actual micrograph. Close agreement between the real and computed micrograph demonstrates the viability of this rapid computational method. Turning to a different class of objects where the atomic positions have been determined by a icosohedral tiling procedures [19] (quasi crystal), the result of an image computation is displayed in Fig. 6e and f. Here the computed micrograph has several five and ten fold symmetry image features characteristic of a penrose pattern. It is interesting to note that close agreement was found between this image and real micrographs of Al_6Mn structure. Additional computer experiments have demonstrated that a wide range of image features, from amorphous to crystal like, can be generated which depend critically on microscope parameters. In other words,

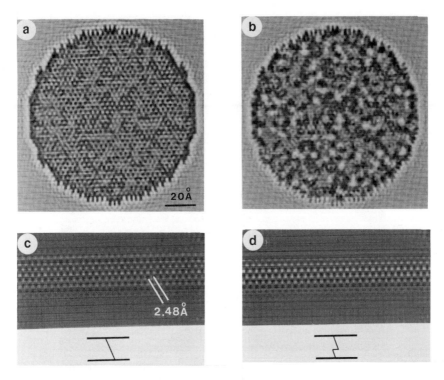

Fig. 5 (a) and (b) Computed images of a (111) Au film with different amounts of surface roughness. (c) Computed image of a double positioning boundary (DPB) and (d) computed image of the DPB with a short (111) twinning plane segment.

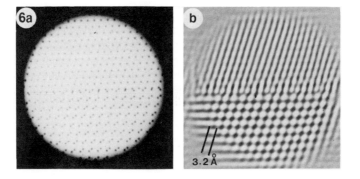

52

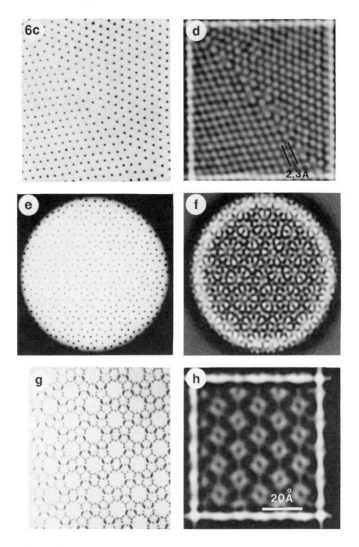

Fig. 6 Models and computed images using real time image processing via a digital television frame store. (a) and (b) Al/Si interface structure. (c) and (d) (110) tilt boundary in Au. (e) and (f) Two dimensional quasi-crystal. (g) and (h) SiO₂ polymorph, silicalite.

great caution must be applied to image interpretation or the correct correlation with atomic structure will not be observed.

Finally, an important area where rapid computer image simulation has great potential is for investigating large unit cell microstructures. An example of a polymorph of SiO_2, silicalite, is displayed in Fig. 6g and h. This zeolite type material has unit cell dimensions of 20.6, 19.8 and 13.36Å and can be described by the space group $Pn2_1a$ as deduced by x-ray structure analysis [20]. The important observation to be made here is that electron microscopy very quickly reveals the pore sizes of the zeolite material and there is close agreement between the simulations and real micrographs.

In all the cases presented in Fig. 6, real time computations are performed in minutes, rather than hours or days. It is interesting to speculate that even faster response times will be possible with direct memory access array processors. Even the extension of this fast technique to dynamical diffraction is possible based upon the proper hardware being available to perform convolution products.

SUMMARY

This paper has shown that a wide variety of electron microscope images of materials structures can be simulated on the computer. Various different levels of simulation of the diffraction processes are possible depending upon the type of information required. The most important aspect of this work is the ability to interpret high resolution electron micrographs and relate the images to structure models.

REFERENCES

1. W. Krakow, A.L.J. Chang and S.L. Sass, Phil. Mag. **35** (1977) 575.
2. W. Krakow, Ultramicrosc. **1** (1976) 203.
3. W. Krakow, IBM J. Res. Devel. **251** (1981) 58.
4. W. Krakow, Ultramicrosc. (1985) in press.
5. F.P. Ottensmeyer, E.E. Schmidt and A.J. Olbrecht, Science **179** (1973) 175.
6. W. Krakow, Ultramicrosc. **1** (1976) 203.
7. R. Benedek and P.S. Ho, J. Phys. F. **3** (1973) 1285.
8. W. Krakow, J. Nucl. Mat. **74** (1978) 314.
9. W. Krakow, Ultramicrosc. **4** (1979) 55.
10. W. Krakow, Ultramicrosc. **5** (1980) 175.
11. T.Y. Tan, H.Föll and W. Krakow, Appl. Phys. Lett. **37** (1981) 1102.
12. W. Krakow, T.Y. Tan and H. Föll, Inst. Phys. Conf. Ser. No. 60 (1981) 23.
13. W. Krakow, T.Y. Tan and H. Föll, Defects in Semiconductors (1981) 185.
14. W. Krakow, Proc. 38th Ann. Electron Microsc. Soc. of Amer., San Francisco, CA (1980) 178.
15. P. Chaudhari and A. Levi, Phys. Rev. Lett. **43** (1979) 1517.
16. W. Krakow, Thin Sol. Films **93** (1982) 109.
17. W. Krakow, Thin Sol. Films **93** (1982) 235.

54

18. F. Legoues, W. Krakow and P. Ho, Mat. Res. Soc. Symp. Proc. **37**, Boston Mass., (1984) 396.

19. A.L. MacKay, Physica **114A** (1982) 609.

20. E.M. Flanigen, J.M. Bennett, R.W. Grose, J.P. Cohen, R.L. Patton, R.M. Kirchner and J.V. Smith, Nature **271** (1978) 512.

DEFECT BEHAVIOUR IN OXIDES

A.N. Cormack*, C.R.A. Catlow**, G.V. Lewis and C. M. Freeman**
Department of Chemistry, University College London, 20 Gordon St., London
WC1H OAJ U.K.

* Present Address: NYS College of Ceramics, Alfred University, Alfred, NY
  14802
** Department of Chemistry, University of Keele, Keele, Staffs, ST5 5BG U.K.

ABSTRACT

     We describe computer-based atomistic simulation studies of defect
behaviour in oxides. In illustrating the value of the technique, contact
with experiment is discussed through comparison with measured energies of
ion migration and other energies. Since the success of these methods is
dependent on the extent to which the interatomic potentials used provide a
realistic model of the material, consideration is given to transferability
of potentials, especially to tenary compounds where the widely adopted
practice of fitting potential parameters to available crystal data is not
feasible. Our principal examples will include ionic conductivity in
gadolinium zirconate, electronic defect properties of $BaTiO_3$ and grossly
nonstoichiometric rutile.

INTRODUCTION

     It is, of course, widely appreciated that defect behaviour controls a
wide range of crytal properties, such as mass transport, for example, and a
considerable amount of effort has gone into elucidating the defect
properties of a wide range of technologically important materials [1]. The
related fact that we appreciate some of the more fundamental factors
governing the energetics of defects is in no small part due to the large
amount of theoretical work, of which computer simulation studies [2,3] have
played a significant part over the last few years. In this brief report we
describe the application of these atomistic modelling methods to behaviour
ranging from electronic defects in $BaTiO_3$, on one hand and to extended

defects in reduced rutile, on the other. At the heart of the technique lies
the validity of the interatomic potentials used to model these materials and
this will receive repeated attention.

CALCULATION OF DEFECT ENERGETICS

Potential Models

     Fairly obviously, in order to evaluate the response of a crystal to the
presence of a defect, one has to have some means of describing the
properties of the material. The two principal ways of doing this at an
atomistic level (i.e. leaving aside continuum theories), by describing the
forces acting on each ion, are through either a prescription for the force
constants or through an interatomic potential model. Whilst force constant
models have been usefully applied in modelling studies, these have been
confined mainly to perfect crystal properties such as calculating phonon
dispersion curves and are not readily applied to defect problems since the
value of the force constant is a function of interatomic separation and
there is no means of evaluating it except at regular crystallographic
distances.
     On the other hand, a potential model description does have this
flexibility, since its derivative is known for all separations and it
remains only to find parameters for the model that allow an adequate
description of the crystal properties. These parameters are mainly derived

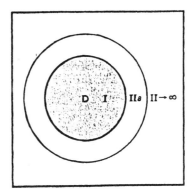

Figure 1.   Schematic representation of division of crystal for defect energy
calculation.

in two ways. Firstly, [16] it is possible to calculate the potential using
either   an   ab   initio   method   based   on   Hartree-Fock   formalism   or   a
semi-empirical method developed by Gordon and Kim which makes use of an
"electron gas" description.   The analytic form is then fitted to the
calculated potential to yield the required parameters. Alternatively, [17]
and this is usually favored when possible, the potential parameters in the
model are found by adjusting them until the model for the material yields
crystal properties in accord with those that have been measured.   This only
works when the experimental data is available, but usually yields more
reliable parameters, since some of the ionocovalent nature of the bonding in
oxide crystals can be accommodated this way.

Defective Crystals

To obtain its energy, the defect (which may be either vacancy,
interstitial, substitutional or, indeed, any combination of these) is
supposed to be at the centre of the crystal which is then divided into two
regions for computational ease [2,3].   In the inner region, close to the
defect and termed Region I, the ions are treated atomistically, the forces
being described in terms of the appropriate interatomic potentials.   This
representation is used because the forces exerted by the defect on the
crystal in this region are strong being both electrostatic and elastic in
nature and considerable distortion of the structure may be expected.
Continuum methods clearly do not provide an adequate means of coping with
these forces.   On the other hand, at large distances from the defect, the
crystal's response is largely electrostatic, being polarized by the
effective charge on the defect; the elastic strain field centred on the
defect has mostly attenuated in this outer Region II, and so a dielectric
continuum treatment is appropriate (providing, of course, that Region I is
chosen sufficiently large).   An interfacial region, IIA, is included to
allow proper consistency between the two regions as indicated in Fig. 1.

Examples

In this section we will describe the application of the techniques to
some contemporary problems in the defect chemistry of oxides.

(i)   Gadolinium Zirconate

Gadolinium   zirconate, $Gd_2Zr_2O_7$,   has   received   detailed   attention recently because it has been found to be a fast oxygen ion conductor [4]. At high temperatures (> 1500°C) it adopts the fluorite structure, although it is anion deficient with respect to the Fm3m in space group stoichiometry. The oxygen vacancies, which, for reasons of charge compensation, comprise the anion deficency have been identified as the mobile species. However, at lower temperatures, a pyrochlore structure is adopted as a result of the ordering of the cations, and the vacancies, into distinct crystallographic sites [4].

Two questions arise: firstly, by which of the possible simple anion to anion sites jumps does migration occur, and secondly, can the ordered vacant sites be considered a source of migrating species (as they are in the fluorite structure) or should they, more properly, be thought of as an integral part of the structure? Calculations [5] have shown that <100> jumps involving 48 (f) ions would have an activation energy of 0.9 eV which is close to the experimental value of $\Delta H = 0.8$ eV for ionic conductivity. The other jumps were considerably higher in energy or could not by themselves form a continuous jump path. An anion Frenkel energy of about 1.9 eV per pair was calculated which suggests, by comparison with experimental data, that the migrating defects (oxygen vacancies) have an extrinsic origin, consistent with the lack of a knee in the Arrhenius plot.

The vacant 8(b) sites were found to be an inherent part of the pyrochlore structure: moving an oxygen from an occupied 48 (f) site to a neighboring (vacant) 8(b) site was calculated to be an unfavourable move. The oxygen relaxed back immediately to the lattice position, thus indicating that no single mechanism exists whereby these vacant positions can contribute to the oxygen migration. The source of vacancies must lie elsewhere, perhaps in incompletely ordered regions of the crystal or in domain boundaries, both of which are known to exist in this material [4].

Barium Titanate

The defect chemistry of this important semi-conductor ceramic has been a controversial area for some time although a concensus is now beginning to appear [6]. Typical of the problems in this material is the donor/acceptor nature of impurities, especially the trivalent ones which may sit on either cation sublattice. A systematic survey [7] of the substitutional energies for a wide range of trivalent dopants yielded complete concordance with the experimental evidence, including the observation that some amphoteric behaviour was to be expected: for example Y^{3+} prefers to self compensate, substituting on both Ba and Ti sites.

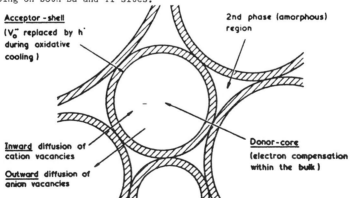

Acceptor - shell
($V_o^{..}$ replaced by $h^.$ during oxidative cooling)

2nd phase (amorphous) region

Inward diffusion of cation vacancies

Outward diffusion of anion vacancies

Donor - core
(electron compensation within the bulk)

Figure 2. Schematic representation of defect chemistry responsible for PTCR effect.

However, the results were affected by the choice of potential; it was found that consistency was necessary and that semi-empirical potentials were less satisfactory than those obtained from structural data.

Calculations of intrinsic defect energies showed that Schottky disorder would dominate and that positive effective charges would be compensated preferentially by Ti vacancies although these are essentially immobile in the lattice. Oxygen vacancies are found to be more mobile than Ba vacancies. These activation energies are relevant to discussions of the PTCR (positive temperature coefficient of resistance) effect as diffusion of donor impurities will be controlled by a Ba vacancy mechanism and diffusion of acceptor impurities by Ti vacancy migration. Hence, acceptor impurities will be confined to regions close to the surface since they will be kinetically prohibited from entering the bulk. (It being assumed that impurities are introduced during processing.)

By calculating the binding energy of electronic defects to substitutional ions (assuming a simple change in the charge state for the electron or hole) ionization energies could be estimated. The prediction of low ionization energies for donors (electron weakly bound) and high ionization energies for acceptors (holes trapped by ~ 0.5 eV) accords with experiment. Taken all together this very extensive study of defect energetics has allowed the development of a model [8] for the PTCR effect in $BaTiO_3$, which, whilst incorporating the relevant features of earlier models, is the most complete explanation so far of this important phenomena (see Figure 2).

Extended Defects in Reduced Rutile

TiO_{2-x} displays an apparently very large range of composition [9], with the nonstoichiometry being accommodated by extended, planar defects known as crystallographic shear (CS) planes, except at compositions very close to stoichiometry, although even then, there is evidence that aggregation of points defects occurs [10]. The CS planes are observed to interact with each other, often over large distances ~100Å, in an attractive manner that leads to the formation of regular arrays of CS planes. These regular arrays have a fixed, if unusual, stoichiometry and thus really constitute a new phase, so the apparent nonstoichiometry range is in fact spanned by a number of discrete stoichiometric line phases, but which have quite similar free energies.

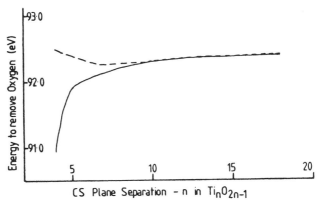

Figure 3. Calculated energy required to remove O^{2-} to infinity as a function of CS plane spacing. The solid line is for $(1\bar{2}1)$ orientation and the dashed line is for $(1\bar{3}2)$.

Two principal questions arise from these observations. Firstly, why are CS planes found in preference to point defects and, secondly what is the nature of the attractive interaction? Answers to these questions have been obtained from computer simulation studies [11].

Arrays of CS planes are treated via a "supercell" approach which makes use of the periodic boundary conditions employed in pefect lattice calculations [12]. The size of the unit cell then reflects the separation between the CS planes. Individual, or isolated CS planes, were modelled using the MIDAS program [13] which is designed to simulate crystal surfaces. The energy required to reduce TiO_2 by removing an O^{2-} ion may be calculated assuming either point defect formation (either oxygen vacancy or cation interstitial) or by extended defect formation (either isolated or arrayed) CS plane formation is found to be preferable and this is basically because considerable structural relaxation occurs at the CS plane thus lowering its energy with respect to the point defect, around which less relaxation is allowed.

Calculation of the energy required to remove oxygen via CS plane formation as a function of separation leads to an attractive interaction energy, the origin of which is due to the interference of elastic strain fields centered on the extruded defects in a similar manner to that identified in ReO_3-structured oxides [14].

We note from the energetics of CS planes with different orientations, ($\{1\bar{2}1\}$ and $\{1\bar{3}2\}$), that the prediction of the formation phases Ti_nO_{2n-1} with CS plane orientation, $\{1\bar{2}1\}$ for 4<n<10, whereas less gross departures from stoichiometry give rise to mixed populations of both orientations, accords well with the experimental structural observations. This is illustrated in Figure 2 which shows the energy required to remove an O^{2-} ion to infinity, as a function of CS plane array spacing. Clearly, for 4<n<10, $(1\bar{2}1)$ arrays will form: these are the well known Magneli phases [15]; for higher n, the energies of both orientations are virtually identical, resulting in intergrowths of these different orientations.

Conclusions

The examples discussed above indicate the range of problems of contemporary importance in defect solid state chemistry that can be addressed using computer simulation techniques. The key to the viability of these methods lies in the choice and validity of the interatomic potentials used in the calculations, but it is clear that adequate representations of these potentials can be found. This is important because not only does it allow one to tackle a wide variety of materials, but also, it enables a deeper understanding of the fundamental nature of interatomic potentials and this symbiosis will encourage greater application of theoretical tools to this technologically vital area.

References

[1] see, for example, "Mass Transport in Solids" eds F. Beniere and C.R.A. Catlow, Plenum Press, NY (1983).
[2] "Computer Simulation of Solids", eds C.R.A. Catlow and W.C. Mackrodt, Lecture Notes in Physics, 116, Springer-Verlag, Berlin (1982).
[3] Special issue of Physica, Vol 131B, (1985): Computer Simulation of Condensed Matter, eds C.R.A. Catlow and W.C. Mackrodt.
[4] M.P. Van Dijk, K.J. DeVries and A.J. Burggraaf (1983), Solid State Ionics, 9/10 913.
[5] M.P. Van Dijk, A.J. Burggraaf, A.N. Cormack and C.R.A. Catlow (1985), Solid State Ionics, 17, 159.
[6] D.M. Smyth (1984), Progress in Solid State Chemistry, 15, 145.
[7] G.V. Lewis and C.R.A. Catlow (1983) Radiation Effects, 73, 307.

[8] G.V. Lewis, C.R.A. Catlow and R.E.W. Castleton, (1985) J. Amer. Ceram. Soc., 68, 555.

[9] L.A. Bursill and B. G. Hyde (1972) Prog. in Solid State Chem. 7, 177.

[10] L.A. Bursill and M. G. Blanchin (1984) J. Solid State Chem. 51, 321.

[11] A.N. Cormack, C.M. Freeman, R.L. Royle and C.R.A. Catlow (1985) in Reactivity of Solids (ed. L.C. Dufour), Elsevier, Amsterdam.

[12] A.N. Cormack in ref [1], Chapter 20.

[13] P.W. Tasker (1979) Philos Mag. A39, 119.

[14] A.N. Cormack, R.M. Jones, P.W. Tasker and C.R.A. Catlow (1982) J. Solid State Chem. 44, 174.

[15] J.S. Anderson (1972) in Surface and Defect Properties of Solids, Vol 1 (Specialist Periodic Reports of the Chemical Society, eds M.W. Roberts and J.M. Thomas), The Chemical Society, London.

[16] R.G. Gordon and Y.S. Kim (1972) J. Chem. Phys. 56, 3122.

[17] G.V. Lewis and C.R.A. Catlow (1985) J. Phys. C: Solid State, 18, 1149.

Simulation of Equilibrium Segregation in Alloys Using the Embedded Atom Method*

STEPHEN M. FOILES
Sandia National Laboratories, Livermore, CA 94550

ABSTRACT

The Embedded Atom Method (EAM) is combined with Monte Carlo simulation techniques to determine the equilibrium segregation at internal defects and surfaces. This approach has been applied in the Ni-Cu alloy system to the calculation of the surface composition profiles and the segregation at an edge dislocation. The surface composition profile of these alloys as a function of distance from the surface is found to vary non-monotonically with the top atomic layer strongly enriched in Cu and the near surface atomic layers enriched in Ni. The compositional variation in the core region of an edge dislocation shows enrichment of Ni in the compressed regions of the partial dislocation core and Cu enrichment in the expanded regions. In addition, the composition changes abruptly at the slip plane of the dislocation.

INTRODUCTION

The Embedded Atom Method (EAM) has been shown in recent work to provide an accurate, *yet computationally simple*, procedure for the calculation of the energetics and structure of metallic systems. This technique was developed by Daw and Baskes [1-2], who showed that it can be used to describe surfaces and defects in pure metals. Other work with pure metals has shown that the EAM also provides a good description of phonons [3] and liquid metal structure [4]. In addition, the EAM has been used to study hydrogen interactions with metals [2,5]. Recently, the EAM has been successfully applied to the description of surface segregation in the Ni-Cu alloy system [6].

In this paper, the study of segregation in the Ni-Cu alloy system will be extended to the equilibrium segregation at internal defects. In the first section of the paper, the EAM and the Monte Carlo simulation techniques will be briefly reviewed. The next section will contain a short review of the results obtained for surface segregation. The third section will describe the equilibrium segregation computed for an edge dislocation in a NiCu(10%) alloy.

Mat. Res. Soc. Symp. Proc. Vol. 63. ©1985 Materials Research Society

COMPUTATIONAL TECHNIQUES

The EAM is an approximate method for calculating the total energy of an arbitrary arrangement of atoms based on the energetics of placing an atom into a background electron density due to the surrounding atoms. The total energy is expressed as

$$E_{tot} = \sum_{i} F_i(\rho_i) + (1/2) \sum_{ij(i \neq j)} \varphi_{ij}(R_{ij}) \qquad (1)$$

where

$$\rho_i = \sum_{j(\neq i)} \rho^a_j(R_{ij}). \qquad (2)$$

In these expressions ρ_i is the electron density at atom i due to the remaining atoms as computed by the superposition of the atomic electron densities, ρ^a_j. $F_i(\rho_i)$ is the energy to embed atom i into the electron density ρ_i and $\varphi_{ij}(R_{ij})$ is a short-range repulsive pair interaction between atoms i and j separated by the distance R_{ij}. The functions F and φ are determined empirically by fitting to the bulk sublimation energies, elastic constants and binary alloy heats of mixing. The details of the fitting procedure are described elsewhere [6]. The advantage of this technique for alloy studies is that for each type of atom the embedding function $F_i(\rho)$ only depends on the local electron density and the chemical species of atom i so that the *same function* may be used in calculations for both pure materials and alloys. The pair interaction term between unlike atoms is also well represented by the geometric mean of the pair interactions for the pure metals.

The use of the superposition of atomic densities to calculate the electron density at each site leads to the computational simplicity of the method. With this approximation, the computation of the electron density, ρ_i, at each atomic site is a simple pair sum as is the evaluation of the pair interaction term in the total energy, (Eq. 1). Once the electron densities are determined, the computation of the total energy just requires a single summation over the atoms to evaluate the embedding energies. Thus, the computational effort involved in the application of the EAM is not significantly different than that required by conventional pair potential models.

The equilibrium compositions are computed here using Monte Carlo simulation techniques. The simulations are performed in a grand canonical ensemble where the chemical potential difference between the two atomic species is held fixed rather than the number of each type of species. The species of an atom can be changed during the simulation. The acceptance of this compositional change is based on the change in the energy relative to the total chemical potential,

$E_{tot} - \Sigma\mu_\alpha N_\alpha$ where N_α is the number of atoms of species α and μ_α is the chemical potential of that species. This procedure produces a rapid convergence of the composition profile since it bypasses the need for diffusion. In addition to the change of species, the simulation allows for the spatial motion of the atoms away from their equilibrium positions. In this way, lattice strain effects on the composition are incorporated naturally in the results. Also, the simulation results incorporate the configurational and vibrational contributions to the entropy. The simulation technique is described in greater detail elsewhere [6].

SURFACE SEGREGATION

In earlier work [6], the segregation of Ni-Cu alloys at low index surfaces was studied in detail. Here we describe the general results and how these results compare with the experimental data. All of the calculations were performed for a

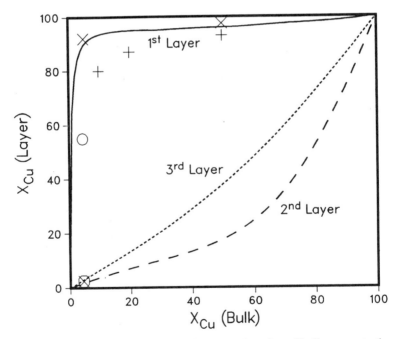

Fig. 1 Calculated Cu concentration by layer as a function of bulk concentration at 800 K for the (111) face of Ni-Cu alloys. The solid curve is the top layer, the long dashed curve is the second layer and the short dashed curve is the third layer. The symbols (x) and (+) are the experimental values for the top layer composition obtained by Auger, reference 8 , and by ion scattering, reference 9, respectively. The symbols (o) and (⊗) are the experimental first and second layer compositions, respectively, determined by field ion microscopy, reference 7.

temperature of 800 K. The calculated compositions for the (100) and (111) surfaces were found to be very similar. The composition of the top three atomic planes are presented as a function of composition for the (111) surface in Figure 1. The top atomic layer is strongly enriched in Cu with the composition of the top layer being >90% for bulk compositions in excess of about 5%. The surprising feature of the results is the composition of the near surface atomic layers. The second layer is enriched in Ni with a composition of about 80% Ni for a bulk composition of 50%. The third layer composition is close to the bulk value but again somewhat enriched in Ni. The oscillations of the composition are unexpected because in the Ni-Cu alloy system the individual elements tends to cluster so that one would have expected Cu to have been enriched near the surface as well as on the top layer. This prediction of composition oscillation is in agreement with the field ion microscopy results of Ng, et al [7]. For the (111) face of a NiCu(5%) alloy, they found strong enrichment of Cu in the surface layer and depletion of Cu from the near surface layer. The results for the (110) face also show a non-monotonic composition variation with depth. There the top two atomic layers are enriched in Cu while the third through fifth atomic layers are somewhat enriched in Ni.

The top layer compositions are compared in Figure 1 with various experimental results. The Auger results of Webber et al.[8] are for a temperature of 850-920K. They studied various low index faces and reported that their results did not depend on the orientation. The ion scattering results of Brongersma et al.[9] are for polycrystalline samples at a temperature of 780K, and the field ion microscopy results of Ng et al.[7] are for (111) surfaces at 820K. The agreement between theory and experiment is quite encouraging.

SEGREGATION AT A DISLOCATION

It is well known that low angle grain boundaries can be decomposed into an array of dislocations. Therefore, the segregation to dislocations is being investigated in order to understand the segregation at low angle grain boundaries. Here the equilibrium composition and structure of the $(a/2)[1\bar{1}0]$ edge dislocation in a Ni-Cu alloy at 800 K are computed. The chemical potentials for this calculation were chosen to correspond to a bulk composition of NiCu(10%). The lattice constant used to determine the periodicities of the calculational cell was determined from constant pressure simulations of the bulk solid solution at the same temperature and chemical potential.

The dislocation line lies along the $[\bar{1}\bar{1}2]$ direction. In an fcc crystal, the dislocation separates into two partial dislocations. The partials are separated along the $[1\bar{1}0]$ direction. For these calculations, periodic boundary conditions are applied along the direction of the dislocation line, $[\bar{1}\bar{1}2]$, with a periodicity of about 13.3Å, and along the direction between the partials, $[1\bar{1}0]$, with a periodicity of about 62.5Å. Free surfaces are used in the third direction, $[111]$. Special treatment was needed for the free surfaces. As shown above, the surface will strongly segregate. However, it was not feasible to make the slab thick enough so that the segregation from the surfaces does not overlap with the region around the dislocation core. Therefore, the atoms in the first three planes of each face of the slab were allowed to move spatially, but are held at the NiCu(10%) bulk composition. The distance in the [111] direction between the fixed composition regions is about 50Å. This simulation involved over 4700 atoms.

The initial atomic configuration of the slab was a fcc lattice with the two additional [1̄10] half planes inserted in the center. Due to the presence of the edge dislocation, the number of [1̄10] planes at the two edges of the slab are different. Thus the *ideal* periodicity in the [1̄10] direction is different for the two edges of the slab. The actual periodicity used is the average of that appropriate for the two edges. Because of this the density is somewhat different on the two edges and so the composition away from the dislocation core on the two sides of the slab is different. During the equilibration period of the simulation, the dislocation split into partials which moved away from each other until they reached the equilibrium separation of 25Å.

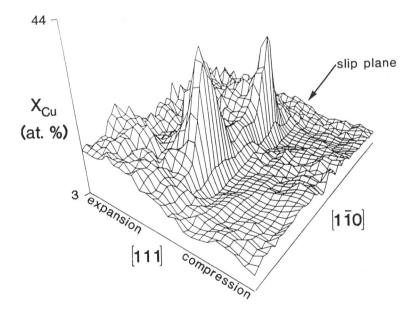

Fig. 2 Calculated Cu concentration of a (a/2) [1̄10] edge dislocation in NiCu(5%) at T = 800 K as a function of position perpendicular to the dislocation core. The length of the [111] axis is 50Å and the length of the [1̄10] axis is 62.5Å. Note that the Cu enrichment occurs predominantly at the partial dislocation cores on the expansive side of the slip plane.

The average composition as a function of the position in the plane perpendicular to the dislocation core is presented in Figure 2. The most dramatic and important feature of these results are the two peaks in the composition. The peaks are located at the centers of the two partial dislocations and have compositions of around 40% Cu. This is a significant enhancement over the bulk Cu content of around 10%. The cliff on the front side of the peaks (as shown in the figure) is located at the slip plane of the dislocation. On the other side of the slip plane, there is a small depletion of Cu. The Cu enhancement occurs on the expanded side of the slip plane and the depletion on the compressed side. This corresponds to the simple picture that the larger lattice constant element, Cu, will segregate to the expanded region of the core. Note that the details of the composition profile away from the cores is not reliable since the boundary conditions do not reflect the correct long-range strain fields. However, the qualitative conclusions about the segregation in the core region of the dislocation should not be affected by the choice of boundary conditions.

SUMMARY

The EAM has been used to determine equilibrium segregation at both the surface and at a dislocation core in Ni-Cu alloys. The agreement between the surface calculations and experimental information on surface segregation is good. Further work on the segregation at internal defects and interfaces of this alloy is in progress. In addition, an empirical parametrization of the embedding functions and pair interactions for the six fcc metals Cu, Ag, Au, Ni, Pd, and Pt as well as the alloys of these elements has been completed [10].

REFERENCES

* Work supported by the U. S. Department of Energy, Office of Basic Energy Sciences.
1. M. S. Daw and M. I. Baskes, Phys. Rev. Lett. 50 1285 (1983).
2. M. S. Daw and M. I. Baskes, Phys. Rev. B29, 6443 (1984).
3. M. S. Daw and R. L. Hatcher, Solid State Comm., in press.
4. S. M. Foiles, Phys. Rev. B32, 3409 (1985).
5. S. M. Foiles and M. S. Daw, J. Vac. Sci. Technol. A3, 1565 (1985).
6. S. M. Foiles, Phys Rev. B32, 7685 (1985).
7. Y. S. Ng, T. T. Tsong, and S. B. McLane, Jr., Phys. Rev. Lett. 42,588 (1979).
8. P. R. Webber, C. E. Rojas, P. J. Dobson, and D. Chadwick, Surf. Sci. 105, 20 (1981).
9. H. H. Brongersma, M. J. Sparnay, and T. M. Buck, Surf Sci. 71, 657 (1978).
10. S. M. Foiles, M. I. Baskes, and M. S. Daw, submitted to Phys. Rev. B.

THE SOLID-LIQUID INTERFACE: THEORY AND COMPUTER SIMULATIONS

B.B. LAIRD AND A.D.J. HAYMET*
Department of Chemistry, University of California
Berkeley, California 94720 U.S.A.

ABSTRACT

We present the results of computer simulations of body centered cubic (bcc)/melt interfaces, with particular emphasis on the "width" of the interface. Both static and dynamic properties of single crystal/liquid interfaces are examined. The implications for crystal growth near equilibrium are discussed. The results of these computer "experiments" are compared with an extended density functional theory of the solid-liquid interface.

INTRODUCTION

The structure and dynamics of an interface between a crystal and its melt are of paramount importance in studies of crystal growth near equilibrium. Such an interface lies between two condensed phases making direct experimental study difficult [1]. Laboratory estimates of the surface tension for a limited number of systems have been obtained both directly and indirectly [2] but experimental data concerning the microscopic structure of the interfacial region is lacking. This experimental difficulty enhances the role of computer simulation in the development of a suitable theory.

Some previous simulations treat the interface as a liquid up against a hard wall or rigid solid face. These include hard sphere models [3] as well as molecular dynamics simulations [4]. The perturbation theory of Abraham and Singh has been used to describe this type of interface [5]. This approach causes the interfacial width to be consistently underestimated because it does not take into account the participation of the crystal.

Several recent simulations have addressed the situation in which both the liquid and solid participate in forming the interfacial region [6-10]. All of these involve fcc crystal faces and all but one [7] involves a Lennard-Jones interaction potential. The density profiles and diffusion constant profiles through the interface, both important in the development of theories of crystal growth dynamics [11], have been calculated for several fcc faces. These profiles show that the interface is quite diffuse with the interfacial region extending over 7 - 10 crystal layers perpendicular to the interface. Estimation of the surface tension has been attempted but has met with limited success to date.

A microscopic theory of diffuse interfaces based on the density functional theory formalism was developed by Haymet and Oxtoby [2]. At present, calculations of interfacial properties have been carried out only for crystals of bcc symmetry, so direct comparison to the simulation results has not been possible. One of the goals of the present work is to simulate bcc solid-liquid interfaces in order to make this comparison.

The current simulation uses the technique of (constant volume) molecular dynamics which has been described elsewhere [12].

Mat. Res. Soc. Symp. Proc. Vol. 63. ©1985 Materials Research Society

SIMULATION RESULTS

The system under study consisted of 2160 particles interacting via a pairwise additive purely repulsive potential

$$v(r) = \varepsilon \left(\frac{\sigma}{r}\right)^6 \qquad (1)$$

This particular form for the potential was chosen for two reasons: (a) the inverse power form of the potential results in scaling relations from which the thermodynamic properties of any point on the solid-liquid coexistence line can be calculated from data at a single point on that line [13], (b) the sixth power is apparently the highest power inverse power potential which freezes into a bcc solid [14].

The preparation of the interface is irrelevant for the results presented below, but may be of interest. The interface was built from five blocks of 432 particles each. Three of the blocks were set up in a bcc solid configuration with the z axis perpendicular to a (100) crystal face. The solid had a density $\rho_s\sigma^3 = 0.7$ with an average temperature of $kT/\varepsilon = 0.1$. The two remaining blocks consisted of liquid equilibrated at the same average temperature but with density, $\rho_L\sigma^3 = 0.687$. The values of the densities and temperature were chosen so that system lies on the phase coexistence line as estimated by Monte Carlo simulations [14]. The blocks were laid end to end in the z direction with the three solid blocks in the middle and a liquid block on each end. Periodic boundary conditions were then applied in all three coordinate directions.

If the simulation were turned on fully at this point the interface would not be stable because the liquid has not been equilibrated next to a solid block. This results in high energy interactions at the interface. In order to create a stable interface the following procedure was adopted. First, the solid particles were held fixed while the liquid particles were allowed to evolve for 5400 time steps of length τ, where

$$\tau = 0.01 \ (m\sigma^2/\varepsilon)^{1/2}.$$

During this phase the liquid temperature was periodically rescaled to the coexistence temperature. The solid atoms were then given their original velocities and allowed to move. The system was equilibrated for 1000 time steps. The system at this point had a flat temperature profile through the interface - a condition of phase coexistence.

Once a stable interface was created by the above procedure the simulation was run for 10,000 more time steps to collect averages.

The density profile for the (100) interface was calculated by dividing the z direction into 1300 bins, counting the number of particles in each bin, and dividing by the volume of the bin. Since the simulation involved two interfaces, one on each side of the solid, a center of symmetry was found so that the profile could be folded over to give an average of the two interfaces. The results of this procedure can be seen in Figure 1.

The variation of the diffusion constant through the interface was calculated by dividing the z direction into 30 bins and calculating the average mean squared displacement as a function of time for each bin. The diffusion constant for a given bin is then calculated from the relation

$$D = \lim_{t \to \infty} \ (6t)^{-1} \ \langle [r(t) - r(o)]^2 \rangle \qquad (2)$$

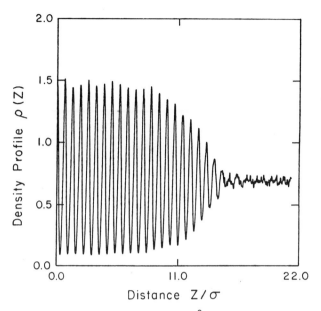

Figure 1. The equilibrium density profile $\sigma^3\rho(z)$, averaged in the perpendicular directions, of the melt/bcc 100 crystal face, as a
function of distance z/σ.

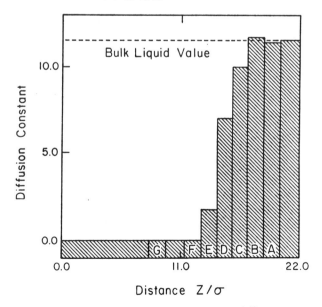

Figure 2. The measured diffusion constant $((\epsilon\sigma^2/m)^{1/2} \times 10^3)$ in
subregions of the interface. Certain subregions are lettered
for reference in Figure 3.

The measured diffusion profile as a function of distance from the solid center is shown in Figure 2. The limiting liquid value of D was obtained in a separate 432 particle simulation of the bulk liquid.

Figure 3 shows the mean squared displacement curves used to calculated the diffusion profile. These curves were calculated using 50 time origins separated by 20 time steps of length τ. Least squares linear regression was used to determine the slope.

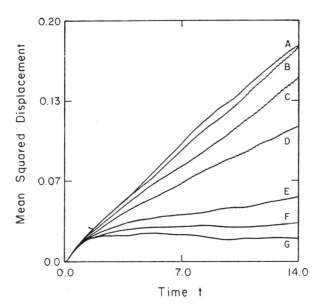

Figure 3. The mean squared displacement (divided by 6) of particles in the interface shown in Figure 1. The letter on each curve refers to a region defined in Figure 2.

COMPARISON TO THEORY

The density functional theory of the interface [2] begins by expanding the local density of the bulk solid in terms of its Fourier components,

$$\rho(\underline{r}) = \rho_L[1 + \eta + \sum_n \mu_n e^{i\underline{k}\cdot\underline{r}}] \tag{3}$$

where ρ_L is the bulk liquid density, η is the fractional density change on freezing, and the $\underline{k}_n$'s are the reciprocal lattice vectors (RLV's) of the solid. The order parameters η and $\{\mu_n\}$ are zero in the liquid phase but assume specific nonzero values in the solid. To account for the transition from solid to liquid as one goes through a planar interface, equation 3 is modified to include spatial variation of the order parameters

$$\rho(\underline{r}) = \rho_L[1 + \eta(z) + \sum_n \mu_n(z) e^{i\underline{k}_n\cdot\underline{r}}] \tag{4}$$

where z is the coordinate direction perpendicular to the interface.

Haymet and Oxtoby then introduce a functional of the order parameters $\gamma[n(z), \{\mu_n(z)\}]$, which is minimized to obtain the equilibrium interfacial density profile. It can be shown that at this minimum value the quantity γ is exactly the surface free energy. At the present level of theory the functional γ is obtained in the square gradient approximation. This functional depends on certain structure coefficients of the bulk liquid related to the liquid structure factor $S(k)$, which in the simplest theory is approximated by unity beyond the magnitude of the nearest neighbor RLV's of the solid.

At the minimum value, the functional becomes

$$\gamma/kT = \int d\underline{z} \left\{ w[\mu_i(z)] + \frac{1}{4} \rho_L \sum_{n=0} c''(k_n)(\hat{\underline{k}}_n \cdot \hat{\underline{z}})^2 \left[\frac{\partial \mu_n(z)}{\partial z}\right]^2 \right\} \qquad (5)$$

where $c(k) = 1 - S^{-1}(k)$ is the Fourier transform of the direct correlation function of the bulk liquid, $c''(k)$ is its second derivative, and we define $\mu_0 = n$, $\underline{k}_0 \equiv 0$ and $(\underline{k}_0 \cdot \underline{z}) = 1$. The grand potential density w is given by [2]

$$w = [c(0) - 1]n(z) + \frac{1}{2} \sum_{n=0} c(k_n)\mu_n^2(z) \qquad (6)$$

The profiles $\{\mu_n(z)\}$ are found by minimizing γ subject to the boundary conditions

$$\mu_n(-\infty) = 0 \ (\text{LIQUID}), \qquad \mu_n(+\infty) = \mu_n^s \ (\text{SOLID}) \qquad (7)$$

Further mathematical details may be found in reference [2].

An estimate of the interfacial width is the 10 - 90 width, ℓ_{10-90}, defined to be the distance between points z_1 and z_2, where $n(z_1) = 0.1n_{SOLID}$ and $n(z_2) = 0.9n_{SOLID}$. The simplest level of density functional theory predicts

$$\ell_{10-90} \approx 4 \left[-c''(k_1)\right]^{1/2} \qquad (8)$$

where $c''(k_1)$ is related to the curvature of the liquid structure factor at the magnitude of the nearest neighbor RLV's of the solid.

For the inverse sixth power bulk liquid, $c''(k_1)$ is estimated to be $-1.17 \ \sigma^2$. This leads to a predicted "density" width of approximately 6 lattice layers. Splitting of the $n(z)$ and $\mu_1(z_1)$ profile, described in Reference 2, leads to an even broader total interfacial profile. At this level the theory appears to be consistent with the simulations. A more detailed comparison is underway.

References

(*) U.S. Presidential Young Investigator 1985 - 1990

1. D.P. Woodruff, The Solid-Liquid Interface (Cambridge University, London, 1973).

2. A.D.J. Haymet and D.W. Oxtoby, J. Chem. Phys. 74, 2559 (1981); 76 6262 (1982).

72

3.   A. Bonissent and B. Mutaftschiev, Phil. Mag. **35**, 65 (1977).

4.   L.F. Rull and S. Toxvaerd, J. Chem. Phys. **78**, 3273 (1983).

5.   A. Bonissent and F.F. Abraham, J. Chem. Phys. **74**, 1306 (1981).

6.   A.J.C. Ladd and L.V. Woodcock, J. Phys. C **11**, 3565 (1978).

7.   J.N. Cape and L.V. Woodcock, J. Chem. Phys. **73**, 2420 (1980).

8.   J.Q. Broughton, A. Bonissent and F.F. Abraham, J. Chem. Phys. **74**, 4029 (1981).

9.   J.Q. Broughton and G.H. Gilmer, J. Chem. Phys. **79**, 5095´ (1983).

10. G. Bushnell-Wye, J.L. Finney and A. Bonissent, Phil. Mag. A **44**, 1053 (1981).

11. T. Munakata, J. Phys. Soc. Japan **43** 1723 (1977); **45**, 749 (1978); B. Bagchi, preprint.

12. A. Rahman, Phys. Rev. **136**, A405 (1964).

13. J. Hansen and I. McDonald, The Theory of Simple Liquids (Academic Press, New York, 1976).

14. W.G. Hoover, M. Ross, K.W. Johnson, D. Henderson, J.A. Barker and B.C. Brown, J. Chem. Phys. **52**, 4931 (1970).

FROM HAMILTONIANS TO PHASE DIAGRAMS

JÜRGEN HAFNER
Institut für Theoretische Physik, Technische Universität Wien,
Karlsplatz 13, A 1040 Wien, Austria

ABSTRACT

   The ab-initio calculation of phase diagrams for metals and alloys using electronic and thermodynamic perturbation theories is discussed.

INTRODUCTION

   One of the central objectives of condensed matter theory is the calculation of the structure and of the thermodynamic properties of materials as a function of the external thermodynamic variables such as temperature, pressure, composition ... , starting from an electronic theory of the chemical bond. The adiabatic theorem teaches us how we have to proceed: The dynamics of the ionic motions can be considered to evolve on a much slower time scale than the electronic excitations. Therefore we can assume that the electrons follow the ions adiabatically, always in their ground state for every configuration of the ions. Hence our problem divides neatly into two steps: (1) The calculation of the electronic ground state properties, first of all of the total energy as a function of the ionic coordinates, but also of the forces acting between the ions. (2) The calculation of the equilibrium structure and of the thermodynamic properties (such as free enthalpy, entropy, compressibility), starting from these interatomic forces and using the methods of solid-state and liquid-state statistical mechanics.

   Recent progress has been most astounding in the first area. Since the mid-sixties, there has been a qualitative change in our ability to predict structures, cohesive energies and even dynamical properties of solids by purely theoretical methods, without any information from experiment. This development has been pioneered by perturbative methods, based on local-density-, pseudopotential-, and linear response theories [1]. With the advent of the linearized band-structure methods and the increase in computer performance, non-perturbative methods began to supercede the older perturbation theories towards the end of the seventies. [2] Today, the perturbation calculations of cohesive energies, structures, phonons, and structural phase transitions of simple metals have been repeated by the non-perturbative methods with greater accuracy , and - what is more important - the calculations have been extended to other materials to which pseudopotential and linear response theories do not apply. However, perturbation theoretical methods continue to be of importance (1) whenever an element of disorder (substitutional or topological) appears and (2) when an explicit expression for the interatomic forces is required - i.e. they are necessary for an extension of the theoretical considerations to substitutionally disordered crystalline alloys at finite temperatures and to the liquid state.

   This brings us to the second part of the problem, the calculation of the thermodynamic properties at finite temperatures. For a crystalline solid, the vibrational contributions to the thermodynamic functions may be calculated using the standard methods of harmonic and anharmonic lattice dynamics [3], the configurational contribution to the entropy can be calculated e.g. using the cluster variation method [4]. For a liquid, the central problem is the calculation of the liquid structure - i.e. of the pair correlation function and of the static structure factor (which are sufficient to express the thermodynamic functions). This might be achieved by (1) computer simulation, (2) by solving one of the integral equations of the statistical-mechanical theory of liquids (hypernetted chain, Percus-Yevick, mean spherical approximation ,...), or (3) by using thermodynamic perturbation theory [5]. Here again thermodynamic perturbation theory continues to be a very attractive - and very often even the only realistic alternative: even today, the computer capacity needed to perform computer simulations with the accuracy and with the speed necessary for phase diagram calculations exceeds that of the existing facilities; integral equations that are suited to handle the rather complex interatomic forces characteristic of metals and alloys are presently not yet available. This is precisely where thermodynamic perturbation theory steps in: the interatomic potentials are separated into a steeply repulsive short-range and an oscillatory long-range part , each part being treated in a convenient perturbation approximation [6,7].

It is the aim of this paper to demonstrate that a judicious application of electronic and thermodynamic perturbation theories renders the ab-initio calculation of the (p,T)-phase diagram of a pure metal or the (c,T)-phase diagram of a binary alloy possible. However, if the theoretical phase diagram calculations would merely serve to reproduce an information already known from experiment, there would be a real danger that the subject degenerates into a pointless amassing of numbers of the kind once qualified by Rutherford as 'stamp-collecting'. The special merit of computational physics in this area is that it is particularly apt, by analyzing trends and by exploring states not accessible to experiment, to deepen and to unify our understanding of the properties of condensed matter. The analysis of the structures of the sp-bonded elements in their crystalline and in their liquid states [8,9] may serve as an example for the way in which computation facilitates the analysis of trends: if one narrows the 92 elements down to this group there are not very many, dotted in a multi-dimensional parameter space (valence, atomic volume, ionic pseudopotential ...) so that clear trends are difficult to grasp. In the computer studies all the parameters - even the valence - may be varied continuously, and only this allowed the main factors to be identified. A good example for the second point is the investigation of supercooled melts at rates of undercooling never attainable in experiment - this has proved to be invaluable in exploring the thermodynamic and electronic criteria for the glass-forming ability of alloys and the relation between liquid and amorphous structures [10].

METHODS

There is certainly no room here for an exhausting review of electronic and thermodynamic perturbation theories - this has been done elsewhere [1,5] and should not be repeated here. However, there are a few non-trivial points which must be emphasized.

The Choice of a Pseudopotential — Optimization versus Transferability

It is well known that the freedom in defining a pseudopotential can be exploited in different ways. One can either optimize the convergence of the perturbation series for the ground-state energy[11] or aim for a better transferability of the ionic pseudopotential from an atomic reference state to a metal or alloy[12-14]. However, it turns out that transferability is not really a useful property within the framework of a perturbation calculation: the pseudopotential is in reality a collective rather then an atomic property, it describes the scattering of electrons by an ionic core placed in a given surrounding effective medium. If this effective medium changes (e.g. by applying an external pressure or by alloying), the pseudopotential changes too [15]. This means that for a fixed pseudo-potential, these changes have to be recovered by adding up higher-order contributions. Within the framework of a self-consistent band-structure calculation, this is no problem and the preference of most band-theorists for the transferable, 'norm-conserving' pseudopotentials is perfectly legitimate. For a perturbation calculation of the total energy on the other hand, it is clearly preferable to optimize the convergence of the perturbation series - even at the prize of constructing a pseudopotential which in addition to its non-locality is also density- and concentration-dependent. For the details of the construction of such optimized pseudopotentials using an orthogonalized-plane-waves construction, see references [15,16].

Thermodynamic Perturbation Theory : Hard − or Soft − core Reference Potentials

Thermodynamic perturbation theory starts from a decomposition of the effective interatomic potential $\Phi(R)$ into a purely repulsive part $\Phi_0(R)$ and an oscillatory long-range part $\Phi_1(R)$[5,6]. In the first step $\Phi_0(R)$ is approximated by a reference potential $\Phi_{ref}(R)$, the following steps correct for the difference $[\Phi_0(R) - \Phi_{ref}(R)]$ and for the effect of $\Phi_1(R)$. Recently, there has been a lot of discussion concerning the optimal choice of a re reference system - the proposed candidates being a hard-sphere[17] or a one-component-plasma[18] reference system (as well as some other systems of intermediate character - charged hard spheres etc.[19], the choice being dictated by the availability of analytical solutions for the structure factor and for the thermodynamic functions). If a hard-core reference is adopted, $\Phi_0(R)$ is conveniently chosen to represent the short-range repulsive part of the interparticle potential up to its first minimum; $\Phi_1(R)$ representing the long-range oscillatory tail (with amplitudes of the order of millirydbergs). If a one-component-plasma (OCP) reference is adopted, the proper separation is to set $\Phi_0(R)$ equal to the direct Coulom repulsion between the ions, while $\Phi_1(R)$ now represents their electron-mediated attraction (which is of the order of rydbergs at the nearest-neighbour distance). At the level of a thermodynamic variational calculation, an OCP reference seems to yield a lower variational upper bound to the exact free energy for the very 'soft' heavy alkali metals, a hard-sphere reference being preferable in all other cases[20]. Corrections to the HS reference potentials are easily (and self-consistently) introduced via the Weeks-Chandler-Andersen 'blip-function' expansion and the optimized random phase approximation (ORPA) [6,21]. In an OCP-framework this would be a much more demanding task[22,23] - note also that now the perturbation Φ_1 is nearly a factor of 10^3 times stronger than in the HS-case. Hence at present the preference must be clearly for the HS-based perturbation theories.

The generalization to binary alloys is straightforward in principle, but cumbersome in practice[25]. This is related to the fact that appropriate reference systems are difficult to find even for moderately non-ideal solutions. Only first steps have been made into that direction. These include a variational approach based on the restricted primitive Yukawa model (hard spheres of equal diameter, but opposite charges and Yukawa interactions, respecting over-all charge neutrality), known from the theory of liquid ionic melts. Combined with pseudopotential-derived interatomic forces, this yields the first microscopic theory of chemical short-range order in liquid alloys.[24]. The variational approach has the particular merit that it requires only an expression for the total energy as a function of the structural parameters (those specifying the reference system) and not an explicit knowledge of the interatomic forces. This allows for an extension of the theory to transition metal alloys where microscopic expressions for the interatomic potentials are not available, but where the total energy is easily calculated in a tight-binding approximation.[26].

The variational methods can also be applied to the crystalline state, where they are equivalent to a self-consistent phonon theory with a simplified reference spectrum.[27] This ability to treat the thermodynamic properties of the solid and of the liquid states on the same level is indeed important for a correct prediction of the solid-liquid coexistence curve.

RESULTS

In the following we discuss a few calculations (out of many others), subjectively chosen to exemplify best the ideas and the capabilities of the techniques which have been described.

The Structures of the sp − Bonded Elements in the Crystalline and in the Liquid States

It is well known that the stable crystal structures of the sp-bonded elements follow characteristic trends in the Periodic Table[1]: (1) A change from close-packed metallic structures in groups I to III to more open covalent structures in groups IV and V with decreasing coordination numbers. (2) An increasing distortion of the lattice in group IIB as the atomic number increases (Zn,Cd), culminating in a unique structure for Hg. (3) A similar phenomenon in group III with a unique structure for Ga, but then a return to a less distorted structure in In and a close-packed structure in Tl. (4) The tendency to more metallic structures and higher coordination numbers in groups IV and V, most pronounced in the series Si-Ge-Sn-Pb. Recent diffraction experiments[28] have confirmed that - as was suggested long ago[29] - these trends persist in the liquid state where the distorted crystalline structures correspond to a distortion of the first few peaks of the static structure factor $S(q)$ and/or the pair correlation function $g(R)$ from a hard-sphere like form.

Hafner and Heine[8] have shown that these structural trends result from a systematic variation of the interatomic potentials with electron density and pseudopotential. The position of the main attractive minimum in $\Phi(R)$ is determined by the radius of the ionic core (if we model the pseudopotential by an empty-core model) and not by the Friedel oscillations, the strength of the minimum is related to the expansion of the repulsive core (and hence via the screening length to the electron density) and to the strenght of the pseudopotential at $q = 2k_F$ setting the amplitude of the oscillations. At low electron densities and large core radii one finds interatomic potentials with a deep minimum at the nearest neighbour distance D_{cp} for close packing, resulting in stable close-packed (regular in the crystal, random in the liquid) structures for the A group elements and Be, Mg, and Al. As the electron density increases and/or the core radius decreases, the minimum is shifted relative to D_{cp} and flattened - this yields distorted structures for Zn, Cd, Hg, Ga, In, Si, Ge, and Sn at first, but then a return to close-packed structures for Tl and Pb as the last trace of an oscillation in $\Phi(R)$ near D_{cp} has disappeared. This is illustrated in Fig.1a where we show the pair potentials of the B group elements. Here the investigation of the liquid structures is particularly convincing - we can calculate the liquid structure from the interatomic force law [9] (we do it using the optimized random phase approximation), whereas in the crystalline state we are restricted to a comparison of the total energies of a finite, arbitrarily chosen set of structures. Fig.1b shows the variation of the pair potentials, the pair correlation functions, and the static structure factors of the tetravalent elements with the ratio (R_c/R_s). We find that the return to hard-sphere like close packed structure in the series Si-Ge-Sn-Pb is indeed triggered by the simultaneous expansion of the repulsive core and the reduction of the amplitudes of the Friedel oscillations.

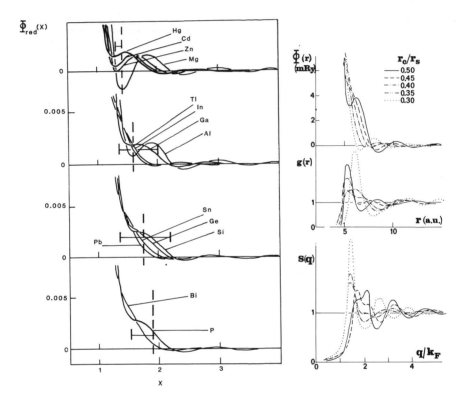

Fig.(1): (a) The interatomic pair potentials $\Phi(R)$ of the B-group elements. The numbers in parentheses indicate the ratio (R_c/R_s) of the core radius R_c to the electron density parameter R_s, the horizontal arrows show the maximum deviation of the nearest neighbour positions from the close-packing distance occuring in each group. Distances are expressed in terms of a scaled variable $x = 2\pi R/2k_F$ - this has the merit of placing the maxima of the ideal Friedel oscillations at integer values of x. (b) Variation of the pair potential $\Phi(R)$, the pair correlation function $g(R)$, and the static structure factor $S(q)$ for valence $Z = 4$ with different values of the ratio (R_c/R_s), demonstrating the return from open structures with low coordination numbers ($N_c = 6.5 - 6.7$ for Si and Ge) to close- packed structures with high coordination number in the series Si-Ge- Sn-Pb ($N_c = 11$ for Pb). After Refs. 8 and 9.

Concentration Dependent Changes in the Chemical and
Topological Short Range Order in Binary Alloys

Over the last decade the outstanding structural, thermodynamic, electrical, and magnetic properties of liquid binary alloys with a strongly non-ideal mixing behaviour, and the interrelation between chemical short-range order in the liquid state with the formation of stable ordered intermetallic compounds has been widely investigated [30]. For simple metals we have been able to show [16,24] that the optimized ab-initio pseudopotentials reflect the competing charge transfer mechanism which, according to Pauling[31], dominate the chemical bond in binary metallic systems. First, electrons are transferred in the direction expected from the electronegativity difference (within the pseudopotential

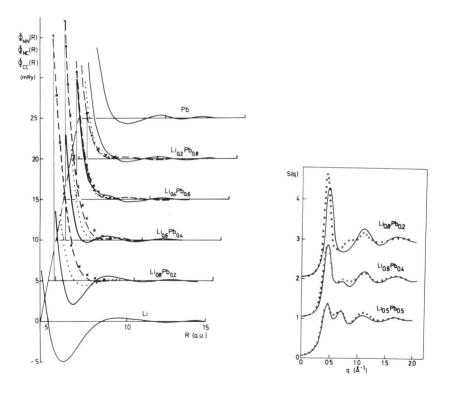

Fig.2: (a) The interatomic potentials $\Phi_{NN}(r)$ (full lines - Φ_{NN} couples to the fluctuations in the mean number density), $\Phi_{CC}(R)$ (dashed line - Φ_{CC} couples to the concentration fluctuations) and $\Phi_{NC}(R)$ (dotted lines) for a series of Li-Pb alloys. (b) The composite (neutron-weighted) static structure factor $S(q)$ for the series of Li-Pb alloys. Note that Li_4Pb is a 'zero-alloy', i.e. in this case the neutrons see only the concentration fluctuations. For higher Pb-concentrations, the composite structure factor is a weighted average over the density- and concentration fluctuation structure factors and their cross-term. Full line - theory, open circles - experiment. After Ref.24.

framework this appears in the form of a variation of the repulsive component of the pseudopotential and of the effective valence with composition). This creates strong potential gradients to which the system reacts by redistributing the electrons accordingly (this happens in the screening of the ionic pseudopotentials) - thus the electroneutrality principle creates a second electron transfer mechanism which goes into the opposite direction.

We have found that indeed these competing charge transfer mechanisms dominate the variation of the ordering potential $\Phi_{CC}(R) = c_A c_B (\Phi_{AA}(R) + \Phi_{BB} - 2\Phi_{AB}(R))$ with the differences in electronegativity and electron density of the two components and with composition. As an example we show in Fig.2 the interatomic potentials for a series of Li-Pb alloys - the strongest ordering effect is correctly predicted for the 'stoichiometric' composition Li_4Pb.

The structure of the liquid alloy may be derived using a thermodynamic variational method with the restricted primitive Yukawa model as a reference system (this is the simplest possible reference system capable of simulating chemical short-range order).[6,24] The comparison with the neutron scattering experiments demonstrates rather concvincingly that we have succeeded in grasping the essential physical effects.[31,32]

As we have already mentioned very briefly, the variational method requires only total energies - this enable us to extend out studies of CSRO to liquid transition metal alloys, by substituting for the pair potential Hamiltonian used for the simple metals a model tight-binding Hamiltonian more appropriate for transition metals. Very recent work of Pasturel and Hafner achieves remarkable agreement with experiment for a series of $Ni - Ti$ and $Ni - Zr$ alloys. [26] The physical effects governing the trends in the ordering interactions are identified again as charge transfer and the distortion of the bands near the Fermi surface.

Changes in the topological short-range order are more difficult to explain: we must push the thermodynamic perturbation theory at least to the level of an optimized random phase approximation. Preliminary results on the Al-Ge system [33] suggest that this is indeed a successful approach to the transition from dense structures with high coordination numbers to more open structures with low coordination.

Freezing and melting

There are two fundamentally different approaches to the solid-liquid transition: (1) The coexistence curve of the crystal and the melt is derived from the equality of the free energies obtained in two separate calculations. (2) Order parameter theories: One can either expand the periodic density of the crystal around the uniform density of the liquid (the order parameters being the fractional density change on freezing and the amplitudes of the Fourier components of the crystalline density function) ,[34,35] or expand the free energy of the harmonically vibrating crystal in terms of a 'disorder-parameter' representing roughly the density of disclination lines.[36]

The total energy approach to the melting curve calculations was pioneered by Stroud and Ashcroft[27] and by Jones[17]. It is clear that neither the calculation of the free energy of the crystal nor that of the liquid can aim individually at an accuracy of the order of a few degrees Kelvin, the estimated absolute error being rather of the order of the latent heat of fusion (about 1 mRy). So it is clear that agreement with experiment will depend largely upon a cancellation of errors. For this reason the simplest realistic approach will again be based on a thermodynamic variational calculation in both phases: the use of the variational method guarantees that the errors have the same sign, and if the refencerence systems are of the same quality, they will also be of the same order of magnitude. Even with the simplest possible reference systems (hard spheres for the liquid, an Einstein- or a Debye-reference system for the solid), qualitative agreement for the melting curves of the alkalis is reached over a substantial range of pressures.[27,37]. However, upon closer inspection it turns out that the fractional volume change on melting is notoriosly overestimated - to correct this it turns out to be necessary to account for the softness of the true interatomic forces by means of the 'blip-function' expansion [38] or by using of soft-sphere reference system [39]. The application of the theory to the polyvalent metals and to large pressures requires a systematic extension of thermodynamic perturbation theory. This has been studied extensively by Pelissier [40,41]. He showed that by pushing the description of the liquid state to the level of the optimized cluster theory (this is essentially the optimized random phase approximation plus some extra diagrams in the free-energy expansion) and that of the crystal to a quasiharmonic phonon calculation plus a perturbation theory for the anharmonic terms, the melting curves of the simple metals form Na to Pb are calculated with good success over a megabar range (see Fig.3), including an accurate predition of the fractional volume change on melting. The discrepancy in the Cs-melting curve appearing at about 40 kbar has to be attributed to the neglected of d-electron effects.

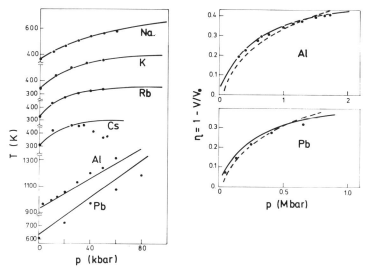

Fig.3: (a) The melting curves of Na, K, Rb, Cs, Al, and Pb. Full curves - theory, dots - experiment. (b) Incipient melting pressure determination along the Hugionot curve: full line - theoretical Hugionot for the solid phase, dashed line - solid volume at melting as a function of pressure. The intersection of both curves marks the beginning of melting along the Hugionot - in good agreement with the experimental estimates. After Pelissier.[40,41]

The disadvantage of the total energy approach is that it does not provide us with an atomistic picture of the solidification or melting process. This is precisely the aim of the order parameter theories. However, the published results refer only to idealized systems (hard spheres, one-component plasma, Lennard-Jones liquid). Any effort to extend these theories to real metals encounters considerable difficulties.

Alloy Phase Diagrams

Ab-initio calculations of binary phase diagrams based on electronic and thermodynamic perturbation theories have now been presented for a variety of different cases: (1) alloys with unrestricted solubility in both the solid and the liquid phases[42], (2) alloys with a melting extremum[43], (3) eutectic systems[44], and (4) compound forming systems[10]. A selected example for each case is given in Fig.4 - note that all these calculations concern homovalent alloys and that all have been performed using thermodynamic variational techniques (and pseudopotentials of various degrees of sophistication and fitting - only the last example is calculated entirely from first principles).

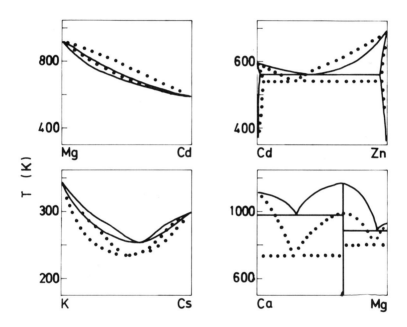

Fig.4: Calculated alloy phase diagrams: full curves - theory, dots - experiment. After Refs. 42 (Mg-Cd), 43 (K-Cs), 44 (Cd-Zn), and 10 (Ca-Mg).

It is clear that any progress for more difficult systems hinges on a progress in the statistical mechanical calculation. According to Bhatia and March[45], the phase boundaries in the (c,T)-plane between two phases α and β are determined by the two coupled differential equations

$$dT/dc^{\alpha} = -(c^{\alpha} - c^{\beta})RT^2/(S_{CC}^{\alpha}(0)L^{\beta})$$

$$dT/dc^{\beta} = -(c^{\alpha} - c^{\beta})RT^2/(S_{CC}^{\beta}(0)L^{\alpha})$$

Here $S_{CC}^{\alpha}(0)$ is the long-wavelength limit of the concentration fluctuation structure factor (which is related to the reciprocal of the Darken excess stability function) and L^{α} is a generalized latent heat. Hence the ability to predict correctly the shape of the liquidus and solidus lines depends on the ability of the theoretical model to describe concentration fluctuations in the liquid and in the solid states. Thus at present we are indeed restricted to systems where these fluctuations are either small (like in the systems demonstrated above - in all those cases a mixture of hard spheres is a rather good reference system) or at least coupled only very weakly to the fluctuations in the mean number density (in that case a restricted primitive Yukawa model could be used as a reference system).

The overestimate of the melting point of the intermetallic compound $CaMg_2$ (a hexagonal Laves phase of the $MgZn_2$-type) in the example (d) clearly stems from the fact that the reference system used in the liquid-state thermodynamic variation calculation (a mixture of hard spheres of different diameters) excludes all concentration fluctuations except those arisng from size effects. The ordering potential derived from pseudopotential calculations points to a preferred A-B coordination in this system (as in the crystalline compound). However, the concentration fluctuations are now strongly coupled to the fluctuations in the number density, as the radius ratio of the two components is rather large. For this case, no reference system with an analytical solution for the correlation functions and for the thermodynamic functions is available. Future calculations will have to rely heavily on numerical work.

CONCLUSIONS

We have shown that, based on electronic and thermodynamic perturbation theories, we are now able to calculate phase diagrams from first principles, with no other input then the Hamiltonian describing the system. This should only be considered as a first step. It should soon be possible to replace the electronic perturbation theory by more accurate methods for the calculation of the ground state energy (has we have indicated at the example of the liquid transition metal alloys), and to replace the simple reference systems with analytical solutions by more realistic ones, for which numerical solutions are required.

ACKNOWLEDGEMENTS

This work has been supported by the Fonds zur Förderung der wissenschaftlichen Forschung in Österreich (Austrian Science Foundation).

The fruitful cooperation of Prof. V.Heine and of Drs. G.Kahl and A.Pasturel is gratefully acknowledged.

REFERENCES

1. V.Heine and D.Weaire, Solid State Physics, ed. by H.Ehrenreich, D.Turnbull, and F.Seitz (Academic Press, New York 1970), vol.24, p.247.

2. See, e.g. the articles by R.M.Martin and S.G.Louie in Electronic Structure, Dynamics, and the Quantum Structural Properties of Condensed Matter, ed. by J.T.Devreese and P. van Camp (Plenum, New York 1985), NATO-ASI Series B, vol. 121, p.175 and p.335.

3. See, e.g. the papers in Ab-initio Calculations of Phonon Spectra, de. by J.T.Devreese, V.E. van Doren, and P. van Camp (Plenum, New York 1983).

4. D.de Fontaine, Solid State Physics. ed. by H.Ehrenreich, D.Turnbull, and F.Seitz (Academic Press, New York 1979). vol. 34, p.73.

5. J.P.Hansen and I.R.McDonald, The Theory of Simple Liquids (New York, Academic Press 1976).

6. H.C.Andersen, D.Chandler, and J.D.Weeks, Adv.Chem.Phys. 34,105(1976).

7. J.Hafner, G.Kahl, and A.Pasturel, in Liquid and Amorphous Materials, ed. by E.Lüscher and G.Fritsch (Sijthoff and Noordhoff, Alpen van Rijn), NATO-ASI Series E, in print.

8. J.Hafner and V.Heine, J.Phys.F (Metal Physics) 13,2479(1983).

9. J.Hafner and G.Kahl, J.Phys.F (Metal Physics) 14,2259(1984).

10. J.Hafner, in Amorphous Metals and Semiconductors , ed. by P.Haasen and R.I.Jaffee (Acta Metallurgica Press, in print); J.Hafner, Phys.Rev. B28,1734(1983).

11. M.H.Cohen and V.Heine, Phys.Rev. 122,1821(1962).

12. D.R.Hamann, M.Sclüter, and C.Chiang, Phys.Rev.Lett. 43, 1494(1979).

13. G.Kerker, J.Phys.C (Solid State) 13,L189(1980).

14. G.B.Bachelet, D.R.Hamann, and M.Schlüter, Phys.Rev. B26, 4199(1982).

15. J.Hafner, J.Phys.F (Metal Physics) 6,1243(1976).

16. J.Hafner, Phys.Rev. B21,406(1980).

17. H.D.Jones, J.Chem.Phys. 55,2640(1971).

18. H.Minoo, C.Deutsch, and J.P.Hansen, J.Physique (Paris) Lett. 38,L191(1981).

19. J.B.Hayter, R.Pynn, and J.B.Suck, J.Phys.F (Metal Physics) 13,L1(1983).

20. K.K.Mon, R.Gann, and D.Stroud, Phys.Rev. A24,2145(1981).

21. G.Kahl and J.Hafner, Phys.Rev. A29,3310(1984).

22. G.Pastore and M.P.Tosi, Physica 124B,383(1984).

23. G.Kahl and J.Hafner, Z.Physik B58,283(1985).

24. A.Pasturel, J.Hafner, and P.Hicter, Phys.Rev. B32,5009(1985).

25. G.Kahl and J.Hafner, J.Phys.F (Metal Physics) 15,1629(1985).

26. A.Pasturel and J.Hafner, in Proc.3rd.Int.Conference on the Structure of Non-Crystalline Materials, ed. by Chr.Janot (Les Editions de Physique, in print) and Phys.Rev.B, submitted.

27. See, e.g. D.Stroud and N.W.Ashcroft, Phys.Rev. B5,371(1972).

28. Y.Waseda, The Structure of Non-Crystalline Materials - Liquids and Amorphous Solids (McGraw Hill, New York 1980).

29. V.Heine and D.Weaire, Phys.Rev. 152,603(1966).

30. See, e.g. W. van der Lugt and W.Geertsma, J.Non-Cryst. Solids 61+62,187(1984).

31. L.Pauling, The Nature of the Chemical Bond (Cornell University Press, Ithaca 1952), Sec. 12.5.

32. These studies have also been supplemented by an investigation of the difference electron densities in intermetallic compounds, see J.Hafner, J.Phys.F (Metal Physics) 15,1879(1985).

84

33. G.Kahl and J.Hafner, to be published.

34. T.V.Ramakrishnan and M.Youssouff, Phys.Rev. B19,2775(1979); A.D.J.Haymet and D.W.Oxtoby, J.Chem.Phys. 74,2559(1981); A.D.J.Haymet, J.Chem.Phys. 78,4641(1983).

35. M.Baus and J.L.Colot, J.Phys.C (Solid State) 18,L365(1985).

36. V.N.Novikov, Phys.Lett. A104,103(1984).

37. A.Angelie and J.L.Pelissier, Physica 121A,207(1983).

38. A.M.Bratkovsky, V.G.Vaks, and A.V.Trefilov, Phys.Lett. A103,75(1984).

39. J.A.Moriarty, D.A.Young, and M.Ross, Phys.Rev. B30,578(1984).

40. J.L.Pelissier, Physica 121A,217(1983); 126A,271(1984); 126A,474(1984); 128A,363(1984).

41. J.L.Pelissier, Phys.Lett. A103,345(1984).

42. M.Hasegawa and W.H.Young, J.Phys.F (Metal Physics) 7,2271(1977).

43. T.Soma, H.Matsuo, and M.Funaki, phys.stat.solidi (b) 108,221 108,221(1981); Z.A.Gurskii and W.I.Baranitskii, Preprint ITF-82-3R, Institute of Theoretical Physics, Ukrainian Academy of Sciences.

44. A.I.Landa, V.E.Panin, and M.F.Zhorovkov, phys.stat.solidi (b), 108,113(1981).

45. A.B.Bhatia and N.H.March, Phys.Lett. A51,401(1975); see also W.Geertsma, Physica 132B,337(1985).

DYNAMICS OF COMPRESSED AND STRETCHED LIQUID SiO$_2$, AND THE GLASS TRANSITION

C. A. Angell, P. A. Cheeseman and C. C. Phifer
Department of Chemistry
Purdue University
West Lafayette, IN 47907

ABSTRACT

Ion dynamics simulations are presented to model the system SiO$_2$ over wide ranges of temperature and density as it passes from liquid into glassy states. Essential features of the diffusive mechanism by means of which the system configuration space is explored are examined in relation to density and structure. The presence of a water-like density maximum in the liquid state of this system is established by using the negative pressure regime to stretch the system and thereby sharpen the phenomenon.

INTRODUCTION

The study of the glassy state, and the process by which ergodicity is "broken"[1] during the cooling of a liquid into the glassy state, is currently a subject of broad interest among physicists and materials scientists. The reasons range from purely fundamental to very practical: for instance, the basic physics of relaxation processes in dense amorphous systems, the kinetics of which determine whether a system is in the liquid or glassy states, are not well understood[1-7] and stand as a challenge to the theoretician. At the other extreme, the role of SiO$_2$ - like glassy grain boundary phases in the binding of high strength ceramic materials during high and low-temperature processing is a much researched problem. In the present paper we use computer simulation methods to explore some aspects of the dynamic and thermodynamic behavior of SiO$_2$ as it passes from the liquid to the glassy state via diffusive slowdown.

It is clear that the long-range structure of a material which qualifies for the description "glass" is frozen on long timescales, although many modes of motion which dissipate energy, and so may be described as "relaxations," are still active in the glassy state.[8] In our view, the particle motions by which a system explores its configuration space and finds its equilibrium state are essentially diffusive in nature. The process of diffusion and the cessation of diffusion by means of which the liquid structure is arrested, are well suited for study by computer simulation methods. This paper is thus devoted to an examination of the slowing-down of diffusion, and the consequent freezing-in of structure. While we focus attention on the best-known if not the prototypical glass-forming substance, SiO$_2$, we carry the study far beyond the range of conditions in which SiO$_2$ is normally investigated in the laboratory.

Although it is generally considered that SiO$_2$ is a covalently bonded network material it has proven possible, via many computer simulation studies in recent years,[9-12] to achieve surprising accuracy in the reproduction of dynamic, structural, and thermodynamic properties of the laboratory substance using simple two-term spherically symmetrical rigid-ion potentials of the form

$$V(r) = z_i z_j / r + A \exp\left[\frac{\sigma_i + \sigma_j - r}{\rho}\right], \qquad (1)$$

where z_i, z_j are ion charges, σ_i, σ_j are ion size parameters, A is a constant containing the Pauling factor and ρ is a constant. The inclusion of van der Waals and multipole terms in the potential has not seemed warranted in view of the simplification involved in the use of potentials of such simple form, and this approximation will be continued in the present work.

As seen in the previous studies[9,10] appropriate choice of σ_i And σ_j parameters in Eq. 1 leads to open tetrahedral network structures, and this often causes surprise in view of the common belief that such structures are a consequence of directional covalent bonding. The reason that a tetrahedral network structure may be obtained by using spherically symmetrical potentials is that when the 4+ cation is small enough it cannot coordinate more than four 2- anions before the anion -anion repulsions become dominant. Thus the four-coordinated state is the one which minimizes the Coulomb energy. Increase of high charged cation size leads immediately to the reorganization of the structure to higher coordination numbers, as was shown in some of the first simulation studies of this subject area.[10]

One of the great advantages of computer simulation experiments is the ability to investigate properties under conditions which are difficult or impossible to reproduce in the laboratory. In this work we will take particular advantage of simulation to investigate the properties of SiO_2 under conditions of negative pressure, i.e. when the average bond in the system is in a stretched condition rather than in a compressed condition as at high pressures. We will also extend the study to exceptionally high pressures to confirm and clarify the anomalous diffusion behavior described in earlier papers[9] on the properties of silica in the liquid state.

We see our first task in this paper as one of illustrating the diffusion mechanism by means of which configuration space is explored in this system, and the manner in which decrease in temperature causes diffusion to stop, locking in the structure at some characteristic, "glass transition," temperature.[13]

PROCEDURE

The computational procedure is that used in previous studies. An initial configuration containing 162 ions with the stoichiometry SiO_2 is obtained by randomization at high temperatures of an initially crystalline configuration, after which a long equilibration run at the normal density and at a temperature sufficient to produce typical liquid-like diffusion (6,000 K in the present instance) was carried out. As usual, the absence of surface effects is guaranteed by the use of periodic boundary conditions. We show, by sample calculations on a system twice the size, that all of the characteristic modes of motion and configurational fluctuations are sampled in a box containing 162 particles and that the objectives of this study are not furthered by the use of larger systems. Indeed, it is important to keep the system size at a minimum because of the particularly slow diffusion which is encountered in some of the more interesting conditions we have explored in this study and which in turn demands very long production runs.

The study was initiated using a Fortran 77 version of the original (largely machine language) Cheeseman multicomponent dynamics program, Dynamo,[14] running on Sun and VAX computers. However, the work was completed (on the eve of the conference) using an optimally vectorized version of the program running on the Purdue Cyber 205 in 1/40th the previous execution time with the CDC 6600. Three thousand time steps now are completed in 300 sec. of CPU time.

Once initial results at the normal density had been shown to reproduce those of earlier work using the original program, detailed diffusion-temperature studies were performed at two different volumes, the first at a volume 12.5% expanded beyond the normal volume ($1.125 V_o$) to increase tetrahedrality and associated structural and behavioral anomalies (e.g. density maximum)[15] and the second at 60% of the normal volume ($0.6V_o$) to observe the high-pressure behavior and to demonstrate the water-like depression of the anomalous characteristics.

RESULTS AND DISCUSSION

To gain insight into the diffusion process in the liquid, and vitrification we depict in Fig. 1 the motions of oxide ions in the vicinity of a randomly chosen silicon ion as a function of time during two picoseconds of simulations at 8,000 and 4,000 K. The process at 8,000 K is characterized by anharmonic oscillations of the oxygens about a mean separation distance from silicon of 1.59 A. The oxygen movements are punctuated by occasional large oscillations and less frequently by sudden jumps out of the coordination shell to which they formerly belonged into a new shell. This event is usually coincident with the arrival of a second oxygen from a nearby silicon, the reference silicon being temporarily five-coordinated. Brawer[16] has identified the five-coordinated network centers as defects of the "glass" structure at which place exchanges characteristically occur, though their essential transience in most cases is indicated by the trajectory plot of Fig. 1. The mechanism of diffusion may thus be described as a "rattle and correlated jump" mechanism. During long runs it is possible to locate the Si to which the jumping oxygen has moved, observe the new oxygen displaced from the receiving Si when the initial jumper arrived and so follow the correlated movements to their stopping point. The latter appears to be some non-transient higher coordinated Si of the type described by Brawer.[16] Thus two types of higher-coordinated Si would appear to exist, the more common one being truly transient with a lifetime less than a vibration period (see Fig. 1), implying that no activated complex in the sense of transition state theory exists. Note in Fig. 1 that it is not the ion with the temporary large displacement (+) which escapes the coordination shell in the first exchange, but a second one whose motion up to the point of departure was quite unexceptional. It should be borne in mind that the reference Si, which is stationary as far as the oxygen motions in Fig. 1 are concerned, is itself in motion within the box though it is generally diffusing less rapidly than the oxygens, as shown below.

The distinction between the 8,000 and 4,000 K behavior is obviously, from Fig. 1, the absence of the neighbor exchange process. At 4,000 K, diffusion has essentially stopped, as shown in Fig. 1(b). Thus we see, in Fig. 1(b), the ion dynamics characteristic of the *glassy* state of the material. Appropriate analysis of the type of motions seen in Fig. 1(b) will lead to the vibrational density of states of the glassy material in a highly thermally excited condition (for example, see Brawer, ref. 17).

From Fig. 1 it seems clear that although no "activated state" (with assignable thermodynamic properties as in transition state theory) is in evidence, the process under observation is an activated process in the sense that an unusual local energy fluctuation, reflected in the large amplitude of local oscillations, is required. The energy involved should therefore be determinable from an Arrhenius law analysis of the appropriate diffusion coefficients. The diffusivities of oxygen and silicon may be obtained by taking the slope of the mean square displacement versus time plot for the system (exemplified in Fig. 2) for several temperatures. Fig. 2 shows the consequences of diminishing temperature in a different manner from Fig. 1. The initial "vibrational" displacement is preserved but the long time slope tends systematically to zero. At 4,000 K, the plot of l^2 versus T is flat within the uncertainty of the computation over the entire period simulated.

For the timescale of this simulation, then, the glass transition for SiO_2 occurs between 4,000 and 5,000 K, enormously higher than the laboratory transition which is found at 1,500 K. The difference, of course, lies in the great difference in timescales between laboratory and computer simulation experiments. A cooling rate appropriate to the present investigation may be assessed by noting that the total of 5,000 timesteps (each 2×10^{-15} seconds in length) is allowed for equilibration after each 2,000 K downward temperature step. The cooling rate q ($q = dT/dt$) thus equals $2,000/5,000 \times 2 \times 10^{-15} = 2 \times 10^{14}$ degrees sec^{-1}, many orders of magnitude faster than in any laboratory quenching experiments. The thermodynamic manifestation of the transition (the falling out of equilibrium or the "breaking of ergodicity") is seen in Fig. 3, where the internal energy E is plotted versus T from 300 K to 10,000 K. Below 5,000 K

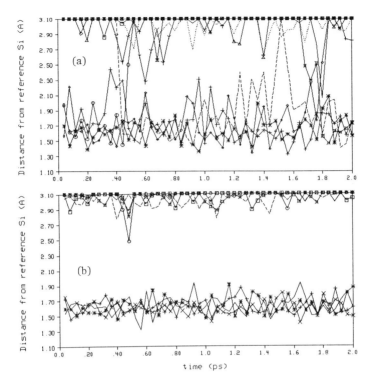

Figure 1. Motion of neighbors (mostly oxygen ions) in the vicinity of a randomly chosen silicon ion, during 2ps of elapsed time in simulated SiO_2 at 8,000 K (part a) and 4,000 K (part b) at 1.125 V_o. V_o is the ambient temperature molar volume.

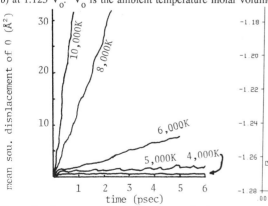

Figure 2. System average mean square displacement of oxide ions from their original coordinates during 3,000 time steps (6 ps) of a simulation (after 4 ps allowed for equilibration) at each of the temperatures indicated. Glass transition occurs between 5,000 and 4,000K for this cooling schedule.

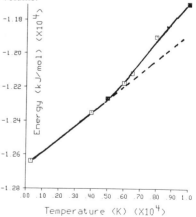

Figure 3. Thermodynamic manifestation of the glass transition. E versus T at $V/V_o = 1.125$ through the temperature of configurational arrest T_g.

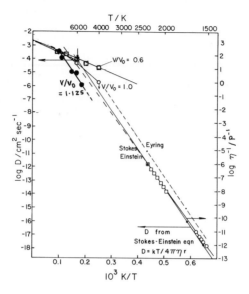

Figure 4. Arrhenius plot of Si and O diffusivities for simulations at three different system volumes as marked.

the slope corresponds to the classical vibrational heat capacity 3Rn per mole of substance where n is the number of atoms per formula unit. The two filled symbols correspond to equal pressures on either side of the minimum indicating the slope observed also corresponds to the constant pressure heat capacity. At this temperature and volume, C_p for the liquid is large compared with its experimental value near T_g. We cannot tell whether this reflects an inadequacy in the potentials used, or the high temperature at which C_p is being assessed.

To obtain the activation energy for the diffusive jumps, the diffusivities ($D = l^2/6t$) obtained from the slopes of the Fig. 2 plots are shown in Arrhenius form (see Fig. 4). Included in Fig. 4 are data assessed from experimental viscosity measurements using conversion formulae due to Stokes-Einstein, and to Eyring,[18] respectively

$$D = k_B T/6\pi\eta r, \tag{2}$$

where η = viscosity in poise and r = radius of diffusing ion, and

$$D = k_B T/\lambda\eta, \tag{3}$$

where λ is the jump distance. The lowest temperature simulation data obtained at the experimental volume V_0 ($\rho = 2.20$ g ml^{-1}) appear to be consistent with the extrapolation of experimental data assessed using the Eyring equation, although the average activation energy is considerably lower and a curvature is evident. On the other hand, the diffusion data at a volume 12.5% greater than the experimental volume exhibit the same activation energy, 114 kcal/mole, which is by far the largest activation energy yet observed for a simulated liquid system The distinction between the behavior at the two volumes is a manifestation of the more perfect tetrahedrality of the configuration forced on the system by the volume *increase* and an accompanying reduction in the number of five coordinated Si or defect sites which serve as mobility centers.[17]

Conversely, decrease in volume, with increase of pressure and increase in particle mobility, is consistent with the observation that increase of pressure decreases the viscosity (hence also T_g) of tetrahedral liquids;[20] in fact, this effect has a limit, and a diffusivity maximum appears to be reached at about 250 kbar, after which the normal decrease in D with increasing pressure is observed. This highly water-like feature of liquid silica and framework aluminosilicates has been discussed at length elsewhere,[21] where it was pointed out, for the case of open network aluminosilicates, that the maximum in diffusivity corresponded to a structure in which the *average* Si coordination number was five. Fig. 4 shows that the activation energy for diffusion at this higher density, $0.6V_o$, is much smaller than in the expanded open network state. Note that the glass transition temperature, at which diffusivity is no longer detectable in the simulations above the noise level ($D < 1 \times 10^{-6}$ for our 20 psec runs), now falls at <3,000 K, so there is, in this system, a strong inversion of the normal effect of pressure on the glass transition temperature. We return to this point in the next section.

In Fig. 5 we show the local motions around a reference Si at 8,000 K at $0.6\ V_o$ to show the change in character of the motion from that seen earlier at $1.125\ V_o$ in Fig. 1(a). Note that the diffusion at $0.6\ V_o$ seems to be a much more continuous process. With multiple particle departures (at 1.2 ps) and delayed replacement it is reasonable to suppose it is the latter which is more characteristic of the normal ionic liquid (or molecular liquid for that matter - we have not seen this type of plot in reports of molecular liquid simulation studies so cannot make comparisons). The occurrence of a common limiting diffusion coefficient at $1/T = 0$ irrespective of the behavior at lower temperatures (see Fig. 4) may be rationalized simplistically in terms of the common frequency of oscillation irrespective of volume (see Fig. 5). When all oscillations become jumps (i.e. each attempt at the barrier is successful) the time between jumps becomes about 0.1 ps from Fig. 1. If the distance moved is the distance between first and second Si - O peaks in the RDF, ≈2.5 A, then we may expect the limiting D $= l^2/6t$ to be 1.04×10^{-3}. We now attempt a partial synthesis of dynamic and thermodynamic behavior.

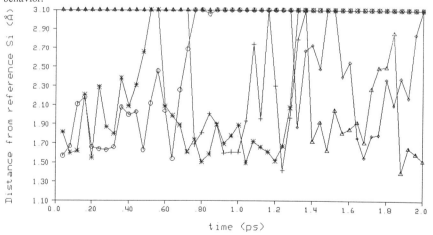

Figure 5. Local motions of a fraction of the nearest neighbors of a randomly chosen Si in compressed SiO_2 at O.6 V_o amd 8,000 K, showing change in character in diffusive modes from that observed at 1.125 V_o in Fig. 1. Other neighbors are omitted to avoid confusion. Average coordination number at this volume is five.

In Fig. 6 we show the variation of the oxygen and silicon diffusivities at 6,000 K over a volume range extending from the high density extreme of 0.4V_o (where the pressure at 6,000 K reaches 2.2 Mbar) to a highly stretched and mechanically unstable state at 1.5 V_o. We deal in turn with (1) the thermodynamic connection to the anomalous negative slope of the diffusivity plot between 0.6 V_o and 1.3 V_o, and (2) with the tendency of D_i to vanish at the spinodal (mechanical stability limit).

1. Anomalous Pressure Dependence Diffusivity and Negative Expansivity.

We noted above that the diffusivity temperature dependences shown in Fig. 4 require that T_g decrease(s) with increasing pressure. Normally, imposition of pressure on a liquid causes T_g to increase at a rate of 3 - 20 K/kbar.[22] The present data, though obtained in a very uncharacteristic time scale range, imply a variation of T_g with P of $\approx$ -3 K/kbar. Only one case of a negative dT_g/dP has been reported in the literature, that of a system containing 91 mole % H_2O (a lithium acetate aqueous solution).[23] In that report it was emphasised that pseudo-second order thermodynamic relations derived for relaxing systems by Davies and Jones[24] required that a negative dT_g/dP be associated with a negative configurational expansivity, $\Delta\alpha$, according to

$$dT_g/dP = VT\frac{\Delta\alpha}{\Delta C_p}, \qquad (4)$$

where $\Delta\alpha$ and ΔC_p are the changes in constant pressure expansivity and heat capacity which occur when the system passes from the glassy (frozen) to the liquid (mobile) state around T_g. ΔC_P can only have positive values since it is proportional to the *square* of the change, upon unfreezing, of the average entropy fluctuation[25]. $\Delta\alpha$ can have negative values, however, since it is determined by the cross product of changes in average entropy and average volume fluctuation ΔV. If unfreezing permits fluctuations to denser packings, as in the case of open network structures then ΔV and hence $\Delta\alpha$ can have negative values. It is this feature which is responsible for the density maximum, in water at 4°C, below which the expansivity, assumes negative values. We arrive at the conclusion that a negative configurational expansivity should be a feature of our simulated SiO_2. If large enough, this should outweigh any positive vibrational (glassy state) expansivity to give a total expansivity which is negative at low temperatures and a density maximum at higher temperatures.

In a constant volume experiment such as is being conducted in the present study, the occurrence of a density maximum i.e. a change in sign of the expansivity will be manifested as a change in the sign of $(\partial P/\partial T)_V$. Since

$$\left[\frac{\partial P}{\partial T}\right]_V = \left[\frac{\partial P}{\partial V}\right]_T \left[\frac{\partial V}{\partial T}\right]_P = \alpha/\kappa_T,$$

where κ_T is the isothermal compressibility. It has been noted in previous simulations of SiO_2[9] that the pressure hardly changes with temperature, but no clear minimum has been reported -- indeed, it would not be expected on the basis of laboratory measurements since, although a density maximum is observable, it is only found at $\approx$ 1,800 K in a viscosity range inaccessible to computer simulation studies. Furthermore, it was shown to be a much weaker maximum than that in water[26]. To enhance this feature we may take advantage of the power of computer simulation studies to explore normally inaccessible ranges of PVT space. In the present case we search for a minimum in P versus T under expanded volume conditions, viz. at V/V_o = 1.125, so that tetrahedrality, which is the root cause of the anomalies, will be enhanced.

The existence of a $(\partial P/\partial T)_V$ minimum, hence a density maximum, at $\approx$ 7,000 K under these "stretched" conditions is witnessed by the data in Fig. 7. The fact that the total expansivity is negative below 7,000 K implies that the *configurational* part will be negative to higher temperatures and also at considerably smaller volumes. This would be consistent with

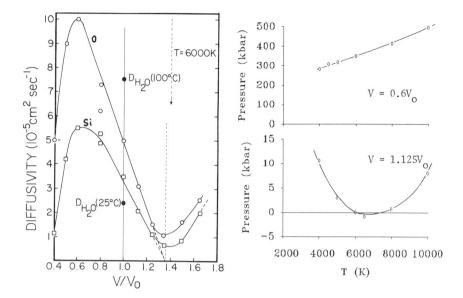

Figure 6. Variation of D_O and D_{Si} at 6,000 K with system volume. V_o is at the ambient temperature molar volume. Comparison is made with diffusivity of H_2O at 298 and 373 K, at is normal volume. Note diffusivity maximum at $0.6V_o$ and tendency of diffusivity to vanish at 1.35 V_o.

Figure 7. Demonstration of the existance of a density maximum in SiO_2 at 7,000 K via the occurrence of a minimum in P versus T hence a change in sign of $(\partial P/\partial T)_V$ $[(\partial P/\partial T)_V = \alpha/\kappa_T$, see text].

the anomalous pressure dependence of diffusion seen in Fig. 6.

2. Negative Pressure Stability Limit - the Spinodal

In Fig. 8 we combine thermodynamic and diffusivity data from preliminary studies carried out some time ago on the CDC 6500[27] to show that the volume at which the diffusion coefficients are tending to vanish is the volume at which the pressure on stretching tends to a minimum value of∠-50 kbar, and a condition of mechanical instability. This instability is identical with that predicted by simple equations of state of the van der Waals type. Although T. H. Soules has shown that simulated silica "rods" break at 50 kbar of uniaxial tension[28] (a result in accord with the known tensile stress of silica glass), it seems that Fig. 8 may represent the first time that the existence of the negative pressure limit of stability for liquid configurations has been specifically demonstrated by a computer simulation. At the stability limit the compressibility passes through infinitely large values and becomes negative, a condition incompatible with mechanical stability, and in principle the system must "fall apart", i.e. change phase. Indeed, the slow growth of large holes in the structure could be observed in simulations beyond the PV minimum of Fig. 8(a). The variation of the compressibility with volume through V_o is shown in Fig. 8(c). The level of agreement with experiment is noted. The comparison is appropriate since compressibility at fixed volume seems to change relatively little with temperature.

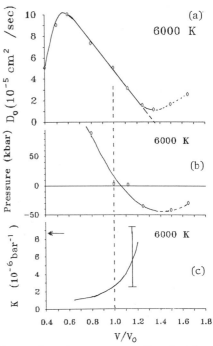

Figure 8(a).

(a) Diffusivity of oxygen at 6,000K as a function of reduced volume.

(b) System pressure versus reduced volume showing extremum where diffusivity tends to vanish.

(c) Compressibility versus reduced volume showing comparison with experimental value at V_0 and 1,500K, and divergence near spinodal.

By obtaining PV isotherms like Fig. 8(b) for a variety of temperatures, a line of mechanical stability limits, the "spinodal" line, may be constructed and connected to the critical temperature for the system. The critical point for SiO_2 occurs at temperatures well outside the range of this study, and we have not sought to identify it at this time. The form of this spinodal line for tetrahedral liquids is of great interest in connection with instabilities in supercooled tetrahedral liquids but will not be pursued further in the present article. We note, however, that these observations are used elsewhere[29] to establish a close relationship between the behavior of water and liquid silica and, by implication, other tetrahedrally coordinated network systems.

ACKNOWLEDGMENTS

This work has been supported by the National Science Foundation under Solid State Chemistry Grant No. DMR 8304887.

REFERENCES

1. R. G. Palmer and D. L. Stein in "Relaxation in Complex Systems" edited by K. Ngai and G. B. Wright (National Technical Information Service, U.S. Department of Commerce, Springfield, VA 1984) p 253.

2. R. G. Palmer, Adv. Phys. **31**, 669 (1982).

3. M. F. Schlesinger and J. T. Bendler, Ann. N.Y. Acad. Sci., 1986 (in press).

4. C. T. Moynihan et al, Ann. N.Y. Acad. Sci. **279**, 15 (1976).

5. I. M. Hodge in "Relaxation in Complex Systems" edited by K. Ngai and G. B. Wright (National Technical Information Service, U.S. Department of Commerce, Springfield, VA 1984) p 65.

6. G. W. Scherer, J. Amer. Ceram. Soc., **61**, 504 (1984).

7. V. Bengtzelius, W. Götze and A. Sjölander, J. Phys. C. Solid State Phys. **17**, 5915, (1984): Ann. N.Y. Acad. Sci. 1986 (in press).

8(a). G. P. Johari and M. Goldstein, J. Chem. Phys. **53**, 2372 (1970); **55** 4245 (1971).

  (b). G. P. Johari in "Relaxation in Complex Systems" edited by K. Ngai and G. B Wright (National Technical Information Service, U.S. Department of Commerce, Springfield, VA 1984) p 17.

9. L. V. Woodcock, C. A. Angell and P. Cheeseman, J. Chem. Phys., **65**, 1565 (1976).

10. C. A. Angell, P.A. Cheeseman, L.V. Woodcock and J.H.R. Clarke in "The Structure of Non-Crystalline Materials," Edited by P. Gaskell (Taylor Publishing Company, Cambridge 1977) pp. 191-194.

11. T. F. Soules, J. Chem. Phys., **71** 4570 (1979).

12. S. K. Mitra, M. Amini, D. Fincham and R. W. Hockney, Phil. Mag. **B, 43**, 365 (1981).

13. C. A. Angell, J.H.R. Clarke and L. V. Woodcock, Adv. Chem. Phys., **48**, 397 (1981).

14. P. A. Cheeseman, Ph.D thesis, Purdue University, 198?.

15(a).R. W. Douglas and J. O. Isard, J. Soc. Glass Tech. **35**, 206 (1971).

  (b).R. W. Bruckner, J. Non-Cryst. sol. **5**, 281 (1971).

16(a).S. A. Brawer and M. J. Weber, J. Non-Cryst. Sol. **38&39**, 9 (1980).

  (b).S. A. Brawer, J. Chem. Phys. **75** 3516 (1981).

17. S. A. Brawer, J. Chem. Phys. **79**, 4539 (1983).

18. H. Yinnon and A. R. Cooper, Phys. Chem. Glasses, **21**, 204 (1980).

19. T. F. Soules and R. F. Busbey, J. Chem. Phys. **75**, 969 (1981).

20. I. Kushiro, J. Geophys. Res., **81** 6347 (1976).

21. C. A. Angell, P. A. Cheeseman and S. Tamaddon, Bull. Mineralogie **1-2**, 87 (1983).

22. T. Atake and C. A. Angell, J. Phys. Chem., **83**, 3281 (1979).

23. C. A. Angell, J. Polymer Science, Polymer Letters, **11, 383 (1973)**.

24. R. O. Davies and G. O. Jones, Adv. Phys., **2**, 370 (1953).

25. L. Landau and E. M. Lifschitz, Stat. Phys. (Pergamon, London and Addison-Wesley, Reading, MA 1958), Sec, 111, p. 350.

26. C. A. Angell and H. Kanno, Science, **193**, 1121 (1976).

27. S. Tamaddon and C. A. Angell (1984) unpublished work.

28. T. F. Soules, J. Non-Cryst. Sol. (Kreidl Symposium) **73**, (1985)

29. C. A. Angell and C. C. Phifer and S. Tamaddon (to be published).

CRYSTAL, LIQUID AND GLASS IN 2 DIMENSIONS.
ANALYSIS OF THE GLASS TRANSITION

Frédéric Lançon* and Praveen Chaudhari
IBM Thomas J. Watson Research Center, Yorktown Heights, NY 10598
*Present address : Centre d'Etudes Nucléaires de Grenoble, DRF/SPh/MP, 85 X,
38041 Grenoble Cédex, France

ABSTRACT

A molecular dynamics technique has been used to simulate the melting of a 2-dimensional diatomic crystal and to quench the liquid phase to a solid phase. We demonstrate that a 2-dimensional dense amorphous structure can be obtained and that a 2-dimensional glass transition does exist. Furthermore, atomic vibrations in the liquid can be separated from motion produced by diffusion. The relaxation time during which atoms have a vibratory motion but do not diffuse, diverges to infinity near the observed glass transition. Because of the 2-dimensionality, we are able to display the microscopic processes associated with the glass transition.

INTRODUCTION

There is now considerable evidence that three-dimensional computer models undergo a glass transition as a result of rapid cooling that can be done by means of the Monte Carlo method [1,2] or the molecular dynamics method [3-11]. Nevertheless, complex atomic arrangements in liquids and glasses make the microscopic description of the glass transition difficult, even though all the required data are, in theory, contained in the atomic trajectories calculated and stored by the computer. Thus, to simplify the problem of analyzing the atomic configurations, we have investigated a two-dimensional system.

Close packings of equal spheres (or atoms with isotropic interactions) tend to maximize the local density by forming tetrahedron units in 3-D systems and triangular units in 2-D systems. But in three dimensions regular tetrahedra cannot fill exactly the space and different locally dense packings of distorted tetrahedra are possible, corresponding either to crystalline or amorphous structures. Unlike 3-D systems, the triangular lattice in two dimensions maximizes the local density and is the only way to pack six atoms around a central one. Polyatomic systems are therefore required to obtain stable two-dimensional glasses.

Randomly packed planar arrays of ball bearings [12,13] and amorphous soap bubble raft [14,15] with two different particle sizes have been used to study amorphous models. We report in this paper the results of a numerical simulation of a two-dimensional system with pair potentials. Using the static relaxation method we have computed a stable diatomic crystal. Then we have used the molecular dynamics method at constant pressure and constant temperature (or with a linear temperature variation) to melt the crystal and quench the liquid. We believe that we have identified a two-dimensional glass transition.

INTERATOMIC POTENTIALS

We chose the modified Johnson potential which has been widely used to simulate interactions in three-dimensional glasses both in monoatomic cases [16,17] and in diatomic cases [18]. There are three kinds of interatomic pairs ij (i,j = A or B) in a two-component system of, say, atoms A and B. For each type of pairs we made an homothetic transformation of the Johnson potential to set the position r_{0ij} and the depth ϵ_{ij} of the minimum such that

$$2\epsilon_{AA} = 2\epsilon_{BB} = \epsilon_{AB} = \epsilon_o$$

$$r_{oAA} = 2\sin(\pi/5)r_o = 1.176 \; r_o$$

$$r_{oBB} = 2\sin(\pi/7)r_o = 0.868 \; r_o$$

$$r_{oAB} = r_o$$

where ϵ_o is the unit of energy and r_o is the unit of length. For simplicity we let the mass of the A and B atoms be equal to the unit of mass m_o. With this set of units, the unit of time t_o is $(m_o r_o^2/\epsilon_o)^{1/2}$, the unit of (two-dimensional) pressure P_o is ϵ_o/r_o^2, and the unit of temperature T_o is ϵ_o/k, where k is the Boltzmann's constant.

The values of ϵ_{ij} have been chosen to avoid a phase separation and the values of r_{oij} are such that 5 atoms of A fit around a B atom and 7 atoms of B fit around a A atom. Note that none of these two units has a crystallographic symmetry. Because in two dimensions the mean number of geometric neighbors defined with the Voronoï construction is exactly six, we assume that the maximum numbers of 5-fold and 7-fold symmetries occur when concentrations of A and B atoms are equal. The total number of atoms in the simulations reported here was 704 divided equally between A and B atoms. The usual periodic boundary conditions were used with a rectangular cell.

SIMULATIONS

Crystal

The triangular lattice is not stable with respect to the previously defined potentials and thus it cannot be used as a reference lattice. We have found a stable crystal structure shown on fig. 1a. This is presumably the ground state of the system although we cannot rule out the possibility that another structure may have a lower potential energy. The static relaxation method is one way to obtain this lattice. We started by setting all the atoms on a triangular lattice with alternate rows of A and B atoms. Then we used a special relaxation program which minimizes the potential energy with respect to the atomic coordinates but also with respect to the length of both sides of the cell defining the periodic conditions. By this procedure, the periodic cell can shrink in one direction and

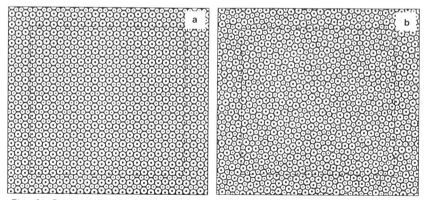

Fig. 1. Crystal (a) and glass (b) states of a two-dimensional system. Circles are centered on the atomic positions and have diameters equal to r_{oAA} and r_{oBB} around A and B atoms respectively. The dotted line indicates the periodic boundaries.

extend in the other. Therefore, the resulting lattice is stable, and its hydrostatic pressure as well as both longitudinal pressures are equal to zero.

Liquid

To introduce the temperature and the pressure in the model, we used the molecular dynamics method. We employed constrained equations of motions [19] for which the temperature and pressure are constants of the motion [20,21]. These equations can be generalized to produce a given variation $\dot{T}$ of temperature with time [22] or a given variation $\dot{P}$ of pressure. Let q_i be the coordinates of atom i, p_i be the momentum of i divided by its mass m_i, V be the volume and n be the space dimension. The constrained equations of motions are :

$$\dot{q}_i = p_i + \dot{\epsilon}q_i \; ; \; \dot{p}_i = F_i/m_i - \beta p_i \; ; \; \dot{V} = nV\dot{\epsilon}$$

For given derivatives of the temperature and pressure, $\dot{T}$ and $\dot{P}$, the damping coefficient β and the volume variation coefficient $\dot{\epsilon}$ have the value :

$$\beta = \left[\sum_i (p_i \cdot F_i) - \frac{n}{2}Nk\dot{T} \right] / \left[\sum_i (m_i \, p_i^2) \right]$$

$$\dot{\epsilon} = \frac{nNk\dot{T} - n\dot{P}V - \frac{1}{2}\sum_{i,j(i\neq j)} \left[\left[\frac{\phi'_{ij}}{q_{ij}} + \phi''_{ij} \right](q_{ij}\cdot P_{ij}) \right]}{n^2 PV + \frac{1}{2}\sum_{i,j(i\neq j)} q_{ij}^2 \left[\frac{\phi'_{ij}}{q_{ij}} + \phi''_{ij} \right]}$$

where ϕ_{ij} is the two-body potential and is a function of the distance q_{ij} between atoms i and j. We used the previous crystal configuration as the starting point for the molecular dynamics. The numerical integration of the dynamical equations was done with a predictor-corrector algorithm [23] and with a time step equal to 0.001 t_0.

The system was heated at a constant heating rate of 0.01 T_0/t_0 up to the temperature 0.70 T_0. The pressure in all the simulations reported here was kept constant at P_0. During this run the positions and velocities obtained at successive temperatures were recorded and used as starting points for later runs at constant temperature. To approach equilibrium and compute time-averaged quantities at constant temperature T, we ran our molecular dynamics for times ranging from 60 t_0 to 990t_0 (60000 to 990000 steps) according to the value of T.

We show the enthalpy, the potential energy and the volume per atom of our model as a function of temperature in Fig. 2. Melting occurs at $T_m = 0.43T_0$ and appears to be a first order transition.

Glass

From the well-equilibrated fluid at temperature 0.7 T_0, we quenched the system to 0.01 T_0 at two different cooling rates of 0.01 and 0.0005 T_0/t_0. Then, as previously, we ran constant temperature simulations from the configurations at intermediate temperature recorded during the cooling. The results are also shown in Fig. 2.

Above the melting temperature T_m, thermodynamic quantities do not depend on the history of the system, i.e. heating or cooling. The glass transition is defined by the change in slope of the potential energy, enthalpy and volume and occurs at $T_g = 0.28 \, T_0$. Between T_g and T_m the system stays during the numerical experiment time in a supercooled liquid state and, unlike below Tg,

98

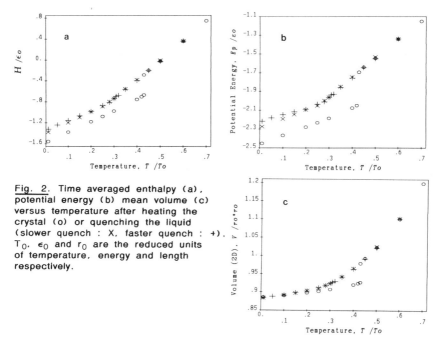

Fig. 2. Time averaged enthalpy (a), potential energy (b) mean volume (c) versus temperature after heating the crystal (o) or quenching the liquid (slower quench : X, faster quench : +). T_0, ϵ_0 and r_0 are the reduced units of temperature, energy and length respectively.

thermodynamic quantities are independent of the cooling rate. As expected, the enthalpy of the amorphous state increases with the cooling rate but the data are not accurate enough to detect a dependency of Tg. The density of the glass is comparable to that of the crystalline lattice and this compact amorphous packing is shown in Fig. 1b.

ANALYSIS OF THE GLASS TRANSITION

In Fig. 3 we show the atomic trajectories just above the glass transition for a time longer than the typical vibration time, t_0. There appears to be clusters of particles that vibrate about a mean position, while others diffuse with correlated motions. This is reminiscent of the free volume approaches of Cohen and Grest [24] where particles are set into two types : liquid-like and solid-like particles.

To have a more quantitative definition of the vibration part of atomic motions in the liquid state, we have introduced the function $f(r,t)$ defined as follows : if at time τ two neighbor particles are at a distance $r_{ij}(\tau)$, we compute the variation in length of the interatomic pair at time $\tau+t$, i.e., $r = r_{ij}(\tau+t) - r_{ij}(\tau)$. $f(r,t)$ is the distribution of the length variations, r, after the time elapsed, t, for all values of the time origin, τ, and for all pairs ij. The criterion which defines neighbors is not critical for our purpose, and we have merely chosen particles separated by less than $1.26\ r_{o\alpha\beta}$ for the pairs of type $\alpha\beta$. The distribution $f(r,t)$ for particles that are only vibrating is a Gaussian with a width independent of time (for t larger than a typical vibration time). On the other hand, if particles diffuse randomly the interatomic distances r_{ij} increase like $t^{1/2}$ and $f(r,t)$ broadens when t increases. In the liquid state $f(r,t)$ decomposes into a vibrating part and a diffusing part. In fig. 4 we show that the vibrating part can be described by a Gaussian with a width σ_T and an area $K_T(t)$. The quantity σ_T represents the amplitude of the interatomic pair vibrations in the liquid state and $K_T(t)$ is the fraction of neighbor pairs which have not diffused away after time t. As expected $\sigma_T(t)$ is independent of time,

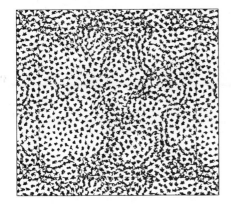

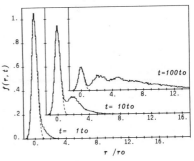

Fig. 4. Distribution $f(r,t)$ of the interatomic pair variation, r, after

Fig. 3. Atomic trajectories during 30 t_0 at T = 0.35 T_0 where t_0 and T_0 are the reduced units of time and temperature respectively.

a time elapsed, t, equals to 1 t_0, 10 t_0 and 100 t_0, in the liquid state at the temperature 0.43 T_0 (t_0, T_0 and r_0 : reduced units).

while $K_T(t)$ decreases with it.

The function $K_T(t)$ versus time could be fitted with a stretched exponential form $K_T(t) = \exp[-(t/\tau_T)^\nu]$. The value of ν was found to be a function of T over the range 0.5 to 0.7. The relaxation time τ_T could be described by a power law : $a(T - T_{op})^{-\mu}$ with $\mu = 2.16$ and $T_{op} = 0.24 T_0$. Note that the temperature T_{op} at which the relaxation time is infinite is close to the value of T_g given by the discontinuity in slope of the enthalpy variation.

We show in fig. 5 the amplitude of vibration σ_T versus temperature. σ_T is larger in the glass or the supercooled liquid than in the crystal at the same temperature and this difference increases rapidly near Tg. A movie has also been made to show that local fluctuations in the supercooled liquid can create voids large enough to allow an atom to jump in. This is illustrated in fig. 6.

CONCLUSION

We have shown in this paper that particles interacting with realistic pair potentials can form a two-dimensional glass. We have proved by means of the molecular dynamics method that this system goes from the liquid state to the

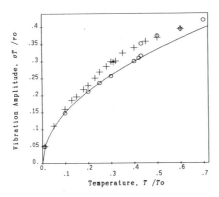

Fig. 5. Amplitude of pair vibrations, σ_T, versus T. See Fig. 2 for the meaning of the symbols.

100

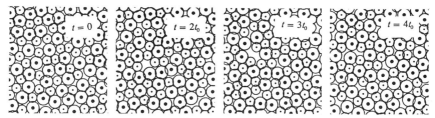

Fig. 6. Atomic configurations at Tg.

amorphous state through a glass transition. This glass transition presents the same characteristics as the three-dimensional glass transition does and thus we expect that the fundamental processes could be extrapolated from two to three dimensions. We have proposed a method to extract the vibration part of the atomic motions in the liquid state and have shown that the relaxation time during which atoms are neighbors before diffusing away, diverges to infinity when we approach the glass transitions temperature.

To yield a complete description of the microscopic mechanisms of the glass transition, we have still to study more quantitatively the repartitions and fluctuations of the free volume in the supercooled liquid and to explain why the amplitude of vibration in the glass increases rapidly near T_g.

REFERENCES

[1] H. R. Wendt and F. F. Abraham, Phys. Rev. Letters, 41, 1244-1246 (1978)
[2] F. F. Abraham, J. Chem. Phys. 72, 359-365 (1980)
[3] A. Rahman, M. J. Mandell and J. P. McTague, J. Chem. Phys. 64, 1564-1568 (1976)
[4] L. V. Woodcock, C. A. Angell and P. Cheeseman, J. Chem. Phys. 65, 1565-1577 (1976)
[5] J. H. R. Clarke, J. Chem. Soc. Faraday Trans. 2, 75, 1371-1387 (1979)
[6] J. N. Cape and L. V. Woodcock, J. Chem. Phys. 72, 976-985 (1980)
[7] Y. Hiwatari, J. Phys. C : Solid St. Phys. 13, 5899-5910 (1980)
[8] C. A. Angell, J. H. R. Clarke and L. V. Woodcock, Adv. Chem. Phys. 48, 397-453 (1981)
[9] S. K. Mitra, Phil. Mag. B 45, 529-548 (1982)
[10] M. Kimura and F. Yonezawa, in Topological Disorder in Condensed Matter, edited by F. Yonezawa and T. Ninomiya (Springer-Verlag Berlin Heidelberg 1983) pp. 80-110
[11] J. R. Fox and H. C. Andersen, J. Phys. Chem. 88, 4019-4027 (1984)
[12] F. Spaepen, J. Non-cryst. Sol. 31, 207-221 (1978)
[13] D. R. Nelson, M. Rubinstein and F. Spaepen, Phil. Mag. A46, 105-126 (1982)
[14] A. S. Argon and H. Y. Kuo, Mater. Sci. & Eng. 39, 101-109(1979)
[15] A. S. Argon and L. T. Shi, Phil. Mag. A 46, 275-294 (1982)
[16] D. Srolovitz, K. Maeda, V. Vitek and T. Egami, Phil. Mag. A 44, 847-866 (1981)
[17] F. Lançon, L. Billard and A. Chamberod, J. Phys. F : Met. Phys. 14, 579-591 (1984)
[18] F. Lançon, L. Billard, J. Laugier and A. Chamberod, J. Physique 46, 235-241 (1985)
[19] W. G. Hoover, A. J. C. Ladd and B. Moran, Phys. Rev. Letts. 48, 1818-1920 (1982)
[20] D. J. Evans and G. P. Morris, Chemical Physics 77, 63-66 (1983)
[21] F. F. Abraham, J. Vac. Sci. Technol. B 2, 534-549 (1984)
[22] F. F. Abraham, Private Communication
[23] A. Nordsieck, Math. comp. 16, 22-49 (1962)
[24] M. H. Cohen and G. S. Grest, Phys. Rev. B, 20, 1077-1098 (1979)

DYNAMIC PROPERTIES OF COMPOSITE MATERIALS
USING A T-MATRIX TO DESCRIBE MICROSTRUCTURE

V. V. VARADAN AND V. K. VARADAN
Department of Engineering Science and Mechanics and
The Materials Research Laboratory
The Pennsylvania State University Park
University Park, PA 16802

ABSTRACT

A dispersion equation is obtained for plane wave propagation in a discrete random medium. The effect of multiple scattering between the discrete inhomogeneities, statistical correlation in the position of the scatterers, details of the geometry, size and properties of the inhomogeneity via the T-matrix are considered. The resulting effective wavenumber for the average or composite medium depends on the above parameters and the frequency. The complex effective wavenumber in turn can be related to the effective properties of the composite material. The formalism is extremely well suited for numerical computation and can thus yield results suited for engineering applications. The other advantage is that the generality of the T-matrix description makes it convenient for describing acoustic, electromagnetic and elastodynamic problems.

INTRODUCTION

The average or effective properties of a composite material containing inclusions of one material or voids distributed in some fashion in a second material called the host or matrix material can be conveniently studied by analyzing the propagation of plane waves in such materials and solving the resulting dispersion equations. Since waves propagating in such a two phase system will undergo multiple interactions with the scatterer phase, it becomes natural to consider multiple scattering theory and ensemble averaging techniques if the distribution of the inclusion phase is random. Multiple scattering theory relies on the existence of a host phase which is assumed to be connected and a scatterer phase which is assumed to be discrete. As the volume fraction of the scatterer phase increases and is greater than or equal to the volume fraction of the host phase, then it is expected that a multiple scattering description will become unreliable. This is in contrast to other approaches such as those of Bruggeman [1] for two phase composites and Biot [2] for porous media where both phases are treated on an equal footing.

In a wave propagation description of the discrete composite material, the effective wavenumber which is complex, frequency and microstructure dependent describes the propagation of the average or coherent field in a particular direction in the medium. The effective wavenumber can in turn be related to the effective material properties of the medium in the usual fashion. This definition of the effective material properties differs formally from other methods in which the effective properties are obtained directly without recourse to an assumption of the effective wavenumber for the coherent field (see for example Bedeaux and Mazur [3]). No formal comparison has been made of the two approaches. It is safe to state that in most cases, the direct method is limited to long wavelengths and the static limit whereas, wave propagation studies enable one to consider wavelengths comparable to the size and geometry of the microstructure and detailed modelling of the interaction of the relevant fields with the microstructure. In the quasi-static theories the scatterer is simply modelled as a point scatterer or by a dipole moment.

In this paper, a multiple scattering theory is presented that utilizes a T-matrix to describe the response of each scatterer to an incident field. The T-matrix is simply a representation of the Green's function for a single scatterer in a basis of spherical or cylindrical functions. In this definition, it simply relates the expansion coefficients of the field that is incident on or excites a scatterer to the expansion coefficients of the field scattered when both fields are expanded in the same spherical wave basis [4]. In theory the T-matrix is infinite, but in practice the T-matrix is truncated at some size that depends on the ratio of size of scatterer to the wavelength and the complexity of the geometry. Formally the T-matrix includes a multipole description of the field scattered by the inclusion and this requires a propagator for multipole fields to describe the propagation from one scatterer to the next. Finally, the technique presented here is for a random distribution of scatterers which requires an ensemble average over the position of the scatterers and requires a knowledge of the positional correlation functions.

The formalism presented is generally applicable to acoustic electromagnetic and elastic waves. Numerical results for the effective properties of dielectric/dielectric or dielectric/metal composites [5], [6], [7] fiber reinforced structural composites [8], ·sound absorbing composites [9], [10], etc., have been obtained. Good agreement has been obtained with available experimental results for all three types of waves for a wide range of wavelengths, scatterer concentration and properties. The theory presented here most closely resembles the work of Twersky [11], [12].

EFFECTIVE WAVENUMBER FOR THE AVERAGE FIELD IN A DISCRETE RANDOM MEDIUM

In this section, the average field in the random medium is written as a partial summation of a multiple scattering series. By assuming that the average field is a plane wave with an effective wavenumber K, the resulting dispersion equation is solved. The formalism is general and applicable to all three types of waves. Then in separate sections applications to specific problems of engineering interest in the three different areas are discussed briefly.

Let the random medium contain N scatterers in a volume V such that $N \to \infty$, $V \to \infty$ but $n_o = N/V$ the number density of scatterers is finite. Let u, u^o, u^e_i, u^s_i be respectively the total field, the incident or primary plane, harmonic wave of frequency ω, the field incident or exciting the i-th scatterer and the field which is in turn scattered by the i-th scatterer. These fields are defined at a point r which is not occupied by one of the scatterers. In general, these fields or potentials which can be used to describe them satisfy the scalar or vector wave equation. Let $Re\ \phi_n$ and $Ou\ \phi_n$ denote the basis of orthogonal functions which are eigenfunctions of the Helmholtz equation. The qualifiers Re and Ou denote functions which are regular at the origin and outgoing at infinity which are respectively appropriate for expanding the field which is incident on a scatterer and that which it scatters which in turn must satisfy outgoing or radiation conditions. Thus, we can write the following set of self-consistent equations:

$$u \quad = \quad u^o + \sum_{i=i}^{N} u^s_i \quad = \quad u^e_i + u^s_i \quad \equiv \quad u^o + \sum_{i \neq j} u^s_j + u^s_i \tag{1}$$

$$u^o(r) = p \exp(ik k_o \cdot r) = \sum_n \alpha^i_n\ Re\ \phi_n\ (r-r_i) \tag{2}$$

$$u^e_i = \sum_n \alpha^i_n \, \text{Re} \, \phi_n(\mathbf{r} - \mathbf{r}_i); \qquad\qquad a < |\mathbf{r} - \mathbf{r}_i| < 2a \qquad\qquad (3)$$

$$u^s_i = \sum_n f^i_n \, \text{Ou} \, \phi_n(\mathbf{r} - \mathbf{r}_i); \qquad\qquad |r - r_i| > a \qquad\qquad (4)$$

where α^i_n and f^i_n are unknown expansion coefficients. We observe in Eqs. (3) and (4) that "a" is the radius of the sphere or cylinder (for 2-D problems) circumscribing the scatterer and that all expansions are with respect to a coordinate origin located in a particular scatterer.

The T-matrix by definition simply relates the expansion coefficients of u^e_i and u^s_i provided $u^e_i + u^s_i$ is the total field which is consistent with the definitions in Eq. (1). Thus [4],

$$f^i_n = \sum_{n'} T^i_{nn'} \, \alpha^i_{n'} \qquad\qquad (5)$$

and the following addition theorem for the basis functions is invoked

$$\text{Ou} \, \phi_n \, (\mathbf{r} - \mathbf{r}_j) = \sum_{n'} \sigma_{nn'} \, (\mathbf{r}_i - \mathbf{r}_j) \, \text{Re} \, \phi_{n'} \, (\mathbf{r} - \mathbf{r}_i) \qquad\qquad (6)$$

Substituting Eqs. (2) - (6) in Eq. (1), and using the orthogonality of the basis functions we obtain

$$\alpha^i = a^i + \sum_{j \neq i} \sigma \, (\mathbf{r}_i - \mathbf{r}_j) \, \alpha^j \qquad\qquad (7)$$

This is a set of coupled algebraic equations for the exciting field coefficients which can be iterated and leads to a multiple scattering series.

For randomly distributed scatterers, an ensemble average can be performed on Eq. (7) leading to

$$\langle \alpha^i \rangle_i = a^i + \langle \sigma \, (\mathbf{r}_i - \mathbf{r}_j) \, T^j \langle \alpha^j \rangle_{ij} \rangle_j \qquad\qquad (8)$$

where $\langle \; \rangle$ ijk... denotes a conditional average and Eq. (8) is an infinite hierarchy involving higher and higher conditional expectations of the exciting field coefficients. In actual engineering applications, a knowledge of higher order correlation functions is difficult to obtain, usually the hierarchy is truncated so that at most only the two body positional correlation function is required.

To achieve this simplification the quasi-crystalline approximation (QCA), first introduced by Lax [13] is involved, which is stated as

$$\langle \alpha^j \rangle_{ij} \approx \langle \alpha^j \rangle_j \qquad\qquad (9)$$

Then, Eq. (8) simplifies to

$$\langle \alpha^i \rangle_i = a^i + \langle \sigma \, (\mathbf{r}_i - \mathbf{r}_j) \, T^j \langle \alpha^j \rangle_j \rangle_i \; ; \qquad\qquad (10)$$

an integral equation for $\langle \alpha^i \rangle_i$ which in principle can be solved. We observe that the ensemble average in Eq. (10) only requires $P(\mathbf{r}_j | \mathbf{r}_i)$, the joint probability distribution function. In particular, the homogeneous solution of Eq. (10) leads to a dispersion equation for the effective medium in the quasi-crystalline approximation. Defining the spatial Fourier transform of $\langle \alpha^i \rangle_i$ as

$$\langle \alpha^i \rangle_i = \int e^{i\mathbf{K} \cdot \mathbf{r}_i} \, X^i(\mathbf{K}) \, d\mathbf{K} \qquad\qquad (11)$$

and substituting in Eq. (10), we obtain for the homogeneous solution

$$X^i(K) = \sum_{j \neq i} \int \sigma (r_i - r_j) \; T^j \; P(r_j | r_i) \; e^{i \; K \cdot (r_i - r_j)} \; dr_j \; X^j(K) \tag{12}$$

If the scatterers are identical

$$X^i(K) = X^j(K) = X(K) \tag{13}$$

and thus for a non-trivial solution to $\langle \alpha^i \rangle_i$, we require

$$\left| I - \sum_{j \neq i} \int \sigma (r_i - r_j) \; T^j \; P(r_j | r_i) \; e^{i \; K \cdot (r_i - r_j)} \; dr_j \right| = 0 \tag{14}$$

In Eqs. (12) and (14), $P(r_j | r_i)$ is the joint probability distribution function. For isotropic statistics,

$$P(r_j | r_i) = 0; \quad |r_i - r_j| < 2a$$

$$g(|r_i - r_j|) / V ; \quad |r_i - r_j| > 2a \tag{15}$$

where we have assumed that the scatterers are impenetrable with a minimum separation between the centers being the diameter $2a$ of the circumscribing sphere in 3-D and circle in 2-D. Equation (14) can hence be simplified to

$$\left| I - n_o \int \sigma (r_1 - r_2) \; T \; g \; (|r_1 - r_2|) \; e^{i \; K \cdot (r_1 - r_2)} \; d(r_1 - r_2) \right| = 0 \tag{16}$$

where $(1/V) \sum_{j \neq i} = (N-1)/V \approx n_o$. The integral in Eq. (16) is simply the spatial Fourier transform of $\sigma \, T \, g$. The zeroes of the determinant as expressed by Eq. (16), yield the allowed values of K as a function of the microstructure as determined by the T-matrix, the number density n_o and the statistics of the distribution as determined by the pair correlation function. In general K, the effective wavenumber is complex and frequency dependent.

It is interesting to examine what type of multiple scattering terms contribute to the quasi-crystalline approximation. If Eq. (8) is iterated, we obtain

$$\langle \alpha^i \rangle_i = a^i + \langle \Sigma_{j \neq i} \; \sigma^{ij} \; T^j \; a^j \rangle_i$$

$$+ \langle \Sigma_{j \neq i} \; \sigma^{ij} \; T^j \; \langle \Sigma_{k \neq j} \; \sigma^{jk} \; T^k \; a^k \rangle_{ij} \rangle_i + \cdots \tag{17}$$

here the abbreviation $\sigma^{ij} = \sigma (r_i - r_j)$ has been used, and matrix multiplication is implied throughout. We recall that σ, T are suitably truncated matrices and α, a are suitably truncated vectors. Suppose the QCA is invoked for each term in Eq. (17), i.e.

$$\langle \sigma^{jk} \; T^k \; a^k \rangle_{ij} \approx \langle \sigma^{jk} \; T^k \; a^k \rangle_j \tag{18}$$

Then we note that only two body correlations are required and the multiple scattering series in Eq. (17) can be easily summed by spatial Fourier transform techniques using the convolution theorem. Symbolically, the multiple scattering series in Eq. (17) may be represented as

$$\langle \alpha^i \rangle_i = a^i + \frac{\qquad}{j} + \frac{\qquad}{j \quad k} + \frac{\qquad}{j \quad k \quad l} + \cdots \tag{17'}$$

where ———— denotes σ, $_j$ $_k$ denotes $p(\mathbf{r}_j|\mathbf{r}_k)$ $_j$ denotes T^j and $\leftarrow_j$ denotes a^j. Eq. (17)$_j$ or its alternate form^j (17') can be summed and written as

$$\langle\alpha^i\rangle_i = \int[1 - n_n \int \sigma(\mathbf{x})\ T\ g(\mathbf{x})\ e^{-i\mathbf{k}\cdot\mathbf{x}}d\mathbf{x}]^{-1} \exp i\mathbf{K}\cdot\mathbf{r}_i\ a^i\ d\mathbf{K} \qquad (19)$$

In Eq. (19), the matrix inverse is the spatial Fourier transform of the Green's function or propagator for the effective medium in the QCA as given in Eq. (18). The dispersion equation for the medium is given by the zeroes of the spatial Fourier transform of the Green's function. Thus, the dispersion equation resulting from Eq. (19) is identical to the one obtained from Eq. (15).

Numerical Results

The dispersion equation as given in Eq. (15) is very well suited for computation. Using appropriate forms of the basis functions ϕ_n which are solutions of the field equations, the T-matrix of the single scatterer can be computed; for example, see Varadan and Varadan [4]. The translation matrix σ, although complicated in form for cylindrical and spherical functions, can nevertheless be computed in a straight forward manner. The spatial Fourier transform of $\sigma\ T\ g$ is fairly easy to compute because the integrand is well behaved for large values of the interparticle distance. In the results presented for different types of wavefields, tabulated Monte Carlo values for inpenetrable spheres [14] were substituted for g at various values of the concentration. The roots of the resulting determinant were found using Muller's method by giving initial guesses using the analytic expressions for K which can be obtained from Eq. (15) in the long wavelength limit. The real and imaginary parts of the effective wavenumber can be related respectively to the phase velocity and attenuation in the effective medium. The attenuation is due to geometric dispersion or scattering which may be further enhanced if there are losses associated with the material properties of the scatterer and/or the host.

(a) **Acoustic properties of a phase fluid.** The scatterers in this case can be penetrable elastic solid or fluid particles, acoustically hard or acoustically soft spheres. The adiabatic compressibility of the effective medium can be obtained from the definition

$$\frac{K^2(\omega)}{k^2} = \frac{\langle\chi\rangle\ \langle\rho\rangle}{\chi_0\ \rho_0} \qquad (20)$$

where k is the wavenumber in the host medium, $\langle\rho\rangle = (1-c)\rho_0 + c\rho_s$, the average mass density, ρ_0 and ρ_s being the density of the host scatterer materials, χ_0 is the adiabatic compressibility of the host material, $\langle\chi\rangle$ is the effective compressibility of the composite fluid and c the volume fraction of scatterers.

(b) **Effective properties of structural composites.** In a fiber reinforced composite, for P – (longitudinal), SV – (transversely polarized shear wave) and SH – (shear waves polarized parallel to the fibers) wave propagation. If the fibers are circular and parallely-oriented, the effective medium is transversely isotropic and will be characterized by five elastic constants. By calculating K_p, K_{sv} and K_{sh}, three of the five elastic constants can be found as a function of frequency, fiber geometry and concentration. If wave propagation along the fibers is considered, then the remaining constants can also be found. In the long wavelength limit analytical expressions for the static values agree with the well known Hashin-Shtrikman bounds [8].

(c) <u>Effective elastic properties of particulate composites</u>. For spherical particles randomly distributed in a host material, the composite material is effectively isotropic and hence characterized by two effective elastic constants which can be obtained by solving the dispersion equations for K_p and K_s, see Refs. 9 and 10.

(d) <u>Effective dielectric properties of dielectric composites</u>. For a random distribution of dielectric or metal spheres distributed in a dielectric host material, the medium is effectively isotropic and characterized by just one dielectric function which can be obtained from the effective wavenumber K via $<\varepsilon> = \varepsilon<K>^2/ k^2$, where ε is the dielectric function of the host material, see Refs. 5-7.

REFERENCES

1. D. A. Bruggeman, Ann. Phys. Lpz. <u>24</u>, 636 (1935).
2. M. A. Biot, J. Acoust. Soc. Am, 179 (1956).
3. D. Bedeaux and P. Mazur, Physica <u>67</u>, 63 (1973).
4. V. K. Varadan and V. V. Varadan, Eds. Acoustic, *Electromagnetic and Elastic Wave Scattering - Focus on the T-matrix Approach*, 103, Perganon Press, New York (1980).
5. V. N. Bringi, V. K. Varadan and V. V. Varadan, IEEE, <u>AP-31</u>, 371 (1983).
6. V. K. Varadan, Y. Ma and V. V. Varadan, Radio Science <u>19</u>, 1445 (1984).
7. V. K. Varadan, V. V. Varadan and Y. Ma, IEEE-MTT <u>33</u>, No. 5 (1986).
8. V. K. Varadan, V. V. Varadan and Y. Ma, J. Acoust. Soc. Am., in press.
9. V. K. Varadan, V. V. Varadan and Y. Ma, J. Acoust. Soc. Am., <u>76</u>, 296 (1984).
10. V. K. Varadan, Y. Ma and V. V. Varadan, J. Acoust. Soc. Am. <u>77</u>, 375 (1985).
11. V. Twersky, J. Math. Phys. <u>19</u>, 215 (1978).
12. V. Twersky, J. Acoust. Soc. Am., <u>64</u>, 1710 (1978).
13. M. Lax, Phys. Rev. <u>88</u>, 621 (1952).
14. J. A. Barker and D. Henderson, Mol. Phys., <u>21</u>, 187, (1971).

PSEUDOPOTENTIAL CALCULATIONS OF STRUCTURAL PROPERTIES OF SOLIDS

MARVIN L. COHEN
Department of Physics, University of California, and Materials and Molecular
Research Division, Lawrence Berkeley Laboratory, Berkeley, CA 94720

ABSTRACT

Through the development of a total energy pseudopotential approach,
it has become possible to compute structural, electronic, and vibrational
properties of solids using only the atomic numbers and atomic masses of
the constituent atoms as input. The method has been applied to semiconductors, insulators, and metals; and agreement with experiment for most properties is usually within a few percent.
Applications include the determination of lattice constants for specific
structural phases, properties of structural phase transitions, cohesive
energies, bulk moduli, lattice vibrational spectra, electron-lattice interaction parameters, and electronic and superconducting properties. A recent
example is the prediction of superconductivity in highly condensed hexagonal
silicon which was subsequently found experimentally.

INTRODUCTION

Since the introduction of the pseudopotential approach by Fermi [1,2],
applications of this method have been varied. Originally, the scheme was
constructed to be appropriate for atoms; however, recent applications have
been made to study molecules and clusters, and the largest effort has been
in the study of solids.
A fundamental property of the pseudopotential is its separation of
core electron properties from those of valence electrons. The cores or
ions composed of nuclei plus core electrons are considered to be inert
and unchanged in going from a gas of isolated atoms to a solid composed
of strongly interacting atoms. It is assumed that the valence electrons
are largely responsible for the bonding and most of the electronic properties
commonly studied in solid state physics and chemistry.
Because the valence electron wavefunctions are orthogonal to the core
electron states, they are repelled from the core region. This is a consequence of the Pauli exclusion principle. Phillips and Kleinman [3] demonstrated that this repulsive potential cancels a large part of the attractive
Coulomb potential from the nucleus leaving a net weak pseudopotential.
It is possible to compute the orthogonality terms and the net potential,
but the development of this area proceeded along a different path. A major
development was the introduction of model potentials and empirical potentials
[4] using experimental input. These potentials were used to explain Fermi
surfaces of metals, optical properties of semiconductors and insulators,
and a large number of other properties of solids.
One of the most widely applied empirical schemes is the Empirical
Pseudopotential Method (EPM) [4,5,6] which involved the fitting of a few
form factors or Fourier coefficients of the potential to experimental data.
Usually, optical measurements were used. The pseudopotential was overdeter-
mined even with one set of data; hence it could be applied to explain and
predict other solid state properties. Some structural studies were done
through the use of electronic charge densities and calculations of bond
charges.
Most modern calculations employing pseudopotentials for structural
calculations use so called ab initio or first-principles potentials. These
potentials are usually generated from atomic wavefunctions [7-12]. The

108

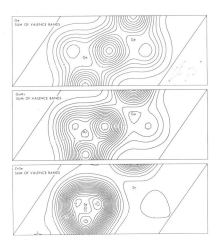

Fig. 1. Total calculated charge density for Ge, GaAs, and ZnSe using the EPM. The contours are in units of charge e per unit cell volume.

total energy of a model system is computed for a series of structures, and the lowest energy structure is assumed to correspond to the observed phase. It is convenient to compute the total energy in momentum-space [13] and to use a local density approach [14] for the electron-electron interactions.

Once the structural energy for a specific material in a given structure is computed as a function of volume, the results can be used to determine the lattice constants, bulk mdoulus, and other structural properties. Since a local density approach is normally used for the electron-electron interactions, the calculated ground-state properties are expected to be reliable. For excited states, recent studies suggest possible modifications [15,16].

The pseudopotential-total energy approach has been used to study a variety of solids, and reviews are available [17]. In this paper, some specific applications to semiconductors, insulators, and metals are described. In addition, some high pressure studies of superconductivity in silicon are discussed. These studies resulted from investigations of the structural properties of this material.

An important application which will not be dealt with here is the study of surfaces and interfaces [18]. Electronic structure and reconstruction geometries have been studied in detail.

SEMICONDUCTORS

In the EPM described above, it was determined that electronic charge density plots [19] give an accurate picture of electron bonding trends. Phillips and Van Vechten developed an empirical theory [20] to explain the structural transition between 4-fold and 6-fold coordinated systems. An interesting material series to study this trend is Ge, GaAs, and ZnSe since the elements involved have the same core configurations. All three semiconductors have eight valence electrons per cell and almost identical lattice constants.

Figure 1 illustrates the charge density and the transition from covalent bonding in Ge to mostly ionic bonding in ZnSe. Using the Phillips-Van Vechten scale of ionicity, it is expected that at an ionicity $f_i \approx 0.8$ a transition from the zincblende to the rocksalt structure is expected. By integrating the charge in Fig. 1 in the covalent bond region, the bond charge Z_b can be computed

$$Z_b = \int_\Omega (\rho - \rho_0) d^3 r \qquad (1)$$

where ρ and ρ_0 represent the charge density and background charge density and Ω is the cell volume. Figure 2 shows the dependence of Z_b on ionicity for Ge, GaAs, and ZnSe and the next row down in the periodic table Sn, InSb, and CdTe. One observes that the bond charge goes to zero at the critical ionicity for 4-fold to 6-fold coordination. This can be interpreted

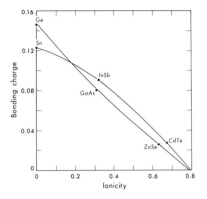

Fig. 2. Bond charge Z_b for two iso-electronic series of diamond and zinc-blende semiconductors as a function of Phillips ionicity.

as a loss of the covalent bond charge which stabilized the tetrahedral co-ordinated structure.

The EPM was used in ways similar to the example used above, and some attempts to calculate the total energy were made. However, most applications of total energy approaches involved <u>ab initio</u> potentials as described above. The first application was to Si [21]. Structural properties and lattice vi-brational properties were computed. These calculations required only the atomic number and atomic mass as input.

Using Si as an example, the energy-volume curve, E(V), is given in Fig. 3 for eight possible structures [22]. The diamond structure is lowest in en-ergy, and the position and curvature of the minimum of the E(V) curve yields the lattice constant and bulk modulus. A calculation of the total energy for very large volumes gives a measure of the energy of an isolated atomic system. Subtraction of this energy from the diamond structure equilibrium energy gives a measure of the cohesive energy of the system. Results for Si and Ge are given in Table I.

The application of pressure decreases the volume, and at smaller volumes, other structures have lower energies. By drawing a common tangent between the E(V) curves for different structures, it is possible to determine [17] the volumes for the transitions and the transition pressures. Typically, results for lattice constants are accurate to $\lesssim$ 1% while bulk moduli, cohe-sive energies, transition volumes, and transition pressures are accurate to $\lesssim$ 6%. Phonon frequencies can also be computed to this level of accuracy [23]. The method was also applied to III-V zincblende semiconductors [24,25] with similar success.

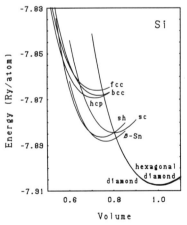

Fig. 3. The total structural energy of Si versus volume normalized by the calculated equilibrium volume in the diamond structure.

INSULATORS

One of the first applications to an insulating system was the case of carbon. In contrast to Si and Ge, C was not found to convert to the ß-tin struc-ture at high pressures. This appears to be a result of the lack of p-states in the C core. Since there is no p-repul-sive potential, the bonding structure differs from Si and Ge. It was suggest-ed [21] that C in the diamond structure would transform to the simple cubic structure at 23 Mbars. It was later de-termined [28,29] that a body-centered cubic structure with eight atoms in the unit cell should be stable at a lower pressure of about 12 Mbar.

Since most insulators are ionic, an interesting series of increasing ionicity is BeO [30], MgO [31], and NaCl [32]. For BeO, the wurtzite structure

Table I. Comparison of calculated and measured static properties of Si and Ge.

	Lattice constant (A)	Cohesive energy (eV/atom)	Bulk modulus (Mbar)
Si			
calculation	5.451	4.84	0.98
experiment	5.429	4.63	0.99
Ge			
calculation	5.655	4.26	0.73
experiment	5.652	3.85	0.77

is found lowest at normal pressures (Fig. 4) with a sizable gap in energy to the zincblende phase. At high pressures (≈ 22 GPa), a transition to rocksalt is predicted. The cases of MgO and NaCl are of particular interest because these compounds exist in the rocksalt structure at normal pressures and are considered to be prototypical ionic insulators.

Following the ideas of Phillips and Van Vechten discussed earlier, it is expected that MgO is more ionic than BeO even though both are II-VI compounds. This suggests that MgO has an ionicity which is larger than the critical value of 0.8. In addition, the question has been raised about the possibility of a transformation of MgO from the rocksalt (B1) structure to the CsCl (B2) structure at pressures found in the earth. The E(V) curves for MgO in these structures are given in Fig. 5. Using these curves, it was found that a transition pressure of around 10 Mbar would be necessary to cause a B1 to B2 structural transition. Hence, it is unlikely that a B2 phase of MgO exists even in the lower mantle of the earth.

The results for the lattice constant and other structural properties of BeO, MgO, and NaCl are all within the range of accuracy found for semiconductor materials. This supports the claim that the pseudopotential approach is applicable to ionic as well as covalent systems. The results are particularly impressive for the case of NaCl since this material has been considered to be the prototype for ionic compounds. Systems of this type are often modelled with a localized combination of atomic orbitals (LCAO). In the present case, the calculations employed a plane wave basis set and pseudopotentials. As indicated above, the results for the structural properties and the B1 to B2 transition pressure obtained from the E(V) curve (Fig. 6) are

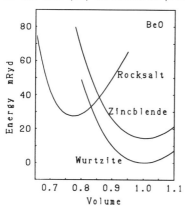

Fig. 4. Calculated total energies per BeO molecule as a function of volume for the wurtzite, zincblende, and rocksalt structures. The volumes are normalized to the calculated equilibrium volume in the wurtzite structure.

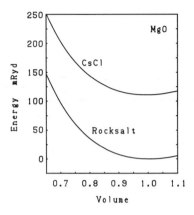

Fig. 5. Total structural energies per MgO molecule as a function of volume for the NaCl (B1) and CsCl (B2) structures. The volumes are normalized to the calculated equilibrium volume in the rocksalt structure.

in good agreement with experiment. A calculation of the transverse optical frequency is also consistent with the measured values.

For localized electronic p-states as found in ionic insulators and similarly for localized d-states, a plane wave basis set often requires a large matrix to represent accurately the secular equation. For these cases, it is more convenient to use a localized basis or a mixture of localized functions and plane waves. A pseudopotential scheme [33] of this kind has been introduced recently, and some applications have been made. The use of the approach for diamond yields structural properties in agreement with the plane wave results. A calculation [34] of the optical phonon frequency as a function of pressure is shown in Fig. 7. This result may have applications in diamond anvil technology as a method for measuring pressure. Raman scattering at high pressures [34,35] can be used to measure the optical phonon frequency and hence be used to estimate the pressure.

METALS AND HIGH PRESSURE SUPERCONDUCTIVITY

Because of its proximity to Si in the periodic table, Al was one of the first metals studied with the methods described. High pressure [36] and lattice vibrational [37] studies yielded results which were comparable in accuracy to the semiconductor and insulator cases. Similar results were obtained for Be [38, 39] and Mg [40]. The latter cases are hexagonal, and a comparison of their structural properties (including the Poisson ratio) reveals the influence of the simpler core of Be.

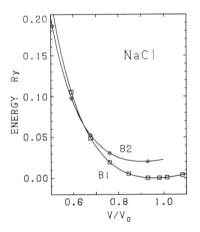

Fig. 6. Total energies per unit cell for NaCl as a function of volume in the B1 and B2 structures. The normalizing volume corresponds to the calculated value at equilibrium for the B1 phase. The solid lines are fits to the calculated points.

The alkali metals are often considered to be prototype metals which are free-electron-like and subject to simple approximations. It is ironic that in the two cases studied, Na [11] and Rb [41], the results are more acutely affected by the approximations used than for any other materials tested. In particular, the total energy curve $E(V)$ depends sensitively on the choice of the local density functional used to model exchange and correlation. Using Na as an example, this solid has a low cohesive energy, and the treatment of exchange among the valence electrons and between valence and core electrons is important. The core in Na occupies a significant fraction of space, and the usual approximations used require modification.

112

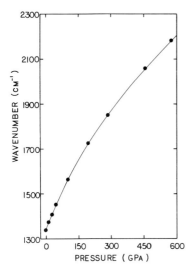

Fig. 7. Calculated frequency of the diamond zone-center optical phonon versus pressure. The solid line is a fit to the calculated points.

Returning to Al where the standard method works very well, it is now possible to compute the phonon spectrum, electronic bands, and electron-phonon matrix elements entirely from first principles (i.e. with just the atomic number and mass as input). Results [42] for the phonon dispersion curves, the wavevector dependent phonon linewidths, and the electron-phonon parameter are obtained without resorting to standard approximations such as the rigid ion approach. The electron-phonon coupling calculation is discussed below.

Assuming a phonon distortion in the crystal of wavevector $\vec{q}$, frequency ω, and branch ν, the electron-phonon matrix element is given by

$$g(\vec{q},\omega) = (\frac{\hbar}{2M\omega})^{1/2} \delta(\vec{k}-\vec{k}'-\vec{q})\langle n,\vec{k}|\hat{\epsilon}\cdot\overline{\nabla V}|n',\vec{k}'\rangle \qquad (2)$$

where $\langle n,\vec{k}|$ and $|n',\vec{k}'\rangle$ are Bloch electron states for wavevectors $\vec{k}$ and $\vec{k}'$ and bands n and n'. The unit vector $\hat{\epsilon}$ is the phonon polarization vector, M is the ionic mass, and $\overline{\nabla V}$ is the change in the potential arising from the phonon distortion. The potential gradient is computed self-consistently, and Eq. (2) can be used to evaluate a $\vec{q}$-dependent electron-phonon parameter

$$\lambda_\nu(\vec{q}) = 2N(E_F) \frac{\langle\langle g(\vec{q},\omega)\rangle\rangle}{\hbar\omega} \qquad (3)$$

where $N(E_F)$ is the density of states at the Fermi energy, and $\langle\langle \rangle\rangle$ denotes a Fermi surface average.

The standard McMillan [42] coupling constant is given by

$$\lambda = \sum_\nu \int_{\Omega_{BZ}} d^3q \lambda_\nu(\vec{q}) \qquad (4)$$

where Ω_{BZ} is the volume of the Brillouin zone. By defining a wavevector dependent coupling, it is possible to examine the phonon line width $\gamma_\nu(\vec{q})$ which arises from electron-phonon scattering. This function is related to $\gamma_\nu(\vec{q})$ through the expression [43]

$$\lambda(\vec{q}) = \sum_\nu \lambda_\nu(\vec{q}) = \sum_\nu \frac{1}{\pi N(E_F)} \frac{\gamma_\nu(\vec{q})}{\hbar\omega^2} \qquad (5)$$

The results for Al appear in reference [42].

It is interesting to apply the above approach to Si at high pressures. Three structures have been studied in detail--β-Sn, primitive (or simple) hexagonal ph (or sh), and hcp. All three are metals and are shown on the E(V) plot in Fig. 3. The β-Sn modification was found to be superconducting [44], and the ph and hcp phases were recently predicted [45,46] to be superconducting at relatively high temperatures.

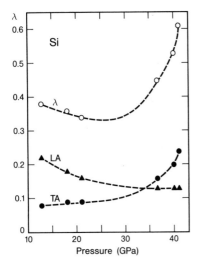

Fig. 8. Pressure dependence of the electron-phonon parameter for the ph phase of Si. The dashed lines represent fits to the calculated points. Contributions from LA and TA phonons are plotted separately.

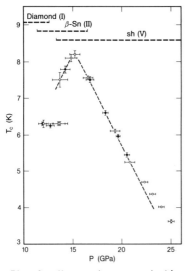

Fig. 9. Measured superconducting transition temperature as a function of pressure for high pressure phases of Si. The two types of points represent data taken for two different samples.

Early pseudopotential calculations [17,21] indicated that hcp Si should be stable in the 40 GPa pressure range. Two experimental groups [47,48] verified this prediction and discovered the ph phase in the 12 GPa range of pressure. Theoretical studies of this phase [22, 49,50] revealed that it is a metal with a relatively large density of states at the Fermi energy. The charge density variation and phonon modes [22] added to the attractiveness of this material as a potential superconductor. In particular, the charge density plots indicated that ph Si is a metal with covalent-like bonds, and this is favorable for superconductivity [51]. It was also suggested that the hcp phase be investigated for superconductivity.

These preliminary suggestions received substantial support by calculations of the electron-phonon parameters. As shown in Fig. 8, the pressure dependence of λ is dominated by LA phonons at lower pressures and by TA phonons at high pressures. The dividing point is roughly 25 GPa. This study was limited to a consideration of the phonons only in the [001] direction; hence it is considered to be only representative of a full Brillouin zone study. The calculation predicted superconductivity in the 5-10 K range and suggested that the transition temperature should decrease with pressure at lower pressures (e.g. see $\lambda(p)$ in Fig. 8). As shown in Fig. 9, superconductivity was observed [45] with the predicted properties. At present, the experimental tests have been made up to pressures of about 25 GPa. Because of the limitations of the one-phonon branch approximation, even a saturation of the transition pressure curve at high pressures would help confirm the theory. However, it should be stated that the theory does predict rather high λ's and hence relatively high transition temperatures.

CONCLUSIONS

In summary, the total energy pseudopotential approach has been applied successfully to explain and predict a variety of properties of semiconductors, insulators, and metals. Current research has focused on structural properties, and this work has led to interesting predictions about high pressure phases of solids and their properties. The

114

results have been found to be accurate in cases where experimental tests are possible, and it is hoped that this method and others like it will lead to the discovery of new materials with desirable properties.

ACKNOWLEDGEMENTS

This work was supported by National Science Foundation Grant No. DMR83-19024 and by the Director, Office of Energy Research, Office of Basic Energy Sciences, Materials Sciences Division of the U.S. Department of Energy under Contract No. DE-AC03-76SF00098.

REFERENCES

1. E. Fermi, Nuovo Cimento 11, 157 (1934).
2. M. L. Cohen, Am J. Phys. 52, 695 (1984).
3. J. C. Phillips and L. Kleinman, Phys. Rev. 116, 287 (1959).
4. M. L. Cohen and V. Heine, Solid State Phys. 24, 37 (1970).
5. J. R. Chelikowsky and M. L. Cohen, Handbook on Semiconductors, ed. W. Paul, Vol. 1 (Amsterdam, 1982), p. 219.
6. T. K. Bergstresser and M. L. Cohen, Phys. Rev. 141, 789 (1966).
7. T. Starkloff and J. D. Joannopoulos, Phys. Rev. B 16, 5212 (1977).
8. A. Zunger and M. L. Cohen, Phys. Rev. B 18, 5449 (1978).
9. G. Kerker, J. Phys. C 13, L189 (1980).
10. G. B. Bachelet, D. R. Hamann, and M. Schluter, Phys. Rev. B 26, 4199 (1982).
11. S. G. Louie, S. Froyen, and M. L. Cohen, Phys. Rev. B 26, 1738 (1982).
12. M. T. Yin and M. L. Cohen, Phys. Rev. B 25, 7403 (1982).
13. J. Ihm, A. Zunger, and M. L. Cohen, J. Phys. C 12, 4401 (1979).
14. W. Kohn and L. J. Sham, Phys. Rev. 140, A1333 (1965).
15. C. S. Wang and W. E. Pickett, Phys. Rev. Lett. 51, 597 (1983).
16. M. S. Hybertsen and S. G. Louie, Phys. Rev. Lett. 55, 1418 (1985).
17. M. L. Cohen, Phys. Reports 110, 293 (1984); Proceedings of the International School of Physics "Enrico Fermi," Course LXXXIX, (North Holland, Amsterdam, 1985), p. 16.
18. M. L. Cohen and S. G. Louie, Ann. Rev. Phys. Chem., 35, 537 (1984); M. L. Cohen, Advances in Electronics and Electron Physics, ed. L. Marton and C. Marton (New York, 1980), p. 1.
19. J. P. Walter and M. L. Cohen, Phys. Rev. B 4, 1877 (1971).
20. J. C. Phillips, Bonds and Bands in Semiconductors (New York, Academic, 1973).
21. M. T. Yin and M. L. Cohen, Phys. Rev. Lett 45, 1004 (1980).
22. K. J. Chang and M. L. Cohen, Phys. Rev. B 31, 7819 (1985).
23. M. T. Yin and M. L. Cohen, Phys. Rev. B 26, 3259 (1982).
24. S. Froyen and M. L. Cohen, Phys. Rev. B 28, 3258 (1983).
25. K. Kunc and R. M. Martin, Phys. Rev. Lett. 48, 406 (1982).
26. M. T. Yin and M. L. Cohen, Phys. Rev. B 24, 6121 (1981).
27. M. T. Yin and M. L. Cohen, Phys. Rev. Lett. 50, 2006 (1983).
28. R. Biswas, R. Martin, R. J. Needs, O. H. Nielsen, Phys. Rev. B 30, 3210 (1984).
29. M. T. Yin, Proceedings of the 17th International Conference on the Physics of Semiconductors, 1984 (Springer, New York, 1985), p. 927.
30. K. J. Chang and M. L. Cohen, Solid State Comm. 50, 487 (1984).
31. K. J. Chang and M. L. Cohen, Phys. Rev. B 30, 4774 (1984).
32. S. Froyen and M. L. Cohen, Phys. Rev. B 29, 3770 (1984).
33. J. R. Chelikowsky and S. G. Louie, Phys. Rev. B 29, 3470 (1984).
34. M. Hanfland, K. Syassen, S. Fahy, S. G. Louie, and M. L. Cohen, Phys. Rev. B 31, 6896 (1985).
35. A. F. Gronharov, I. N. Makarenko, and S. M. Stishov, J.E.T.P. Lett. 41, 184 (1985).

36. P. K. Lam and M. L. Cohen, Phys. Rev. B 27, 5986 (1983).
37. P. K. Lam and M. L. Cohen, Phys. Rev. B 25, 6139 (1982).
38. M. Y. Chou, P. K. Lam, and M. L. Cohen, Solid State Comm. 42, 861 (1982).
39. P. K. Lam, M. Y. Chou, and M. L. Cohen, J. Phys. C 17, 2065 (1984).
40. M. Y. Chou and M. L. Cohen, to be published.
41. W. Maysenholder, S. G. Louie, and M. L. Cohen, Phys. Rev. B 31, 1817 (1985).
42. M. M. Dacorogna, P. K. Lam, and M. L. Cohen, Phys. Rev. Lett. 55, 837 (1985).
42. W. L. McMillan, Phys. Rev. 167, 331 (1968).
43. P. B. Allen and B. Mikovic, Solid State Phys. 32, 1 (1982).
44. J. Wittig, Z. Phys. 195, 215 (1966).
45. K. J. Chang, M. M. Dacorogna, M. L. Cohen, J. M. Mignot, G. Chouteau, and G. Martinez, Phys. Rev. Lett. 54, 2375 (1985).
46. M. M. Dacorogna, K. J. Chang, and M. L. Cohen, Phys. Rev. B 32, 1853 (1985).
47. H. Olijnyk, S. K. Sikka, and W. B. Holzapfel, Phys. Lett. 103A, 137 (1984).
48. J. Z. Hu and I. L. Spain, Solid State Comm. 51, 263 (1984).
49. K. J. Chang and M. L. Cohen, Phys. Rev. B 30, 5376 (1984).
50. R. Needs and R. M. Martin, Phys. Rev. B 30, 5390 (1984).
51. M. L. Cohen and P. W. Anderson, Superconductivity in d- and f-band Metals, ed. D. H. Douglass (AIP, New York, 1972), p. 17.

HUNTING MAGNETIC PHASES WITH TOTAL-ENERGY SPIN-POLARIZED BAND CALCULATIONS

P. M. MARCUS*, V. L. MORUZZI* AND K. SCHWARZ**
*IBM Research Center, Yorktown Heights, N.Y. 10598
**Institut für Technische Elektrochemie, Technische Universität, A-1060 Vienna, Austria

ABSTRACT

Comments are made on total energy band calculations as tools for exploring properties of solids; the importance of fixed spin moment calculations is noted. Use of energy - magnetisation curves to locate magnetic phases is described. Detailed results for fcc and bcc Co and Ni and phase diagrams on the magnetisation - volume plane exhibit two new phases for each metal and show that ferromagnetic fcc Co and bcc Ni break down at small volumes and make first order transitions to nonmagnetic phases in a metamagnetic volume range.

INTRODUCTION

Modern total-energy band-structure programs are remarkable tools for the exploration of basic properties of crystalline solids. Such calculations can discriminate among proposed hypothetical structures or explore the effects of vast ranges of applied fields, such as stress fields or magnetic fields. It is now possible to solve the wave equation for all electrons at a dense net of points in k-space, integrate over the occupied states to fix the Fermi level and find the corresponding electron density in position space, recalculate the potential and iterate to self-consistency with adequate accuracy for total energies in times of a few minutes. Hence it is feasible to repeat the calculation many times in searching parameter space for special phenomenology. The physical conclusions about structure are not drawn from the *total* energy computed in this way, but from the more reliable *cohesive* energy, which is the total energy of the crystal minus the total energy of separated atoms, computed in the *same* way.

The efficiency of such a calculation is made possible by three approximations: 1) the local-density approximation for electron exchange and correlation, which reduces the many-electron wave equation to a set of one-electron equations, coupled through the total density, 2) the muffin-tin form of the potential, which separates the scatterers (ion cores) in a homogeneous background and greatly simplifies the multiple scattering problem; it is especially suitable for metals, 3) linearization of the secular equation for the eigenvalues and vectors of individual electron states, which permits fast diagonalization procedures to be used, as is the case for the augmented spherical wave method used here.[1]

In this work we have systematically studied the cohesive energies of the ferromagnetic elements as functions of volume per atom V and spin magnetic moment per atom M (using the spin-polarized form of the local density expression[2]), searching for minima in the energy that indicate the presence of magnetic phases of the elements. The search has been made possible by the procedure of fixing M as well as V when iterating to self-consistency. This fixed-moment method[3] will be discussed further below, where its advantages when several magnetic phases exist at the same V will be noted. We find clear evidence in several cases for volume ranges in which two magnetic phases coexist and cohesive energy curves cross at a particular volume, indicating occurrence of a first-order phase transition. Our calculations so far are restricted to static cubic ferromagnetic phases, without relativistic effects such as spin-orbit interactions. We give detailed results for Co and Ni in this paper; the more complicated behavior of Fe will be published later.

IDENTIFYING MAGNETIC PHASES

The basic tool for locating magnetic phases consists of a set of curves of energy per atom E as functions of M at a series of values of V (or, equivalently, of the Wigner-Seitz radius r_{WS} of the sphere with volume V). Figure 1 shows such curves for fcc Co over the interesting range of V in which Co makes a first-order transition from a ferromagnetic (FM) phase to a nonmagnetic (NM) phase. Thus at $r_{WS}=2.52$ a.u. a single minimum in E at finite M corresponds to a FM phase (its value is taken as the energy origin). All points along this curve correspond to states of the system in an applied magnetic field $H=(\partial E/\partial M)_V$, (the states with $H<0$, i.e., opposite to M, are thermodynamically unstable), and under an applied pressure $p=-(\partial E/\partial V)_M$. At the minimum the applied field vanishes, although p does not, and it is appropriate to refer to the corresponding state as a magnetic phase since it sustains a magnetic moment without an applied field when maintained at a lattice volume corresponding to $r_{WS}=2.52$. The minima are located more precisely by means of the derivative curve, which goes through zero at these points. Figure 2 shows M vs. $(\partial E/\partial M)_V$ along each of the curves of Fig. 1 and shows clearly that at M=0 the maximum at $r_{WS}=2.52$ is converted to a minimum for $r_{WS}\leq2.50$. The FM phase at 2.52 is continuously connected by similar minimum states to the FM state at p=0, as shown in Fig. 3 at $r_{WS}=2.58$ for fcc Co; we refer to the connecting line on the M - V plane as a phase line. The fcc Co phase line ends at $r_{WS}=2.46$, where Figs. 1 and 2 show the finite moment minimum no longer exists; however we continue the line as dashed down to 2.42 to indicate a range in which the system shows a sudden jump in properties as a magnetic field is applied, i.e., the M vs. H curves in Fig. 2 are multivalued up to that point and as H is increased the system jumps over the unstable ranges of M. Such jumps are part of metamagnetic behavior,[4] which includes both the two-phase region, where two minima are present (with a maximum between) starting around 2.50 (Fig. 2) and continuing to about 2.46, and the multivalued range of M vs. H down to 2.42.

The fixed-spin-moment calculations give every point along the E(M) curve, including the unstable segments, and effectively solve the problem of finding states of the system in an applied magnetic field. Thus Fig. 2 gives the magnetisation curves of the system. Direct calculation of such states with a magnetic field introduced explicitly in the wave equation, while M is allowed to float to minimize E is much more difficult, since several solutions with different M can exist at the same H; however H is always a single-valued function of M (Fig. 2). Applying density functional theory to a system with fixed M is equivalent to adding a constraint to the

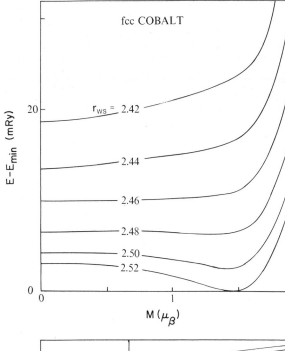

Fig. 1. Set of energy vs. magnetic moment curves at six values of r_{WS} for fcc Co. Energy in mRy is measured from ferromagnetic minimum at $r_{WS}=2.52$ a.u.

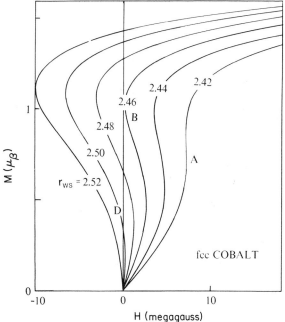

Fig. 2. Magnetic moment vs. applied magnetic field for fcc Co at the same six r_{WS} values as in Fig. 1; H is obtained by differentiating the curves of Fig. 1. Curve D begins the two-phase region, curve B ends the two-phase region, curve A ends the metamagnetic region.

minimization problem that leads to the one-electron differential equations,[5] which are now separate equations for the two spin directions, coupled through the relation to M. Thus the solution of the density-functional equations for a given M is the ground state of a constrained system. Since density functional theory is justified primarily for ground states, and the theory is not as reliable for excited states, it is useful to note that we have ground state reliability in all our calculated values. We are thus able to discuss thermodynamic metastability of magnetic states and the energies and free energies of states of different M and V with the same reliability that we have for the ground state at zero pressure and applied field.

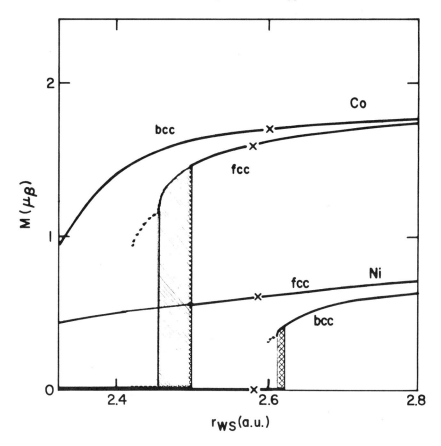

Fig. 3. Magnetic phase diagram on the M - V plane for Co and Ni in bcc and fcc structures; shaded areas are metamagnetic ranges; dashed sections indicate absence of second minimum in E(M) curve; ×'s mark positions of zero pressure. The nonmagnetic phases of the fcc Co and bcc Ni extend to the right-hand edges of the respective shaded areas.

RESULTS FOR Co AND Ni

The phase diagram on the M - V plane (Fig. 3) summarizes the results for the magnetic phases of Co and Ni in both the fcc and bcc structures. Thus fcc Co is FM at zero pressure[6] (marked by ×), and is in fact a well-known phase, but we see from Fig. 3 that in fact this phase breaks down amd terminates under pressure. Before it terminates, an overlapping NM phase appears and continues as a stable phase when the FM phase has disappeared. By examining the E values along these phase lines. (Fig. 4) we see that a first-order phase transition occurs at about $r_{WS}=2.48$ inside a two-phase region existing below 2.50. Surprisingly, FM bcc Co, which is seen from Fig. 4 to be about 5 mRy less stable than FM fcc Co, does not break down under pressure over the range studied, although M decreases as V decreases. This bcc phase is now known through epitaxial growth[7]; however the NM fcc Co phase has not been made, although Fig. 4 shows that where it exists it is more stable than bcc Co.

Figure 3 shows that Ni has everywhere a substantially smaller moment than Co, and that there is an interesting reversal of the roles of fcc and bcc phases compared to Co. Now it is the well-known FM *fcc* Ni phase that is stable over the entire range studied, whereas FM *bcc* Ni, which only exists under negative pressure, breaks down to an overlapping NM bcc phase. The bcc Ni phases have not been made, but Fig. 5 shows that they are at most 4 mRy less stable than fcc Ni, which is closer than bcc Co is to fcc Co.

CONCLUSIONS

The goal of this paper is to demonstrate that new and interesting properties of crystalline materials can be found by exploring system behavior for large ranges of applied fields by first-principles total-energy band calculations; the goal is achieved by finding two new magnetic phases of both Co and Ni and predicting their ranges of existence and degrees of stability, and their lattice constants and magnetic moments over their ranges of existence. An important byproduct of this demonstration is a change in the basic description of magnetic behavior of the ferromagnetic elements at $0°K$, i.e., M is found in some cases to be a discontinuous function of V, as each of a sequence of different phases becomes in turn the most stable phase, rather than a continuous function as is sometimes stated[8]. The necessary sensitivity to pick up the details of the transition is provided by the fixed spin moment calculation. It is noteworthy that the point of view developed here is very similar to the description of metamagnetism[4] within the collective electron theory of Stoner. The results given here for the behavior of the magnetisation curve and the transitions between magnetic phases also correspond to metamagnetism, and give metamagnetism a wider and more fundamental validity. In fact we may consider these calculations to provide a first-principles approach to Stoner theory; it is an approach which gives quantitative results without phenomenological constants. Finally we note that the new phases found here may well be experimentally approachable by the technique of epitaxial growth, which clamps the crystal at a particular lattice constant, as was used for bcc Co[7], thereby exerting large positive or negative stresses on the epitaxially grown crystalline film.

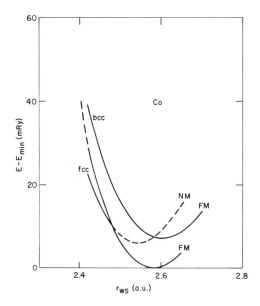

Fig. 4. Energy vs. r_{WS} along the phase lines of Fig. 3 for fcc and bcc Co; the dashed sections of the curves indicate absence of an energy minimum, but the derivative vanishes or has a minimum along those sections.

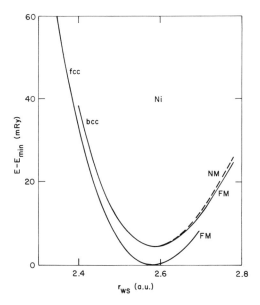

Fig. 5. Same curves for Ni that are given in Fig. 4 for Co.

REFERENCES AND FOOTNOTES

1. A. R. Williams, J. Kübler, and C. D. Gelatt, Jr., *Phys. Rev.* **B19**, 6094 (1979).
2. U. von Barth and L. Hedin, *J. Phys.* **C5**, 1629 (1972); J. F. Janak, *Solid State Commun.* **25** 53 (1978).
3. A. R. Williams, V. L. Moruzzi, J. Kübler and K. Schwarz *Bull. Am. Phys. Soc.* **29**, 278 (1984); K. Schwarz and P. Mohn, *J. Phys. F: Met. Phys.* **14**, L129(1984).
4. E. P. Wohlfarth and P. Rhodes, *Phil. Mag.* **7**, 1817 (1962); M. Shimizu, *J. Physique* **43**, 155 (1982)). The curves of Fig. 2 are obtained from the energy points of Fig. 1 by fitting with a Landau expansion (powers of r_{WS}^2 to r_{WS}^8); the curves are not quantitatively accurate because the Landau expansion does not fit well.
5. P. H. Dederichs, S. Blügel, R. Zeller, and H. Akai, *Phys. Rev. Lett.* **53**, 2512 (1984).
6. The r_{WS} value for p=0 is at the minimum of the $E(r_{WS})$ curve along the phase lines(pl), as given in Figs. 4 and 5, since $(\partial E/\partial M)_V=0$ on the phase line and $(dE/dV)_{pl} = (\partial E/\partial V)_M + (\partial E/\partial M)_V(dM/dV)_{pl}=-p$.
7. G. A. Prinz, *Phys. Rev. Lett.* **54**, 1051 (1985), describes the growth of this phase; P. M. Marcus and V. L. Moruzzi, *Solid State Commun.* **55**, 971 (1985) show how well the theory given here fits the experimental lattice constant and magnetic moment; V. L. Moruzzi, P. M. Marcus, K. Schwarz and P. Mohn, *J. Magn. Magn. Mater.* **54-57** (1986) compare fcc and bcc Co over a wide range of lattice constants, which includes the breakdown of fcc Co.
8. D. Bagayoko and J. Callaway, *Phys. Rev.* **B28**, 5419 (1983), show such behavior in $M(r_{WS})$ curves for fcc and bcc Fe, although calculations like those for fcc Co show that fcc Fe is also multiphased.

FRACTURE AND FLOW VIA NONEQUILIBRIUM MOLECULAR DYNAMICS[*]

W. G. HOOVER, G. DE LORENZI, B. MORAN, J. A. MORIARTY & A. J. C. LADD
University of California & Lawrence Livermore National Laboratory
Livermore, California 94550 U. S. A.

ABSTRACT

The scope of molecular dynamics problems designed to simulate materials properties is described, focussing on the limits computation imposes on space and time scales, as well as the limits theoretical understanding imposes on our knowledge of interatomic forces. Five strategies for improving the efficiency of the simulations are described. Shock-induced solid-solid phase transformations are discussed to illustrate these ideas.

1. SCOPE OF NONEQUILIBRIUM MOLECULAR DYNAMICS SIMULATIONS

Experimental values of fracture toughness and yield strength vary over about six orders of magnitude. The corresponding laboratory fracture and plasticity experiments typically use specimens with dimensions measured in centimeters and deformation times ranging from a microsecond to a year. Computer experiments, designed to simulate laboratory experiments or to illustrate physical principles, although registering steady gains in size and duration, are far from being able to match these length and time scales. Molecular dynamics simulations have gained five orders of magnitude in complexity and speed since the Fermi-Pasta-Ulam calculations carried out at Los Alamos 30 years ago[1]. Presentday size and duration limits, which are now improving relatively slowly, sharply restrict the maximum scope of computer simulations of physical processes. The largest computer simulations have involved 161,604 particles[2] and the longest simulations correspond to physical times of order 1 microsecond.

Both the laboratory-sized compact tension specimens used to measure fracture toughness and the Hopkinson-Bar specimens used to measure dynamic yield strengths, although "small," are still much too large for full-scale modelling at the atomistic level. Thus real laboratory creep experiments, in which metals flow very slowly under relatively small applied loads, are carried out with laboratory time scales ranging from seconds to years. The relatively much faster deformations caused by impacting small plastic sample disks with relatively-massive flying elastic bars, " Hopkinson-Bar experiments," with strain rates which can exceed 100,000 hertz, are still too slow for an accurate simulation of atomistic trajectories.

The computational limitations on space and time do not affect the accuracy of a simulation. Provided that we are satisfied with six-figure accuracy there is no difficulty in using Adams or Runge-Kutta finite-difference methods to integrate the equations of motion[3]. Despite this impressive numerical accuracy computer calculations can only be caricatures of the behavior of real materials. This is because our knowledge of interatomic forces remains primitive.

Current models[4,5] used for the interaction of relatively simple atoms, such as sodium and magnesium, are based on highly-intricate and detailed models for the electronic and nuclear structure of a metal. The complexity of this work defies exact description or reproducibility, and is a subtle mixture of art and science. Thus it is highly unlikely that a

typical published calculation incorporating these electronic pseudopotential models could be accurately reproduced, even with years of effort.
The figure shows two recent versions of the effective atomic interaction in
magnesium. The energy, in millirydbergs, is shown as a function of
distance, in Bohr radii. The Moriarty-McMahan calculation[4] involves no
adjustable parameters; the Barnett-Cleveland-Landman calculation[5] is
fitted to experimental data. The difference between the two, around 15% at
the potential minimum, gives a rough estimate as to the reliability of the
theoretical work. Uncertainties of this order in the interaction potential
are sufficient to shift the positions of phase transitions by pressures of
order 10 kilobars and temperatures of order 100 kelvins.

    The fact that density fluctuates rather wildly on an atomic scale
suggests that no single density-dependent potential is likely to be a fully
adequate approximation. In problems involving mechanical deformation one
expects that energies of the order of microrydbergs could be significant.
These energies, about six orders of magnitude below the hydrogen-atom
energy, are three orders of magnitude smaller than the millirydberg scale
of the figures. It seems highly unlikely that quantitative calculations of
interactions will ever become possible on the accuracy scale necessary to
reproduce mechanical behavior. The need for greatly improved accuracy in
interatomic forces is well recognized. Serious efforts to calculate
accurate atomic forces as functions of coordinates are underway. New
emerging techniques may eventually make semiquantitative molecular dynamics
possible[6], even for transition metals and covalently bonded materials.

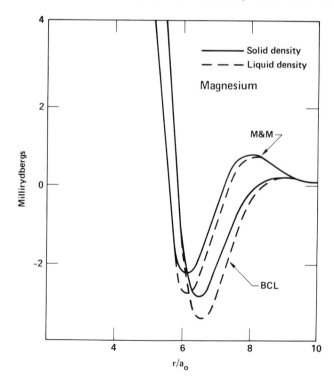

2. STRATEGIES FOR EFFICIENCY IN MOLECULAR DYNAMICS

Molecular dynamics became possible with the development of computers for the weapons calculations.of the second world war. The forces used in the early work were simple, slightly anharmonic springs or hard spheres with additional attractive forces, in keeping with the limits of knowledge and computational capacity. The original aim of molecular dynamics calculations was the understanding of a long-standing problem which fascinated Boltzmann and his colleagues, the consistency of Newton's reversible equations of motion with the irreversible second law of thermodynamics.

Fracture and flow, for instance, are irreversible processes. As a specimen is broken or caused to change shape, in an irreversible way, we nevertheless expect that the underlying dynamics is reversible, described either by Newton's or Schroedinger's equations of motion. It is a rare and unusual circumstance for a crack to heal, and the necked region of a tensile test specimen never reverts to its original diameter and shape if the tension is changed to compression. A detailed understanding of the paradox that reversible equations yield irreversible behavior was given by Boltzmann for dilute gases. Even so, this topic remains under intensive investigation 100 years later.

. When the qualitative thermodynamic features of irreversibility and the existence of gas, liquid, and solid phases had been established, emphasis shifted to the cataloging of equilibrium and linear transport properties for more elaborate potentials. Inverse powers, combinations of powers, exponentials, and "realistic" pseudopotentials began to be studied. This work made it possible to assess the usefulness of the theories, developed in the period 1930-1960, which incorporated well-defined but poorly known distribution functions and made approximations of uncertain value. The computer experiments provided the first accurate distribution functions and were used to test the common truncation and superposition approximations of statistical mechanics.

In the past ten years the emphasis has changed, largely due to the success achieved by hard-sphere perturbation theory in calculating the thermodynamic properties for many-body fluid systems. A good survey of current research can be found in the proceedings of the 1985 Enrico Fermi Summer School on Molecular Dynamics Simulation of Statistical-Mechanical Systems. The present focus of attention has shifted to nonlinear dynamical problems such as those involved in fracture, plastic flow, and shockwaves[7]. The nonlinear work also makes contact with the current focus on the fractals and attractors associated with nonlinear dynamical systems[8].

During this past decade several strategies have been developed which have increased the power of molecular dynamics beyond what could be achieved by straightforward solution of Newton's conservative equations of motion. Here we list five strategies for such improvement:

1. Taking number-dependence of the microscopic results into account systematically in order to make more accurate macroscopic predictions.

2. Considering mesoscopic constitutive simulations with many degrees of freedom, intermediate between microscopic atomistic calculations and macroscopic continuum simulations.

3. Developing constrained dynamics so that independent variables other than energy and volume can be used. This work includes the introduction of

thermostats to remove irreversibly generated heat.

4. Developing corresponding states relationships, analogous to the scale models used in mechanical engineering, linking results for several related problems together.

5. Developing artificial processes to avoid at least a part of the size-dependence which makes full-scale simulation uneconomic.

Let us consider these five strategies in more detail:

2.1 Number Dependence

Molecular dynamics is at its best in elucidating mechanisms on an atomic scale. Such mechanisms dominate problems involving fracture, plastic flow, or chemical reactions. Attempts to deal with these same problems using continuum mechanics are complicated by the presence of singularities in the continuum equations. The singularities in the stress and strain fields at cracks and dislocations are relatively long-ranged. The stress in the neighborhood of a crack tip in an elastic crystal falls off as the square root of L/r, where L is proportional to the size of the crystal.

The stress in the neighborhood of a dislocation also falls off slowly, as d/r, where d is the interatomic spacing. Thus simulations covering 1% of. the sample size or 100 atomic diameters are required to reproduce, accurately, macroscopic stress fields in the vicinity of crystal defects.

The size dependence for many problems of interest in materials science is surprisingly straightforward to analyze. This is illustrated in the figure [9]. Stress near a brittle crack tip--made by cutting bonds linking the top and bottom halves of the crystals for 40% of the crystal length--was determined for a series of triangular-lattice two-dimensional crystals varying in size from 10 rows of 70 atoms to 40 rows of 280 atoms. The Hooke's-law linear force becomes increasingly attractive to a separation of 1.15d, and then is reduced linearly to zero, vanishing at a cutoff of 1.30d.

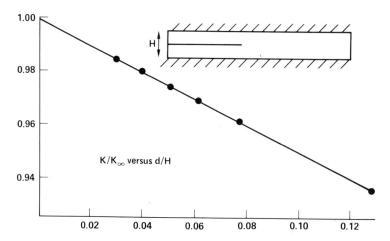

K/K_∞ versus d/H

The top and bottom of the crystal were displaced in order to stress the crack, and the fracture toughness K was measured. The intercept from the empirical straight line shown in the figure is nicely consistent with the square-root dependence mentioned above. This correspondence of continuum and atomistic results on length scales large with respect to the range of interparticle forces makes it possible to combine these points of view in fracture and plasticity problems.

It is interesting to see that accurate data for crystals much too large for direct simulation can be obtained by extrapolating small-crystal data to the large crystal limit. This requires much less computer time than would a straightforward brute-force simulation of the stress field in the largest possible crystal. The results shown in the figure display much less size-dependence than would ductile cracks in crystals containing dislocations, or than would cracks in crystals with long-range, stiff interactions.

2.2 Mesoscopic Simulations

Lattice dislocations provide the mechanism for plastic flow. One part of a crystal flows relative to a contiguous part, through the motion of dislocations on the glide plane separating the two parts. The dislocations themselves can be treated as particles, in two dimensions, or as deformable lines in three dimensions. The dislocations move through a crystal with an equation of motion giving their velocity as a function of orientation and local stress. Dislocations can move at speeds near the sound speed under relatively small stresses. They interact with a tensor force and can react and combine to form other dislocations.

By studying the properties of dislocations, it has proved possible to determine equations of motion directly from computer experiments [10]. By including dislocation interactions and stress-induced motion , "mesoscopic" simulations of plastic flow can be carried out. Such simulations, intermediate in length scale between atomic and continuum calculations, greatly increase the potential of computer experiments in the understanding of plastic flow.

2.3 Thermostats

In any irreversible process, such as fracture or plastic flow, stored mechanical energy is converted to heat. The temperature in a "cold-worked" metal or in the vicinity of a propagating crack can exceed the ambient temperature by hundreds of kelvins. The fundamental source of this heat is the potential energy stored in the solid through the action of external deformation forces. The energy is released as heat in the irreversible process of breaking interatomic bonds. In laboratory experiments this heat can then be carried, by phonons, to the boundary of the system. A typical thermal diffusivity for a metal is one square centimeter per second. Thus, in a microsecond, heat diffuses about 40,000 particle diameters in a simple metal such as room-temperature sodium.

It is difficult to model this motion of heat on an atomistic level. This is because the phonons which carry the heat have free paths which often substantially exceed the size of convenient computational cells. Thus a third strategy for extending the capability of simulations is to introduce thermostats directly into the equations of motion. This eliminates the rapid changes of thermodynamic state which would otherwise accompany small-scale irreversible processes, and also makes it unnecessary to look at systems large with respect to a typical phonon mean free path.

Thermostats based on rescaling atomic velocities have been in use for over a decade. A recent breakthrough in this area was announced by Shuichi Nose[11,12], who found a way to obtain the canonical constant-temperature phase-space distribution from ordinary differential equations of motion slightly modified from Newton's equations:

$$ma = F - zmv ,$$

where m is the atomic mass, v, and a are the velocity and acceleration, F is the force, and z is the friction coefficient. Nose calculated the friction coefficient as a memory function for the kinetic energy, proportional to the time integral of $K - 3NkT/2$.

The same friction-coefficient form of the equations of motion, including the linear thermostatting force had already been used in simulations designed to keep the kinetic energy of selected sets of particles constant, but in the earlier work the friction coefficient had no memory. It was simply chosen to fix the kinetic energy (and has the form [dO/dt]/2K, where O and K are the potential and kinetic energies, respectively).

Several other forms of thermostats and ergostats have been introduced and compared[7,13]. So far each of these choices can be described, in the language of control theory, as differential, proportional, or integral control, but it seems likely that more general forms will emerge in the near future. The canonical Nose thermostat and the isokinetic Gaussian thermostat seem to be the most useful. Evans and Holian made a comparison of six different thermostats applied to the calculation of the nonlinear fluid viscosity. Although all six methods produced essentially the same viscosity the Gaussian thermostat was more efficient than the others by at least a factor of six in computer time[13]. Why it is that the nonlinear viscous response, far from equilibrium, is nearly independent of the type of thermostat used is not yet understood.

2.4 Corresponding States Relationships

In metals the electronic energy is a sensitive function of density. Thus the corresponding pseudopotentials, such as those shown in the first .figure, which impute a part of this energy to the ion cores, depend relatively strongly on density and nearby crystal defects. For this reason it may well be unrealistic in most cases, to imagine a quantitative simulation of a particular material. The many aluminum and iron alloys have very different fracture and flow properties relative to single crystals of the pure metals.

Thus corresponding states relationships which link together simulations with one structure or force law with other simulations can be extremely useful. These relationships identify properties which are insensitive to the form of interparticle forces and which therefore can be predicted, with fair confidence, from computer simulations. Van der Waals' fluid equation of state, which describes gas-liquid coexistence in a semiquantitative way, is a familiar example. By choosing characteristic van der Waals constants, a and b, many different materials can be described by the same equation of state.

Corresponding states relationships away from equilibrium are not so well known. At low rates of strain materials deform in individual ways, depending very much upon crystal structure, inclusions, grain size, radiation history, and a host of such variables. At high strain rates

things become simpler. The figure shows the variation of solid-phase yield
stress with strain rate for isothermal steady shear[14]. The stress is
made dimensionless by dividing by the shear modulus. The strain rate is
made dimensionless by using the interparticle spacing d and the transverse
sound speed characteristic of dislocation motion. Within the width of the
corresponding-states lines the two- and three-dimensional crystals produce
the same results. An extrapolation of the computer data, the solid lines
at the extreme upper right corner of the figure, agrees reasonably well
with data, of uncertain validity, for metals. The temperatures of the
computer simulations, relative to the melting temperature, are shown.

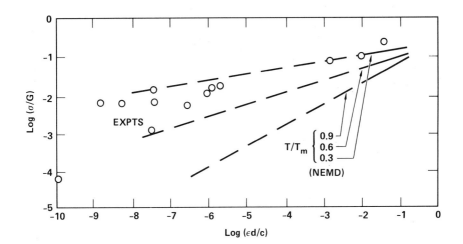

132

2.5 Artificial Processes

Finally, in some cases, it may be possible to replace inhomogeneous time-dependent physical processes by homogeneous artificial processes which span essentially the same thermodynamic states. This idea has been fully explored, for both fluids and solids, in simulating steady viscous flows and heat flows in fully periodic homogeneous systems[14,15].

Viscous flow has been simulated by using the dashed-line velocity field shown in the figure. The basic system extends initially from -L/2 to L/2, and is repeated spatially with periodic boundaries. Then a macroscopic velocity, shown in the figure, is added to the initial microscopic thermal velocities. The smooth dashed line corresponds to simple longitudinal compression. If the macroscopic velocity varies in a series of steps, as shown with the heavy line, then a pair of shockwaves is generated, at -L/2 and L/2. Shear flows, as well as flows combining shear and dilation can be simulated in an analogous way.

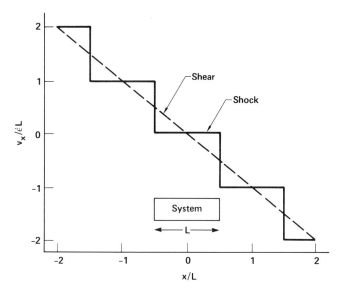

Heat flow has also been simulated in periodic homogeneous systems. This has been accomplished by using an external field, derived from linear response theory, which drives more-energetic particles in one direction and less-energetic particles in the opposite direction. It has to be emphasized that these artificial processes generating shear flows, shockwaves, and heat flows are not the same as those determined in laboratory experiments on the same thermodynamic state. But, as the Evans-Holian calculations[13] demonstrate, the differences are in most cases considerably.less than the statistical uncertainties in the measured transport, and hence negligible. This potentially useful idea of artificially reproducing nonequilibrium states by using external forces is explored more fully in the last section. There we suggest a new method for simulating the polymorphic phase transformations induced by shockwaves.

3. SIMULATION OF SHOCK-INDUCED PHASE TRANSFORMATIONS

Shockwaves can produce high pressures[16] much more cheaply than static experiments and are therefore coming into relatively common use in materials research. Ongoing work at Livermore[17] is directed toward the low-cost shockwave-induced synthesis of superconductors from amorphous metals. The experiments proceed by treating relatively thin layers of metal with shockwaves of controlled pressure and duration.

The shock process is relatively complicated, even in the simplest steady-wave case. In this case, in a frame centered on the shock wave, the flows of mass, momentum, and energy are all constants. Otherwise, mass, momentum, or energy would build up in a part of the wave, which would then not be a steady one. The conservation equations for these three fluxes provide relations between the pressure tensor component in the shockwave direction, the density, and the heat flux. These state variables are connected in a definite way in a shockwave. As shown in the figure, the longitudinal pressure varies linearly with volume. The energy flux has a quadratic variation with volume.

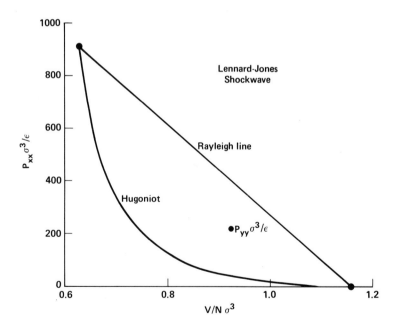

How can these same states, characteristic of a steady · shockwave, be traversed in molecular dynamics simulations in which homogeneous periodic boundaries are used? In the simulations we can follow the Gauss-Nose ideas of introducing a friction coefficient into the equations of motion. Because two constraints must be satisfied the shockwave-simulation friction coefficient is a tensor, with different components parallel and perpendicular to the shock propagation direction.

We are testing this idea in the one case where a strong dense-system shockwave transformation has been well characterized, the 400-kilobar shockwave studied by Holian, Hoover, Moran and Straub[18]. The main point which is presently unclear is the choice of· strain rate which must be imposed on the simulation. When the fluid-phase results have been successfully reproduced we will apply the same idea to the solid. transformations used to synthesize superconducting alloys. In the solid-phase shockwaves studied by Holian and Straub at Los Alamos considerable dependence of the shockwave structure on system width was found. Because the systems studied had to be many times longer than wide in order to study a steady profile, a cross section containing only 32 atoms was the maximum practical size. With the development of a new· technique for homogeneous simulation of the shockwave process, cross sections containing a few hundred atoms should become possible.

REFERENCES

<*>   This work was performed under the auspices of the United States
      Department of Energy at the Lawrence Livermore National Laboratory
      under Contract W-7405-Eng-48

[1] J. L. Tuck and M. T. Menzel, Advan. Math. 9, 399(1972).

[2] F. F. Abraham, W. E. Rudge, D. J. Auerbach, and S. W. Koch, Phys.
      Rev. Letts. 52, 445(1984).

[3] See H. Berendsen's contribution in the Proceedings of the Enrico
      Fermi International Summer School on "Molecular Dynamics Simulation of
      Statistical-Mechanical Systems" (Lake Como 1985), and W. E. Milne,
      Numerical Solution of Differential Equations (Dover, New York, 1970).

[4] A. K. McMahan and J. A. Moriarty, Phys. Rev. B 27, 3235(1983).

[5] R. N. Barnett, C. L. Cleveland, and U. Landman, Phys. Rev. Letts. 54,
      1679(1985).

[6] R. Car and M. Parrinello, Phys. Rev. Letts. 55, 2471(1985)

[7] D. J. Evans and W. G. Hoover, Ann. Rev. Fluid Mechanics 18, 243(1986)

[8] H. A. Posch, W. G. Hoover, and F. J. Vesely, Phys. Rev. A (submitted)

[9] B. Moran, Ph. D. Dissertation, "Crack Initiation and Propagation in
      the Two-Dimensional Triangular Lattice," University of California at
      Davis-Livermore(1983).

[10] A. J. C. Ladd , "Molecular Dynamics Studies of Plastic Flow at High
      Strain Rates," Topical Conference on Shockwaves, Spokane, Washington,
      (July 1985)

[11] S. Nose, J. Chem. Phys. 81, 511(1984).

[12] W. G. Hoover, Phys. Rev. A 31, 1695(1985).

[13] D. J. Evans and B. L. Holian, J. Chem. Phys. 83, 4069(1985).

[14] W. G. Hoover, A. J. C. Ladd, and B. Moran, Phys. Rev. Letts 48,
      1818(1982).

[15] R. Grover, W. G. Hoover, and B. Moran, J. Chem. Phys. 83, 1255(1985)

[16] M. Ross, Repts. Prog. Phys. 48, 1(1985).

[17] W. J. Nellis, W. C. Moss, H. B. Radousky, A. C. Mitchell, L. T.
      Summers, E. N. Dalder, M. B. Maple, and M. McElfresh,
      "Superconducting Critical Temperatures of Niobium Recovered from
      Megabar Dynamic Pressures," Lawrence Livermore Laboratory Preprint
      UCRL-92742(June 1985).

[18] B. L. Holian, W. G. Hoover, B. Moran, and G. K. Straub, Phys. Rev. A
      22, 2798(1980).

INTERATOMIC FORCES AND STRUCTURE OF GRAIN BOUNDARIES

V. VITEK* AND J. Th. M. DE HOSSON
Department of Applied Physics, Materials Science Centre, University of Groningen, Nijenborgh 18, 9747 AG Groningen, The Netherlands
* On leave from the Department of Materials Science and Engineering, University of Pennsylvania, Philadelphia, PA 19104, U.S.A.

ABSTRACT

Over the past two decades, grain boundary structures have been investigated, with increasing frequency, using computer simulation techniques. The accuracy of the potentials used to describe the interaction between atoms is essential to the success of any computer modelling and practically all these studies have been made assuming that the atoms exert pair-wise forces on each other. This paper discusses the applicability and limitations of this assumption and the validity and applicability of the simulation results obtained with pair-potentials. First, it is shown that certain structural features of grain boundaries are independent of the potential used. Such results form, for example, the basis of the structural unit model which relates the structures of boundaries corresponding to different misorientations. Secondly, the multiplicity of boundary structures and its physical implications are discussed. In particular, we report on the complexities of the structure of $\Sigma=5$ [001] twist boundaries and on alternative structures of [111] twist boundaries near the $\Sigma=3$ twin orientation, both found by computer modelling. These results, which are of general nature, are compared with experimental observations by X-ray diffraction and transmission electron microscopy and a very good agreement is found. However, exact positions of atoms in calculated structures depend on the potential used. Hence, to study properties sensitively dependent on these positions requires to go beyond the pair-potentials.

INTRODUCTION

Grain boundaries are the most common interfaces always present in materials which are not in a single crystalline form. These interfaces affect a large variety of material properties because many important physical phenomena, such as, for example, diffusion, segregation of alloying elements, decohesion, occur either preferentially or exclusively at grain boundaries. The region in which these processes take place is very localized, spanning only a few lattice spacings, and it is, therefore, the atomic structure of the "core" of grain boundaries which controls the boundary properties. This is the reason why the atomic structure of grain boundaries has been the subject of research ever since the importance of these interfaces was recognized (see e.g. [1]). In the last fifteen years investigations of the structure of grain boundaries became particularly intensive. Experimental observations have been made using electron and X-ray diffraction (for reviews see [2-4]) and, more recently, the electron microscope lattice imaging techniques (e.g. [5]). These studies have now established beyond any reasonable doubt that the structure of grain boundaries is crystal-like. The first most important steps in the theoretical description of grain boundaries were the dislocation model [7] and the coincidence site lattice and O-lattice theories [8, 9] which, together with the analysis of bicrystal symmetry [10], provide a crystallographic basis for the description of grain boundaries. The atomic structure of crystallographically defined boundaries has then been studied by computer modelling (for reviews see [3, 11, 12]) and results of these calculations form a basis for our present understanding of the general features of the atomic structure of grain boundaries (see e.g. [2]).

Practically all the computer simulations of grain boundaries have been made assuming that the interatomic forces can be described in terms of pair potentials. This assumption clearly cannot be justified in general and the question arises how valid and applicable are the results of these calculations. The purpose of this paper is to discuss this question. We show first, using as an example the structural unit model [13], that there are certain structural features which are independent of the potential used. These features are then likely to have general applicability. The second type of results, which are equally valuable, are those on the basis of which various possible structural alternatives and trends can be established, although these may not be uniquely associated with specific materials. As examples of such results we discuss here the multiplicity of structures. In particular we concentrate upon the complexities of the structure of $\Sigma=5$ [001] twist boundaries and alternative structures of [111] twist boundaries. Finally, we assess the limits of the pair potential approximation and discuss possible alternatives.

PAIR POTENTIALS

The description of the interaction between atoms in terms of pair-potentials corresponds to writing the total energy of the system of N particles as

$$E = U(\rho) + 1/2 \sum_{i,j=1}^{N} \phi(r_{ij})$$

(1)

where ϕ is the pair potential, r_{ij} the separation between atoms i and j and U is the part of the energy which depends only on the average density, ρ, of the material. All the terms corresponding to higher order interactions are either neglected or included, in an average form, into the density dependent term. The cohesive energy is principally given by $U(\rho)$, while the pair potential part describes changes of the energy of the system associated with changes of relative positions of the atoms and it may contribute either positively or negatively to the cohesive energy. This description of the total energy can be fully justified only for simple metals and in this case it can be derived theoretically using the pseudopotential theory [15, 16]. However, it has been shown recently [17, 18] that equation (1) is feasible in all metallic materials but for the same material different pair-potentials are obtained using different reference states. The use of these potentials is then limited to configurations which do not differ substantially from the corresponding reference states, in particular in density.

The potentials constructed for s,p metals on the basis of the pseudopotential theory generally consist of repulsive and attractive parts and possess long range Friedel oscillations. However, it has been shown that these oscillations can be damped [19, 20] and it is reasonable to limit the interactions to a small number of neighbours. Potentials for other metals have frequently been constructed empirically assuming similar forms of the potential functions as in simple metals. Every potential can be expected to be repulsive for atom separations smaller than the nearest neighbour spacing but no general rules governing the behaviour of the potentials beyond the first nearest neighbour spacing exist. However, since the pair potential represents an effective interaction between atoms at a fixed average electron density, its form at large separations of atoms will follow Friedel-type oscillations which then suggest approximate positions of the minima and maxima on the potential curve [21]. In particular it is important to decide whether the first and second nearest neighbours are positioned within different minima of the potential curve or in the same minimum. An example of the former is the potential for aluminium [22], shown in figure 1, and of the latter two empirical potentials for copper [21, 23], shown in fig. 2. The important difference between these two different shapes of the potential curve is that in the former case both the first and second nearest neighbours are positioned at strongly repulsive parts of the potential while in the latter case the first nearest neighbours are at the repulsive but the second nearest neighbours at the attractive part of the potential.

Once the shape of the potential has been chosen the curve can be parametrized and the parameters determined so as to fit certain experimental data for the ideal lattice, e.g. elastic constants, phonon dispersion curves, stacking fault

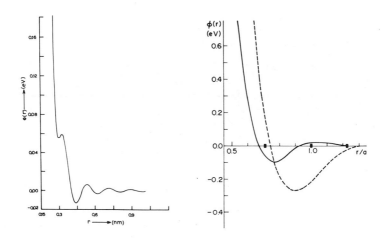

Fig. 1 (Left)   Potential for aluminium [22].
Fig. 2 (Right)   Two empirical potentials for copper constructed in [23] (full
curve) and [21] (broken curve).

energy etc. (It should be noted that while the Cauchy relation $c_{12} = c_{44}$ would be
required in the cubic structure if only pair potentials were used, it is no more
imposed when the density dependent term is included.) An important parameter to
choose at this stage is the cut-off radius. This is usually done so as to include only
second and third nearest neighbour interactions in the ideal lattice which is
justified due to the rapid damping of the potentials in real situations [19, 20].
Finally, every potential must lead to mechanically stable ideal lattice structure
with the correct lattice parameter and it should guarantee even the mechanical
stability with respect to large deformations if it is to be used in defect studies.
Nevertheless, empirical potentials cannot be considered as representing exactly
interatomic forces in specific materials and, therefore, we cannot expect to
evaluate, for example, absolute values of grain boundary energies for a specific
material. As discussed below, their use is most valuable when following structural
features which are insensitive to the potential used, studying possible structural
alternatives or investigating structural trends in dependence on certain parameters
of the potentials.

The computer simulations of defects such as grain boundaries have usually
been made subject to the constraint that the total volume of the relaxed block is
fixed. This condition is consistent with the use of pair potentials and the density
dependent part of the energy need not be considered explicitly. If the volume
associated with the defect is small compared to the volume of the block the
constant volume and constant pressure calculations always lead to the same results,
as explained in [11, 24]. Constant pressure calculations have also been made (e.g.
[25]) but the way how to include explicitly the density dependent part of the energy
into the relaxation calculation is always uncertain. The solution to this problem has
recently been proposed in [26] where it was suggested that the density dependent
part can be replaced by a term $- \sum_i \sqrt{\sum_j \beta(r_{ij})}$, where $\beta(r_{ij})$ is a positive, rapidly

decreasing function of the separation of atoms i and j. The justification for this
choice is provided in the framework of the tight binding approximation when
considering the expansion of the energy up to the second moment of the density of
states. A similar approach, called the embedded atom method has been developed
in [27] and using these approaches constant pressure and constant volume calculations can be done equally easily.

DEPENDENCE OF THE BOUNDARY STRUCTURE ON MISORIENTATION: STRUCTURAL UNIT MODEL

A grain boundary is characterized geometrically by five parameters: the orientation of the rotation axis, the angle of misorientation of the grains and the orientation of the boundary plane. The relative displacements of the grains and the position of the grain boundary (in non-primitive lattices) are sometimes regarded as additional four parameters but these are determined by energetics if the above five parameters are fixed. Interesting general questions are how does the atomic structure of boundaries vary with the geometrical parameters, what is the relationship, if any, between structures of boundaries with different parameters, and how sensitively the answers to these questions depend on the interatomic forces. To investigate the variation of the grain boundary structure with all five parameters is a very complex task and it is usual to vary only one of the parameters while the others are fixed. Detailed computer simulation studies using a number of empirical pair potentials, as well as the potential for aluminium shown in figure 1, were made to investigate the dependence on the misorientation of the grains. These calculations were performed for both tilt [13, 14] and twist [28, 29] boundaries. For a chosen rotation axis the misorientation is the only variable in the latter case but in the former case the orientation of the boundary plane can still be varied. Therefore, in the case of tilt boundaries the study was made for families of boundaries with the same mean boundary plane. This plane is defined such that its normal is parallel to the vector $n_1 + n_2$, where n_1 and n_2 are normals of the boundary plane in the upper and lower grains, respectively. Different boundaries possessing the same mean boundary plane differ only in misorientation. (Symmetrical tilt boundaries are obtained if the mean boundary plane is a mirror plane or if it contains a two-fold rotation axis, such as (001) and (110) planes in the cubic case.)

The general results of these atomistic studies which were found to be entirely independent of the potential used form a basis for the structural unit model which relates the structures of boundaries corresponding to different misorientations. All tilt boundaries possessing the same mean boundary plane as well as the twist boundaries in a certain misorientation range are composed of mixtures of two structural elements identified as 'units' of two small period boundaries that delimit the misorientation range. (Additional filler units are present in twist boundaries.) The structure of each of the delimiting boundaries corresponds to the contiguous sequence of one of these units. If the unit of a delimiting boundary is not composed of units from other boundaries the corresponding delimiting boundary is called 'favoured', and this unit is then a fundamental structural element of nearby boundary structures.

As examples we show in figure 3 the $\Sigma=353$ (1780) 39.60° which is a symmetrical tilt boundary, and in figures 4 and 5 the $\Sigma=689$ [001] 35.49° and the $\Sigma=183$ [111] 7.34° twist boundaries, respectively. The structure of every unit cell of the $\Sigma=353$ boundary is composed of seven units of the $\Sigma=5$ (210) 36.87° boundary (marked by full lines) and one unit of the $\Sigma=5$ (310) 53.13° boundary (marked by broken lines). All other boundaries in the misorientation range 36.87° - 53.13° are composed of these units. The structure of every unit cell of the $\Sigma=689$ boundary is composed of 64 units of the $\Sigma=5$, 36.87° boundary (marked by full lines) and one unit of the ideal crystals, i.e. "$\Sigma=1$ boundary" (marked by broken lines in the corners of the unit cell). The units of the ideal crystals are always linked along the sides of the unit cell by rows of eight filler units. The situation is more complex in the $\Sigma=183$ boundary. The structure of this boundary could be assumed to contain units of the $\Sigma=21$, 21.79° boundary and units of the ideal crystal. The hexagons marked in Fig. 5 are, indeed, the units of the $\Sigma=21$ boundary which are linked by filler units. However, when analyzing the stacking of the planes in the triangular regions, we can see that they correspond alternately to the ideal crystal and the (111) stacking fault. This is related to the dislocation content of this boundary as explained below.

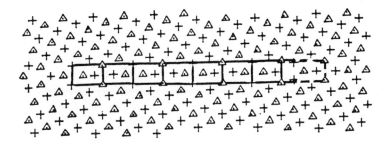

Fig.   3 Structure of the Σ=353 (1780) symmetrical tilt boundary. Triangles and crosses represent in this and all other figures atoms belonging to two different (002) planes within the [001] period.

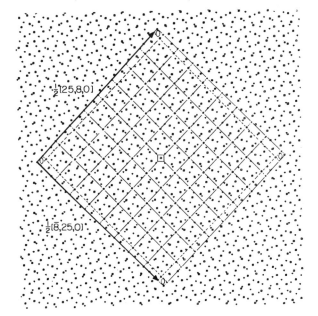

Fig.   4 Structure of the Σ=689 [001] twist boundary.

Geometrically, any large period boundary can, of course, be decomposed into strained units of smaller period boundaries, as first noted in [31]. However, it is not known a priori which boundary units to chose in such decomposition and whether the same units comprise boundaries nearby in the misorientation range or how distorted the units are. Atomistic studies have revealed the continuity of the unit description and showed that relaxations minimized distortions of the units. Further-more, the calculations showed that the minority units represent the regions of misorientation variation and are the sources of the elastic fields of the boundary which can be interpreted as fields of networks of DSC grain boundary dislocations superimposed on the reference structure composed of majority units. The structural unit model thus also provides a link between the dislocation model of grain

142

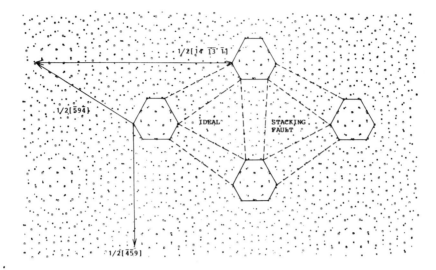

Fig. 5 Structure of the Σ=183 [111] twist boundary. The symbols in the sequence
△ + ◻ * represent atoms from four consecutive (111) planes. The
geometrical boundary plane is positioned between layers marked + and ◻.

boundaries and their atomic structure. In tilt boundaries rows of minority units
parallel to the tilt axis can be identified with the cores of edge DSC dislocations
and in twist boundaries the rows of filler units correspond to the cores of screw
dislocations which then intersect at minority units. In figure 3 the rows of the Σ=5
(210) units are the cores of 1/5 [210] edge dislocations and in figure 4 the filler
units represent the cores of 1/10 <310> screw dislocations. All these dislocations
are the complete DSC dislocations of the Σ=5 coincidence lattice. In the Σ=183
boundary the corresponding "complete DSC dislocations" would be 1/2 <1$\bar{1}$0>
lattice dislocations since the ideal crystal is the reference structure. However, it
can easily be seen using Frank's formula [7], that the filler units represent cores of
1/6 <11$\bar{2}$> partial dislocations which is also consistent with the alternating regions
of the ideal crystal and the stacking fault. Apparently, it is energetically favourable for every other nóde of 1/2 <1$\bar{1}$0> dislocations to dissociate and form thus a
triangular network of 1/6<11$\bar{2}$> partials [32].

The structural unit model is an example of the general rule governing the
dependence of the boundary structure on geometrical parameters which was deduced on the basis of atomistic studies but is independent of interatomic forces
employed. However, which of the delimiting boundaries are favoured may vary with
the change of interatomic forces [13] and, therefore, variation in boundaries
favoured in different materials with the same crystal structure can be expected.
Furthermore, although the structures of the same delimiting boundaries are often
similar for different interatomic forces they are not identical and detailed positions
of atoms in the units of the delimiting boundaries will vary with the potentials used
and, presumably, from material to material [13, 33].

MULTIPLICITY OF BOUNDARY STRUCTURES AND ITS PHYSICAL IMPLICATIONS

In a number of atomistic studies of grain boundaries several different
metastable structures of a boundary characterized by the five geometrical para-

meters were found (e.g. [11, 12, 35, 36]). This structural multiplicity is not surprising and it would be of no particular significance if one of these structures possesses a much lower energy than the others. This structure would be then the only one which is physically important. However, already some of the early studies (e.g. [36]) indicated that several different structures might have very similar energies so that alternative structures may be of comparable importance. In particular a very extensive multiplicity of grain boundary structures may occur in general, large period boundaries. As explained in the previous section, structures of those boundaries can be regarded as composed of units of two short period delimiting boundaries. If the delimiting boundaries possess several alternative structures and units of all these structures participate in the structures of intervening boundaries, a very extensive structural multiplicity of the intervening boundaries follows. In principle, an infinite number of alternative structures may exist if the periodicity of the intervening boundary structures is allowed to become an indefinite multiple of the coincidence site lattice periodicity. This type of structural multiplicity has been investigated recently for symmetrical tilt boundaries in both f.c.c. and b.c.c. metals [14, 37] and it has, indeed, been found to be very common with alternative structures frequently possessing very similar energies. An example of eight possible structures of the periodic $\Sigma=65$ (740)[001], 30.51° symmetrical tilt boundary is shown in fig. 6.

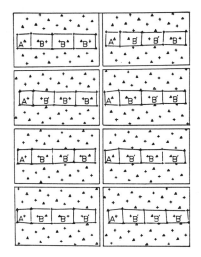

Fig. 6 Eight alternative periodic structures of the $\Sigma=65$ (740) boundary.

Every period of this boundary consists of three units of the $\Sigma=5$ (210) boundary (marked B or B') and one unit of the ideal crystal (marked A). Two alternate structures of the $\Sigma=5$ (210) boundary, B and B', exist and their various combinations are found in different structures of the $\Sigma=65$ boundary. The lowest energy configuration corresponds to the unit sequence ABBB' and the highest energy configuration to the sequence AB'B'B. The energies of these two structures differ by 9% but much smaller differences exist for other alternatives. For example, the energy of the structure ABB'B' differs from that of the lowest energy structure by less than 1% [14].

The existence of alternative structures with similar energies suggests that at high temperatures transformations of the grain boundary structure could occur. These transformations may either be of the order-disorder type involving transition from a periodic structure to a non-periodic multiple structure, or the transitions from one alternative structure to another [37].

Such transformations of the grain boundary structure have been suggested earlier on thermodynamic grounds [39] and several experiments can be interpreted in terms of such transformations [39, 40]. Transitions between alternative multiple structures provide a basic physical model for such transformations the details of which will be different for different boundaries and different materials. They also represent an alternative to the recently proposed grain boundary melting [41, 42] which has not been confirmed experimentally [43].

Another important feature of the multiple structures of tilt boundaries is that they may transform mutually by absorption or emission of vacancies. For example, structural unit B (fig. 6) transforms into unit B' if one of the atoms in its center is removed. Because the energies of these units are similar those transformations will be easy which explains the ability of grain boundaries to act as practically ideal sources and sinks of vacancies. Furthermore, the energetically most favourable structure of a vacancy in, for example, a boundary composed of units B corresponds to the isolated unit B' and vice versa. Hence, the results of the simulation of alternate structures of grain boundaries also provide a general model for the structure of vacancies in grain boundaries [37].

Σ=5 [001] Twist Boundaries

For the Σ=5 [001] twist boundary three alternate structures of very similar energy which differ in symmetry and relative displacements of the grains, were found already in the first atomistic study of [001] twist boundaries [36]. At the same time the structure of this boundary in gold was studied using the X-ray diffraction from the region of the boundary core [4, 44, 45]. This shows reflections located in the reciprocal space at points given by the vector h/5 [310] + k/5 [130] that depend on the relaxation of atomic positions away from the ideal CSL positions. The absence of certain reflections suggests that the symmetry of the structure is p42$_1$'2' [45]; the prime indicates that the symmetry elements relate sites in different crystals. This is the symmetry of one of the structures found, called the CSL structure in [36], and thus a direct comparison of the experimental diffraction intensities with those calculated for the structure found in the computer simulation, can be made. This is shown in figure 7a for the CSL structure calculated using a potential for Au [21, 46]. A pronounced disagreement between the experiment and theory is apparent. Furthermore, detailed positions of the atoms in the CSL structure depend sensitively on the potential used [33, 45] which leads to a significant variation of the calculated intensities with the potential but none of the CSL structures calculated so far agrees with the observations [45].

Another alternative structure of the Σ=5 boundary has been found in recent calculations [46] using potentials for Cu and Au [21, 23]. This structure possesses symmetry p42'2' and the boundary plane coincides with the one of the (002) layers rather than being positioned in between the two (002) planes as in the CSL structure. Furthermore, it has been found that structures composed of quasi-random mixtures of the units of the CSL structure and the structure with the p42'2' symmetry, the boundary plane of which is positioned alternately below and above that of the CSL structure, may possess lower energy than any of the CSL periodic structures. Since randomness is also favoured for entropic reasons, this structure will be even more favoured at finite temperatures. The diffraction intensities calculated for this non-periodic structure composed of two types of units are shown in figure 7b. In this calculation it was also considered that different distortions will occur in different parts of the boundary due to inhomogeneities.

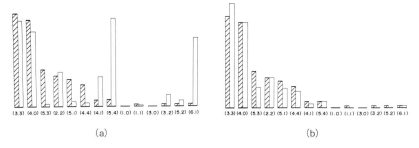

(a)   (b)

Fig.   7   Comparison of experimental diffraction intensities (shaded columns) with calculated ones (blank columns) for various reflections (h,k).
(a) CSL structure. (b) Structure composed of a random mixture of two types of units assuming that the boundary is variously distorted in different regions.

These distortions are slowly varying on the atomic scale and we, therefore, assume that the structure of the real boundary consists of patches each made of a mixture of slightly but differently distorted units. These distortions were simulated by assigning small random displacements to all the atoms in the unit cell. Differently distorted regions of the boundary were then regarded as mixtures of the same distorted units of the CSL and p42'2' boundary. The corresponding structure factors were calculated for each of these regions and the total diffraction intensity evaluated as a sum of the intensities corresponding to these regions. While the effect of these distortions is not negligible the main reason for the big difference in calculated intensities when compared with those for the CSL structure, stems from the mixing of two types of structural units [46].

An excellent agreement between calculations and experiment is now obtained. This result was found to be the same for all the interatomic potentials used in [46] and it is, therefore, likely that the same diffraction pattern for $\Sigma=5$ twist boundaries will be observed in different materials, similarly as found in the case of $\Sigma=13$ boundaries [47]. However, when the calculation has been made using the potential for aluminium [22] the CSL structure is highly favoured, (its energy is 30% lower than that of alternatives, see also [34]) and the structure with the p42'2' symmetry is not stable. Hence, it can be expected that a different diffraction pattern of $\Sigma=5$ twist boundaries will be observed in aluminium.

[111] Twist Boundaries near the $\Sigma=3$ Twin

As an example of the [111] twist boundary the misorientation of which is close to that of the $\Sigma=3$ twin we show in figure 8 the $\Sigma=91$, 53.99° twist boundary calculated using the potential for aluminium [22]. In this figure the structure is shown in the projection along [111] and four (111) planes adjacent to the boundary are shown; atoms belonging to these four planes are marked consecutively as $\Delta + \mathbf{O} *$. In this configuration one finds that some atoms in the lattice planes adjacent to the boundary are at a distance closer than the nearest neighbour spacing in the undistorted f.c.c. lattice. Consequently, these atoms produce a large contribution to the grain boundary energy due to the repulsive part in the interatomic potential. In order to look for a lower energy configuration and to avoid the close approach of atom pairs in the boundary a different starting configuration has been adopted. The lattice plane in the middle of the computational block was divided into two separate sections belonging to the upper and lower crystal, respectively, in such a way that no two atoms in different (111) planes are too close to each other. The initial density of this particular plane was taken to be the same as the density of the other (111) planes. After relaxation, structures different from that depicted in fig. 8 were found depending on the interatomic potential used. Figure 9 shows one structure obtained for the $\Sigma=91$ boundary which resembles closely the six-star pattern observed experimentally for boundaries with misorientation near the twin orientation [48, 49]. The stacking sequences of (111) planes in a cross section along P_1P_3 in fig. 9 is illustrated in fig. 10 where the symbols I and II represent ABC... and CBA... stacking sequence of the (111) planes, respectively. Apparently, the boundary plane has a stepped character such that the double steps only occur at the corners of the triangles in fig. 9, in agreement with the experimentally observed structure of the network. Comparing the structures with the triangular pattern (fig. 8) and the six-star pattern (fig. 9) it is easily seen that a shift of the dislocation lying vertically in fig. 9 by half of the dislocation spacing transforms the configuration of fig. 9 into that of fig. 8. As a result of such a shift, the dashed triangle in fig. 9 disappears. This implies that the triangular structure of fig. 8 consists entirely of single steps in the boundary. Experimentally, a six-star pattern as well as a triangular pattern have been observed in near coherent twin boundaries [33, 48, 49] in gold. This suggests that perhaps the energy difference, at least in Au, between the two structures is not very large, as may have been anticipated from the previously discussed simple transformation of one structure into the other. This transformation preserves the dislocation density in the boundary plane. Both structures consist of

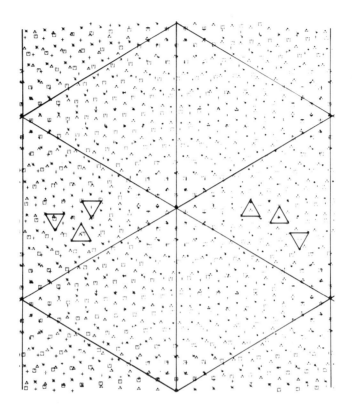

Fig. 8 Structure of the $\Sigma=91$ [111] twist boundary with the triangular pattern. Symbols have the same meaning as in figure 5.

single atomic steps in the boundary, in contrast to a hexagonal network which introduces double steps in the boundary, thereby increasing the boundary energy [32].

The general features of the structures found for various coincidence [111] twist boundaries in the vicinity of the twin orientation did not depend on the exact form of the interatomic potential used. Both triangular and six-star patterns were obtained for $\Sigma=127$, 61, 37, 91 using two completely different interaction function for Al: Dagens-Rasolt-Taylor (fig. 1) and a Lennard-Jones type. However, the computed energy of the triangular configuration is, at least in the case of Al, much lower than the energy of the six-star pattern. It means that detailed agreement with the experimental observations on Au, as far as the relative stability of structures is concerned, cannot be achieved by using 'material independent' potentials.

DISCUSSION

Pair potentials have been used to describe interatomic forces in most computer simulations of grain boundaries and other lattice defects. At the same time a complete description of interatomic forces in solids generally requires

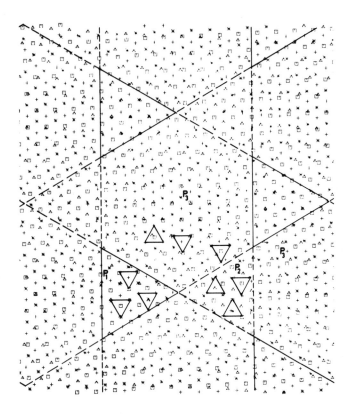

Fig.   9 Structure of the Σ=91 [111] twist boundary with the six-star pattern. The
symbols have the same meaning as in figure 5.

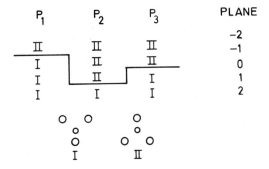

Fig.   10 Stacking of (111) planes in the vicinity of the boundary shown in figure 9
depicted in the cross-section along the line P_1P_3.

148

introduction of non-central many-body forces. The purpose of this paper has been, therefore, to discuss whether useful results regarding the grain boundary structure can be obtained with pair potentials. The positive answer to this question has been demonstrated on several examples. First, the most reliable results are clearly those which are independent of the potential used. The structural unit model which relates structures of boundaries for different misorientations of the grains is an example of such results. Using this model basic dependences of various grain boundary properties on misorientation can be deduced [2, 13, 50] although their detailed studies for specific materials will require a more sophisticated description of interatomic forces. It is likely, however, that pair-potential calculations can similarly reveal dependences of the boundary structure on other geometrical parameters.

Similarly, general features of grain boundaries are the multiplicity of structures and related phenomena such as possible transformations of boundary structure, mechanisms of absorption and emission of vacancies and the structure of vacancies in the boundaries. Again, detailed atomic positions in specific materials and related activation energies for the transformations and formation and diffusion of vacancies cannot be evaluated using only pair potentials. On the other hand, the present calculations are capable to reveal which structural alternatives can be expected. One example shown here are the [111] twist boundaries with misorientations near that of the $\Sigma=3$ twin for which two alternate structures and their relationship, were found. Similarly, several alternate structures of the short period $\Sigma=5$ [001] twist boundaries were found, and, furthermore, it was shown that non-periodic structures composed of two types of units may be most favoured. This is, indeed, an unexpected result which if proved correct brings new concepts into the studies of grain boundary structures.

Ultimately, of course, the validity of the simulation results must be checked experimentally. In this paper we have shown that the predicted non-periodic structure of the $\Sigma=5$ twist boundary leads to a very good agreement with the observed intensities of X-ray diffraction from this boundary. Similarly, the alternative structures of the near-twin [111] twist boundaries have both been observed. Nevertheless, a wider variety of experiments including, for example, the lattice imaging and recently developed convergent beam-electron diffraction which is capable of producing detailed information on the symmetry of the bicrystals [51, 52], are needed.

The exact positions of the atoms in the structures shown in this and other papers quoted here do depend on the pair potential used. Therefore, for any given material, they cannot be regarded as unambiguously predicted in these studies owing to the limitations of the pair potential approximation. Similarly, it has been found, for example, that the relative displacements of the grains across a geometrically defined boundary which are an important relaxation mode [11], vary in dependence on the potential used. Indeed, recent experimental studies of these displacements show that for the $\Sigma=3$ (112) boundary these displacements are different in different materials [53]. To investigate the dependence of these displacement relaxations on the type of material requires, in general, to go beyond pair potentials although the potential for Al derived from the electron theory [22] leads to the displacements at these boundaries which agree very well with the experiments [35]. Similarly, more sophisticated methods of the solid state physics have to be employed when studying cohesion at grain boundaries and any chemical and electronic phenomena at grain boundaries.

ACKNOWLEDGEMENTS

This research is a part of the research program of the Foundation for Fundamental Research on Matter (FOM-Utrecht) and has been made possible by financial support from the Netherlands Organization for the Advancement of Pure Research (ZWO - The Hague). This research was also supported in part by the National Science Foundation through the MRL Program Grant No. DMR-8216718.

149

REFERENCES

[1]   D. Mc Lean, Grain Boundaries in Metals (University Press, Oxford, 1957).
[2]   R. W. Balluffi, Met. Trans. A13, 2069 (1982).
[3]   A. P. Sutton, Int. Metals Reviews 29, 377 (1984).
[4]   K. R. Milkove, P. Lamarre, F. Schmückle, M. D. Vaudin and S. L. Sass, J. Physique 46, C-4-71 (1985).
[5]   A. Bourret, J. Physique 46, C4-27 (1985).
[6]   H. Ichinose and Y. Ishida, J. Physique 46, C4-39 (1985).
[7]   F. C. Frank, in Symp. on the Plastic Deformation of Crystalline Solids (Office of Naval Research, 1950), p. 150.
[8]   D. G. Brandon, B. Ralph, S. Ranganathan and M. S. Wold, Acta Metall. 12, 813 (1964).
[9]   W. Bollmann, Crystal Defects and Crystalline Interfaces (Springer, Berlin, 1970).
[10] R. C. Pond and D. S. Vlachavas, Proc. Roy. Soc. A385, 95 (1983).
[11] V. Vitek, A. P. Sutton, D. A. Smith and R. C. Pond, in Grain Boundary Structure and Kinetics (American Society Metals, Metals Park, Ohio, 1980), p. 115.
[12] V. Vitek, in Dislocations 1984, eds. P. Veyssière, L. Kubin and J. Castaign (CNRS Press, Paris, 1984), p. 435.
[13] A. P. Sutton and V. Vitek, Phil. Trans. Roy. Soc. London A309, 37 (1983).
[14] G.-J. Wang, V. Vitek and A. P. Sutton, Acta Metall. 32, 1093 (1984).
[15] W. A. Harrison, Pseudopotentials in The Theory of Metals (Benjamin/Cummings, Menlo Park, CA, 1966).
[16] R. Taylor, in Interatomic Potentials and Crystalline Defects, ed. J. K. Lee (TMS AIME, Warrendale, PA, 1981), p. 115.
[17] A. E. Carlsson and N. W. Ashcroft, Phys. Rev. B27, 2101 (1983).
[18] A. E. Carlsson, Phys. Rev. B32, to be published (1985).
[19] M. S. Duesberry, G. Jacucci and R. Taylor, J. Phys. F (Metal Phys.) 9, 413 (1979).
[20] D. G. Pettifor and M. A. Ward, Solid State Comm. 49, 291 (1984).
[21] V. Vitek and Y. Minonishi, Surface Sci. 144, 196 (1984).
[22] L. Dagens, M. Rasolt and R. Taylor, Phys. Rev. B11, 2726 (1975).
[23] A. G. Crocker, M. Doneghan and K. W. Ingle, Phil. Mag. A41, 21 (1980).
[24] M. W. Finnis, J. Phys. F (Metal Phys.) 4, 1645 (1974).
[25] A. P. Sutton and V. Vitek, Acta Metall. 30, 2011 (1982).
[26] M. W. Finnis and J. Sinclair, Phil. Mag. A50, 45 (1984).
[27] M. S. Daw and M. I. Baskes, Phys. Rev. B29, 6443 (1984).
[28] G.-J. Wang, A. P. Sutton and V. Vitek, Acta Metall. 32, 1093 (1984).
[29] P. D. Bristowe and R. W. Balluffi, J. Physique 46, C4-155 (1985).
[30] D. Schwartz, V. Vitek and A. P. Sutton, Phil. Mag. A51, 499 (1985).
[31] G. H. Bishop and B. Chalmers, Scripta Metall. 2, 133 (1968), Phil. Mag. 29, 515 (1971).
[32] R. F. Scott and P. J. Goodhew, Phil. Mag. A44, 373 (1981).
[33] D. Wolf, Acta Metall. 32, 245 (1984); ibid 32, 735 (1984).
[34] G. Hasson, J.-Y. Boos, I. Herbeuval, M. Biscondi and G. Goux, Surf. Sci. 31, 115 (1970).
[35] R. C. Pond and V. Vitek, Proc. Roy. Soc. London, A357, 453 (1977).
[36] P. D. Bristowe and A. G. Crocker, Phil. Mag. A38, 487 (1978).
[37] V. Vitek, Y. Minonishi and G.-J. Wang, J. Physique 46, C4-171 (1985).
[38] E. W. Hart, in Nature of Behaviour of Grain Boundaries, ed. Hsun Hu (Plenum Press, New York, 1972), p. 155.
[39] K. T. Aust, Can. Metall. Quart. 8, 155 (1972).
[40] T. Watanabe, S.-I. Kimura and S. Karashima, Phil. Mag. 44, 845 (1984).
[41] R. Kikuchi and J. W. Cahn, Phys. Rev. B21, 1893 (1980).
[42] G. Kalonji, P. Deymier, R. Najafabadi and S. Yip, Surface Science 144, 77 (1984).
[43] Sin-Wan Chan, J. S. Lin and R. W. Balluffi, Scripta Metall. 19, 1251 (1985).
[44] R. W. Balluffi, S. L. Sass and T. Schober, Phil. Mag. 26, 585 (1972).

[45] J. Budai, P. D. Bristowe and S. L. Sass, Acta Metall. 31, 699 (1983).
[46] Yoonsik Oh and V. Vitek, to be published (1986).
[47] J. Budai, A. M. Donald and S. L. Sass, Scripta Metall. 16, 393 (1982).
[48] J. J. C. Hamelink and F. W. Schapink, Phil. Mag. A44, 1229 (1981).
[49] J. Th. M. De Hosson, F. W. Schapink, J. R. Heringa, J. J. C. Hamelink, Acta Metall. 33 (1985) in press.
[50] G.-J. Wang and V. Vitek, Acta Metall., in press (1986).
[51] F. W. Schapink, S. K. E. Forghany and B. F. Buxton, Acta Cryst. A39, 805 (1983).
[52] J. Th. M. De Hosson, J. Heringa, F. W. Schapink, J. H. Evans and A. van Veen, Surface Science 144, 1 (1984).
[53] D. Schwartz, Ph. D. Thesis, University of Pennsylvania (1985).

MONTE CARLO SIMULATION OF GROWTH OF CRYSTALLINE AND AMORPHOUS SILICON

BRIAN W. DODSON AND PAUL A. TAYLOR
Sandia National Laboratories, Albuquerque, NM 87185

ABSTRACT

The authors have previously introduced a method, based on Monte Carlo techniques, for simulation of crystal growth processes in a continuous space. We have applied the method, initially used to simulate growth of two-dimensional Lennard-Jones systems, to treat growth of silicon in three dimensions. The interaction model for silicon is taken to be the recently introduced Stillinger-Weber (S-W) potential, which is a two- and three-body classical potential. Although the early stages of growth seem to be well modelled by the S-W potential, growth of even a single monolayer of epitaxial (111) silicon does not seem to be possible. Modifications to the S-W potential were considered, and found to be unacceptable physically. More accurate treatment of non-ideal atomic configuration energies is necessary to arrive at physically realistic growth simulations.

INTRODUCTION

The discovery and characterization of semiconductor strained layer superlattice (SLS) systems is of great importance for device applications because of the extremely fine control which can be exerted over electronic properties [1]. A good understanding of the processes involved in growth of crystals of covalent semiconducting materials is therefore of great interest. To date, however, despite much empirical information concerning growth of semiconductors, there is little fundamental understanding of the growth processes, especially those which dominate the early (submonolayer to few layer) stages of growth.

The authors have initiated a program to examine these issues through numerical modelling of crystal growth. Our principal tool is a technique, developed by the authors, to simulate crystal growth in continuous space using Monte Carlo methods [2]. The use of Monte Carlo techniques allows the simulation of greater effective time scales than do molecular dynamics techniques, at the cost of an accurate real time scale. This technique has been previously used to study the growth of 2-dimensional Lennard-Jones (L-J) SLS interfaces [3].

We begin with a short description of the Monte Carlo growth routine, and a description of its application to the 2-dimensional L-J system. The Stillinger-Weber classical potential for silicon [4] will then be described, and its strengths and limitations will be discussed. Finally, the early results of our simulations of silicon growth using this potential will be presented, and the improvements necessary for the description of the interaction potential will be discussed.

MONTE CARLO SIMULATION OF CRYSTAL GROWTH

At present, most simulations of crystal growth have depended on the definition of a natural spatial lattice by the substrate crystal [5]. This is usually done in the context of an SOS (solid-on-solid) model, wherein one considers a lattice gas model of the crystal and only very short-range interactions between the particles [6]. Such models can generally be mapped onto some version of an Ising model, thus rendering them quite tractable for computer simulation. However, SOS models ignore all effects which depend on the motion of particles relative to the ideal lattice sites. Such effects include displacements associated with thermal phonons, heteroepitaxy with lattice mismatch, surface reconstructions, and other effects inaccessible to a simulation involving a discrete spatial lattice. Clearly, such restrictions are not acceptable in the simulation of crystal growth from the vapor, where even in homoepitaxy such phenomena as structural freezing of initially metastable configurations and the behavior of the adatoms in the long-range substrate potentials are of crucial importance. We therefore developed a technique [2], based on Monte Carlo principles, to model crystal growth from the vapor in continuous space. The choice of Monte Carlo methods rather than molecular dynamics allows for longer effective diffusion and equilibration times, both of which are important for the accurate simulation of crystal growth.

Our intention is to model the growth of new crystal from the vapor phase onto a substrate. The substrate consists of an atomic lattice whose atoms interact through a continuous classical many-body potential A. The substrate, generally consisting of 50-500 atoms, is subjected to periodic boundary conditions perpendicular to the growth surface so that edge effects are avoided and correlation effects can be treated over longer distances. New atoms are added to the system in the following manner. First a random position above the substrate is chosen. Then a vapor phase atom, interacting with other vapor phase atoms through the potential B and with the substrate atoms through the potential AB (in homoepitaxy A, B, and AB are the same), is placed at that position several interatomic spacings above the substrate. The atom is then lowered towards the surface until it satisfies some criterion for small, but substantial, interaction with the substrate atoms (perhaps 10% of the final binding energy). This is then the initial position for the adatom. During this process the substrate atomic positions are held fixed. The results are insensitive to the details of the proximity criterion.

Once the new adatom is in its initial position, the complete system is equilibrated at the desired substrate temperature using a conventional Monte Carlo technique [7]. The Monte Carlo method generates a representative set of states from the canonical ensemble of the system at a given temperature. This is accomplished through generating new configurations by randomly moving one of the atoms, and accepting or rejecting the new configuration based on a Boltzmann-weighted energy criterion. The extent to which the system must be equilibrated between introduction of new adatoms is an important issue. Clearly, at the rates of growth involved in such techniques as molecular-beam epitaxy (on the order of monolayers/second), perhaps 100 atoms/second would be incident on our model substrates, which is a vastly longer time than can be simulated numerically. Thus, the introduction of subsequent adatoms must appear to be isolated events, that is, adatom N must come into equilibrium with the substrate and move into a statistically characteristic position before adatom N+1 is introduced. This requires a compromise between performing large numbers of Monte Carlo steps between introducing successive adatoms and satisfying the constraints of available computer time. This is to some extent an art, but a useful rule of thumb is that enough Monte Carlo steps must be taken between introduction of new adatoms to allow signifigant surface diffusion near the melting point (at perhaps $0.8T_{melt}$). For the systems under consideration in the current work, we find that this criterion is met with 100-200 Monte Carlo steps per atom between successive adatoms. Recall that in a typical molecular dynamics calculation, 100-200 time steps will represent at most a few atomic vibrations, during which time equilibrated surface diffusion will not take place. This avoidance of `real` dynamics combined with the statistical independence of the adatom introductions results in the efficiency of the MC method over MD. Once the adatom equilibration has been accomplished, another adatom is introduced, and the process is repeated until done.

GROWTH OF 2-D LENNARD-JONES SLS INTERFACES

The procedure described above to simulate crystal growth on an atomic scale was first tested by simulation of homo- and hetero-epitaxial growth of atoms on a 2-dimensional substrate [3]. The interaction potential was chosen to be the Lennard-Jones potential, modified so that materials with different lattice parameters but the same binding energy could be modelled simultaneously. The potential used was

$$\phi(r)=[r^{-12}-\alpha r^{-6}]/\alpha^2 \qquad (1)$$

where the lattice parameter is varied by α, and r and ϕ are in dimensionless units. A cutoff distance of 3 units was adopted (the interatomic distance with $\alpha=1$ is 1.1225.)

The substrate was an 80x4 trigonal 2-D lattice with periodic boundary conditions in the long direction. Several hundred atoms were added to the substrate, representing 3-5 new crystal layers. Using 100 Monte Carlo steps/atom (MCS/atom) between adatom depositions, the growth simulations required roughly 10^7MCS, costing about 40 hours on a VAX780-class minicomputer.

We found that homoepitaxial growth without defects would occur, for this model system ($\alpha=1$), at substrate temperatures from 0.02 to 0.09 units. (The bulk melting point of the Hamiltonian used is 0.12 units.) At higher temperatures melting interfered with the growth process, whereas equilibration at lower temperatures became too slow.

The case of heteroepitaxial growth is more complex and more interesting. We simulated the growth of systems with mismatches of 1,2,3.5, and 5% at substrate temperatures of 0.06 and 0.09 units. The adatoms had the larger lattice parameter. At T=0.06, all systems exhibited imperfect interfaces. The defects took the form of conventional mismatch dislocations occurring roughly every 100/(%mismatch) atoms along the interface. However, at T=0.09, we were successful in growing properly registered SLS interfaces at mismatches of 1 and 2%. In contrast, the 3.5 and 5% mismatch cases resulted in misfit dislocations similar to those described above.

THE STILLINGER-WEBER SILICON INTERACTION POTENTIAL

Having observed that the simulation of growth in the simple model L-J system made clear the importance of mismatch parameter and substrate temperature in the successful growth of defect-free SLS interfaces, we decided to extend our studies to more realistic systems. A superlattice system of great current interest is the Si/SiGe system. Fortunately, this is also, conceptually, one of the easiest for which to develop an interaction potential, since all of the atoms exhibit tetragonal covalent binding, and one of the layers is monoatomic. If one makes the further simplifing assumption that the SiGe layer can be modelled by a monoatomic layer with the same average size and potential, one has a relatively tractable model to work with.

However, one is still faced with the problem of choosing an appropriate description for the interaction of the silicon-like atoms. It is well-known that no reasonable pair potential can stabilize the open tetragonal structure required here. It is possible to express a potential energy function describing interactions between N particles in terms of n-body contributions

$$\phi(1,2,3,\ldots N)=\sum_{i}v_1(i)+\sum_{\substack{i,j \\ i<j}}v_2(i,j)+\sum_{\substack{i,j,k \\ i<j<k}}v_3(i,j,k)+\ldots \qquad (2)$$

The utility of this expansion depends on the rapid convergence of the energy as the order of the interaction terms increases. The standard phenomenological implementation of Eq.2 is the valence force field [8]. This method involves expansion of the binding energy in terms of small displacements from the atomic equilibrium positions. Such a description cannot be used to model growth from the vapor, since atomic configurations will arise which cannot be mapped naturally onto the underlying geometry assumed in this procedure.

An interaction potential for silicon which allows for arbitrary atomic configurations was recently introduced by Stillinger and Weber [4] to model the melting of silicon. The Stillinger-Weber (S-W) potential is in the form of Eq.2, truncated to include only the 2- and 3-body terms. The pair potential $v_2(i,j)$ has the form

$$v_2(r) = \begin{cases} A(Br^{-p}-r^{-q})\exp[1/(r-a)], & r<a \\ 0, & r>a \end{cases} \qquad (3)$$

where a sets the range of the potential, B sets the lattice parameter, and A controls the size of the binding energy. The 3-body term v_3 is expressed as a symmetrized sum

$$v_3(ijk)=h(ijk)+h(jik)+h(jki) \qquad (4)$$

where h(ijk) depends only on the distances r_{ij}, r_{jk}, and the angle θ_{ijk}. If both radii are less than a, h has the form

$$h(r_{ij},r_{jk},\theta_{ijk})=\lambda\exp[\gamma((r_{ij}-a)^{-1}+(r_{jk}-a)^{-1})][\cos\theta_{ijk}+1/3]^2 \qquad (5)$$

otherwise h is zero. This angular term, which is always repulsive, discriminates in favor of the ideal tetrahedral geometry without excluding the possibility of other configurations.

Stillinger and Weber assigned the adjustable parameters the values

A=7.0496   a=1.8
B=0.60222   λ=21.0
p=4   γ=1.2
q=0

These values were chosen by requiring that the diamond structure be the most stable periodic structure and that the melting point and the liquid structure be in reasonable accord with experiment. It is of interest, especially in the context of structural calculations, to examine the accuracy with which the S-W potential models the mechanical properties of silicon, which were not used as input to the parameter determination. This question was addressed by developing a valence force field model for both the S-W potential and real silicon, and comparing the resulting parameters. The two terms which are appropriate for a 2- and 3-body potential such as the S-W potential are a bond-stretch and a bond-bending term. We find that, for the S-W potential, the bond stretch term is 11% higher, and the bond-bending term is 17% smaller than in real silicon. This fit is remarkably good considering that no mechanical information was used to obtain the S-W parameters.

The S-W potential does a satisfactory job of simulating the mechanical and thermal properties of bulk silicon. However, the simulation of growth from the vapor imposes especially strong constraints on the structure of the interaction potential. The structure of the substrate surface must be accurately modelled, and the interaction of a single adatom with the surface atoms must be accurately described. In both cases the proper description will include the effect of dangling bonds on the interaction potential. At present, this is where most simple representations for covalent binding fail, and the S-W potential is not without flaw in this regime. It is well-known that a clean (111) silicon surface undergoes a reconstruction driven by the degenerate surface states through the Jahn-Teller effect. Unfortunately, the S-W potential shows no surface reconstruction, because the potential is short-ranged and only the 2-body term has an attractive part. The treatment of adatom-surface interactions is also inadequate. However, it is not immediately clear that these flaws will render the S-W potential useless for simulation of growth processes.

MODELLING GROWTH OF (111) SILICON

A (111) silicon substrate 2 monolayers thick and roughly 20x20 A on the surface was generated. Periodic boundary conditions were imposed on the edges perpendicular to the surface to avoid edge effects. We then investigated the interaction energy between a single adatom and the substrate as a function of position. Recall that on an unreconstructed (111) silicon surface, alternating surface atoms will have a dangling bond directed perpendicular to the surface. The equilibrium position for an adatom is not directly above such an atom, but rather is located so as to allow binding with three adjacent dangling bonds, being equidistant from the corresponding surface atoms and above the hole in the surface hexagons. (This is the bonding geometry suggested by Lander and Morrison [9].) This bonding geometry will continue to be energetically favored until 1/3 of a surface layer is added, thus occupying all of the dangling bonds. Thereafter, adatom-adatom interactions will qualitatively change the nature of the growth process.

Having examined the nature of the initial adatom-surface interactions and finding that they are at least qualitatively correct, we attempted to model the epitaxial growth of silicon on the (111) silicon surface. There was no substrate temperature between zero and the melting point where epitaxial growth was obtained. Instead, an essentially amorphous layer was built up. Upon finer investigation, it became clear that the difficulty occurred in the neighborhood of 1/3 monolayer deposited, where the nature of the adsorption process changes. Proper treatment of this regime requires the accurate calculation of small energy differences in non-ideal atomic configurations, and the S-W potential seems unable to model configurations with low coordination number with the required accuracy.

The early stages of (111) silicon growth are so well modelled by the S-W potential that it seems reasonable to investigate possible modifications which would extend its usefulness. It was found that increasing λ from 21 to 60 resulted in growth of perfect (111) silicon. However, the resulting material has very strange properties, including a liquid-crystal-like molten state. The angular interactions are too large to be physically reasonable. It is possible to obtain a similar effect on atomic configurations with grossly non-tetrahedral angles without changing the bulk equilibrium properties by adding a fourth-order angular term to the three-body potential.

$$h(ijk) = \lambda \exp[\gamma((r_{ij}-a)^{-1}+(r_{jk}-a)^{-1})][(\cos\theta_{ijk}+1/3)^2+\beta(\cos\theta_{ijk}+1/3)^4] \quad (6)$$

When β is 10 or larger, we again obtain reasonably good (111) silicon growth. However, the qualitative nature of the early stages of growth has changed, with adatoms bonding directly above surface atoms with dangling bonds. Since the realistic simulation of SLS interfaces depends strongly upon proper treatment of both regimes of crystal growth, the S-W and other closely related potentials are not adequate for such studies.

156

ACKNOWLEDGEMENTS

This work was performed at Sandia National Laboratories and was supported
by DOE contract DE-AC04-76DP00789.

REFERENCES

1. G.C. Osbourn, J Appl Phys 53, 1586 (1982).
2. B.W. Dodson, submitted to Phys Rev B.
3. B.W. Dodson, in Layered Structures, Epitaxy, and Interfaces, J.M.
    Gibson and L.R. Dawson eds. (MRS, Pittsburgh, 1985).
4. F.H. Stillinger and T.A. Weber, Phys Rev B 31, 5262 (1985).
5. See, for example, H. Muller-Krumbhaar, in Monte Carlo Methods in
    Statistical Physics, K. Binder, ed. (Springer, New York, 1979).
6. T.L. Hill, J Chem Phys 15, 761 (1947).
7. K. Binder, op cit.
8. W.A. Harrison, Electronic Structure and the Properties of Solids.
    (Freeman, San Francisco, 1980).
9. J.J. Lander and J. Morrison, Surf Sci 2, 553 (1964).

COMPUTED STRUCTURES OF [001] SYMMETRICAL TILT BOUNDARIES IN COVALENTLY BONDED MATERIALS

J.T. Wetzel, A.A. Levi and D.A. Smith
IBM Thomas J. Watson Research Center, Yorktown Heights, New York 10598

ABSTRACT

The dependence of the structure of (210) and (310) symmetrical [001] tilt boundaries in silicon, germanium and diamond on the Keating covalent force field (potential) has been investigated by computer modelling. We have found that the sensitivity of grain boundary structure to variations of the Keating potential depends on the local atomic arrangement at the grain boundary.

INTRODUCTION

Studies of grain boundary structure in covalently bonded materials have been carried out mainly by ball and stick geometric modelling.[1-3] Relatively few efforts have been made to use computer modelling techniques to study grain boundary structure in covalently bonded materials.[4,5] One difficulty in modelling such materials is the lack of appropriate interatomic potentials which can handle various aspects of covalent bonding in a crystalline solid, such as bond stretching, torsion, bond angle distortions and dangling bonds; however, recent calculations of dislocation structures in silicon have shown that covalent force field potentials developed for molecules may be applicable.[6] Another difficulty is developing appropriate computer algorithms which allow bonds to break and be remade between different atoms so that as many atoms as possible have four-fold coordination without recourse to "doing it by hand", i.e., ball and stick modelling.

We have used the harmonic Keating potential[7] to model grain boundary (gb) structure in silicon, germanium and diamond; thus our initial goal was to ascertain if changing the Keating parameters significantly changed grain boundary structure. We have studied the (210)/[001] and (310)/[001] $\Sigma = 5$ tilt boundaries in Si, Ge and C, and a series of $< 110 >$ symmetrical tilt boundaries in silicon.[5] The harmonic Keating potential is the simplest of the covalent force field models because it considers only bond stretching and bending; it does not include any effects due to dangling bonds. We do not allow dangling bonds in our calculations, but we do allow reconstruction to take place by considering only the four optimal nearest neighbors during relaxation. This algorithm, explained briefly below, is still not sufficient for all reconstructions, so our method is more akin to "computer-aided" ball and stick modelling.

METHOD OF CALCULATION

A bicrystal with a given misorientation was built in the computer. Periodic conditions along the tilt axis direction and a direction normal to the tilt axis but parallel to the grain boundary plane were used to form the bicrystal unit cell and thus reduce the number of atoms needed for a given calculation. Rigid body displacements and local atomic relaxation in all three degrees of freedom were allowed during any relaxation. A conjugate gradient algorithm was used to relax the atoms in a bicrystal towards a low energy configuration. Exit criteria from the conjugate gradient algorithm were no energy changes between successive conjugate gradient iterations and the largest displacement experienced by any atom be less than $10^{-3}\text{\AA}$. The minimum energy structure thus found is a local minimum; tests for structural stability were done by using different starting points and obtaining the same structure and by rigidly displacing one crystal of a relaxed bicrystal with respect to the other crystal and then re-relaxing the bicrystal to obtain the first relaxed structure.

The harmonic Keating potential[7] was used in the present work to calculate the energy of the bicrystal and the forces between atoms. The Keating potential is given as:

$$E = \frac{1}{2}\left\{ \frac{\alpha}{4a^2}\sum_{i=1}^{N}\sum_{j=1}^{4}(x_{ij}^2 - 3a^2)^2 + \frac{\beta}{2a^2}\sum_{i=1}^{N}\sum_{j=1}^{4}\sum_{k \neq j}^{3}(\vec{x}_{ij} \cdot \vec{x}_{ik} + a^2)^2 \right\}$$

where α and β are the bond stretching and bond bending parameters, respectively, a is $a_0/4$, and a_0 is the lattice constant; $\vec{x}$ are vectors connecting atoms ij and ik, respectively, x_{ij} is the distance between atoms i and j, N is the number of atoms in the bicrystal unit cell and the summations of j are always only four atoms. The parameters of the harmonic Keating potential are shown in Table I.

Table I. Parameters for the Harmonic Keating Potential (10^5 dyne/cm)

	α	β	β/α
Silicon	0.485	0.138	0.29
Germanium	0.38	0.12	0.32
Diamond	1.29	0.85	0.66

A cutoff distance must be assumed in order to use the Keating potential, and we found that a distance of $0.5 \times a_0$, which is between first and second neighbors in the perfect crystal, was useful. Only the four optimal nearest neighbors were used to calculate the force on any one atom; these neighbors were chosen by bond distances, that is, how closely an interatomic distance approximates the equilibrium nearest neighbor spacing for all the atoms found within the cutoff distance. Bond angles could also be considered as one of the criteria for choosing optimal neighbors, but experimental evidence concerning the pair correlation function in amorphous silicon shows that first neighbor spacings are preserved as much as possible[8], implying that bond angle distortions are not as important as bond stretching in the total energy.

The energy values reported here are given only as a guide; these values are not particularly significant because the Keating potential gives {111} twin boundary and stacking fault energies of zero. However, the energies can be used to distinguish unlikely structures from preferred, or more probable, structures.

CALCULATED STRUCTURES

The structure of (210)/[001] and (310)/[001] tilt boundaries has been modelled by Hornstra[2] using ball and stick techniques. These boundaries are composed of linear arrays of lattice dislocations with Burger's vector 1/2<110> and line vector parallel to [001]. Our computed structures for (210)/[001] and (310)/[001] boundaries in silicon are shown in Figures 1 and 2. Two stable structures for each boundary have been found thus far.

. For the (210)/[001] boundary, both structures are different from Hornstra's results. The lower energy structure in Figure 1a is characterized by a rigid body translation of $a_0/8[001]$ away from the coincidence position, see Figure 1b.. The structure of the higher energy (210)/[001] boundary (Figure 1c) is composed of lattice dislocations with a core structure that is not identical to that in Figure 1a. A compilation of the rigid body translations and computed grain boundary energies for the stable structures found thus far is given in Table II.

The lower energy (310)/[001] structure shown if Figure 2a is nearly identical to Hornstra's result; the small differences arise due to the relaxation that is allowed in our models. The higher energy structure shown in Figure 2c can be distinguished from Figure 2a because of the different lattice dislocation core structure which constitutes the grain boundary. These structures are henceforth referred to as 210(a), 210(c), 310(a) and 310(c); where (a) and (c) correspond to Figures 1 and 2, and A-type structures are those of lower energy and C-type structures are those of higher energy.

Varying the parameters of the Keating potential to represent germanium and carbon does not result in any new reconstructions; the structures are nominally the same. There are differences in the rigid body translations (RBTs) between the grains of a bicrystal, and these are shown in Table II. The RBTs are referred to the coincidence position in units of a_0, the lattice parameter, and a negative value for Y means a compression. The directions X,Y,Z are three mutually perpendicular vectors that describe the bicrystal unit cell. X is parallel to [$\bar{1}$20] or [$\bar{1}$30], Y (the grain boundary plane normal) is parallel to [210] or [310] and Z (the tilt axis) is parallel to [001].

Table II. RBTs and GB Energy(mJ/m²) for (210) and (310) Tilt Boundaries in Si, Ge and C.

	210(a)			210(c)			310(a)			310(c)		
	X	Y	Z	X	Y	Z	X	Y	Z	X	Y	Z
Si	0.05	-.02	-.12	0.01	-.01	0.0	0.8	0.01	0.0	0.03	0.0	0.0
		730			2210			610			1980	
Ge	0.05	-.02	-.12	0.01	-.02	0.0	0.8	0.01	·0.0	0.03	-.01	0.0
		665			1855			565			1825	
C	0.05	-.02	-.13	0.01	-.04	0.0	0.8	0.01	0.0	0.02	-.03	0.0
		4285			11250			3725			10900	

The data from this comparison show that increasing the β parameter relative to the α parameter (see Table 1 where values are given) results in a compression of the bicrystal and thus the grain boundary for the C-type structures. The A-type grain boundaries are relatively insensitive to changes in β/α; the minor differences are not significant.

160

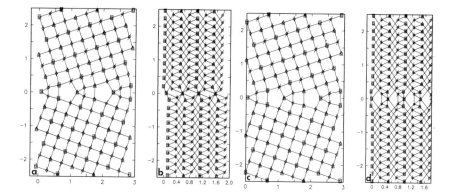

Figure 1. Relaxed, stable grain boundary structures of $\Sigma=5$, $(210)/[001]$ symmetrical tilt boundaries. There are two projections shown for each stable structure: Figures 1a and 1c are [001] projections and Figures 1b and 1d are $[\bar{1}20]$ projections. Circles represent atomic positions; squares and triangles around the circles and straight and slanted lines through the circles show the stacking along $a_0[001]$, which is in the sequence $A\,(\square)\,,B\,(/)\,,C\,(\Delta)\,,D\,(\,|\,)$. Figure 1b shows most clearly the rigid body translation of $a_0/8[001]$. Figures 1a-b and Figures 1c-d correspond to 210(a) and 210(c), respectively, in Table II.

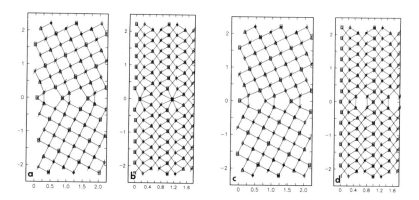

Figure 2. Relaxed, stable grain boundary structures of $\Sigma=5$, $(310)/[001]$ symmetrical tilt boundaries. There are two projections shown for each stable structure: (a) and (c) are [001] projections and (b) and (d) are] and $[\bar{1}30]$, projections. Circles represent atomic positions; squares and triangles around the circles and straight and slanted lines through the circles show the stacking along $a_0[001]$, which is in the sequence $A\,(\square)\,,B\,(/)\,,C\,(\Delta)\,,D\,(\,|\,)$. The structures depicted in Figures 2a-b and c-d correspond to 310(a) and 310(c), respectively, in Table II.

DISCUSSION

The structure of (210) and (310) symmetrical tilt boundaries is not a sensitive function of the Keating potential, though differences do arise when the net volume change of the bicrystal is calculated. The correlation of the ratio β/α with the net volume decrease, or compression that acts on a grain boundary, is a result which may depend largely on the grain boundary structure. We note that the A-type structures do not have systemactically varying RBTs but C-type structures do when β/α is changed to larger values. A careful accounting of the number and strength of the bond stretching and bending interactions across the grain boundary needs to be done to understand this difference. Also other grain boundaries with different misorientations ought to be studied to see if similar behavior is found for all stable structures.

Compressive strain at the grain boundary is not the only behavior found; tensile strain in the 310(a) structures should be compared with compressive strain in 210(a) structures. The fact that different RBTs were found gives some confidence in the use of the harmonic Keating potential for computer modelling of grain boundary structure.

There are not many experimental studies to which we can compare our results. The work of Bacmann, et al.,[9] on the structure of $\Sigma=5$, (310)/[001] does not agree with our models. Their structure has a rigid body translation of $a_0/8[001]$ whereas neither of our structures for the (310) boundary have this translation. However, we have found such a result for the (210) boundary. As was mentioned above, it is possible that there are several stable alternate structures with essentially the same energy for a particular misorientation angle. Indeed this is the case for grain boundary structure in metals[10], and we have found this to be true for symmetrical <110> tilt boundaries in silicon.[5]

CONCLUSIONS

The (210) and (310) symmetrical [001] tilt boundaries have been modelled in silicon, germanium and diamond using the harmonic Keating potential. Two stable structures were found for both tilt boundaries. Changing the parameters of the Keating potential did not cause marked structural differences in grain boundary structure, but different rigid body translations were found that depended on the local atomic structure of the grain boundary and the relative strength of the bond stretching and bending parameters.

REFERENCES

1. J. Hornstra, Physica **25**, 409 (1959)

2. J. Hornstra, Physica **26**, 198 (1960)

3. M. D. Vaudin, B. Cunningham and D. G. Ast, Scripta Metall. **17**, 191 (1983)

4. R. C. Pond, D. J. Bacon and A. M. Bastaweesy, in **Inst. Phys. Conf. Series No. 67**, Section 4, 253 (1983).

5. J.T. Wetzel, A.A. Levi and D.A. Smith, to be published in the Proc. Jap. Inst. of Metals Int. Conf. on Grain Boundary Structure

6. A. Lapiccirella and K.W. Lodge, in **Inst. Phys. Conf. Series No. 60**, Section 1, 51 (1981)

7. P. N. Keating, Phys. Rev. **145**, 637 (1960)

8. M. H. Brodsky, S. Kirkpatrick and D. Weaire, in **Tetrahedrally Bonded Amorphous Semiconductors**, A.I.P. Conf. Proc. No. 20 (1974).

9. J.J. Bacmann, A.M. Papon and M. Petit, J. de Phys. **43**, Coll. C-1, 15 (1982)

10. A.P Sutton and V. Vitek, Phil. Trans. Roy. Soc., **A309**, 1 (1983)

SURFACE RECONSTRUCTION OF SI(100)

THOMAS A. WEBER
AT&T Bell Laboratories, Murray Hill, New Jersey 07974

ABSTRACT

The reconstruction of the Si(100) surface has been studied using a potential consisting of two- and three-body interactions [1]. Minimum energy quenches of a slab of 320 Si atoms 10 layers thick with periodic boundary conditions in the xy plane found the p(2x1) reconstruction lower in energy than the c(2x2) reconstruction. The potential energy contours for the addition of an adatom to the (100) surface were determined for the unreconstructed surface and for the p(2x1) and c(2x2) reconstructions.

INTRODUCTION

Although potentials have been available to model the crystalline phases of silicon [2,3], it is only recently that potentials capable of also modeling the fluid phase of silicon have become available [1,4,5]. Abraham and co-workers [6,7] have studied the p(2x1) reconstruction of the (100) surface and the melting of the (100) and (111) surfaces using the potential of ref. [1]. NoorBatcha, Raff and Thompson have determined the diffusion of silicon atoms on the (111) and (100) surfaces [8,9].

Experimentally the reconstruction of the Si(100) surface has been studied by various techniques, among them LEED [10-13] and He atom diffraction [14,15]. Both a (2x1) and a disordered c(4x2) reconstructions have been observed. The c(4x2) reconstruction is the most stable structure although both the p(2x1) and c(2x2) reconstructions are also observed.

In this paper, the reconstruction of the (100) surface and Si adatom addition to the reconstructed surfaces are studied using the potential of ref. [1] that was constructed to model the melting of bulk Si. The form of the potential and the parameters are give in ref. [1]. The potential is a modified Keating form [2] in that bond stretch is modeled by the two-body interactions while bond bending is modeled by the three-body interactions.

In this study a slab of 320 atoms representing a surface-vacuum interface was constructed with periodic boundary conditions used in the lateral (x and y) directions. The slab consists of 10 layers with vacuum on two faces. Only one surface was allowed to reconstruct. The initial density of the silicon slab was 2.41 g/cm^3. The unit of length used in the simulations is 0.20951 nm; the unit of energy is 2.16 eV; and the unit of mass of ^{28}Si is 4.6457×10^{-23}kg.

To determine the minimum energy structures for the various surface reconstructions, the steepest-descent equations

$$\frac{d\mathbf{r}_j}{ds} = -\nabla_j \Phi \qquad (1)$$

were solved in the limit (s →∞). This approach finds the mechanically stable structures starting from any given initial conditions.

SURFACE RECONSTRUCTION

Starting with an unreconstructed (100) surface, the application of the steepest-descent equations causes the surface to relax, but no dimer formation characteristic of both the p(2x1) and c(2x2) reconstructions is observed. However by displacing the first layer atoms along the (110) direction from their crystalline lattice sites, both the p(2x1) and c(2x2) reconstructions are then obtained. Figure 1 shows a top view of the dimer formation in the first layer of atoms for both reconstructions. The sizes of atoms have been reduced for the layers further below the surface. Each dimer formed in the p(2x1) reconstruction recovers 78% of the energy of a bulk Si-Si bond. The dimer bond formed on the surface is 2% larger than a bulk Si-Si bond. Each dimer formed in the c(2x2) reconstruction recovers 71% of the energy of a bulk Si-Si bond.

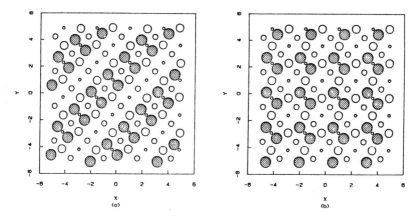

Figure 1. Top view of the p(2x1) reconstruction (a) and the c(2x2) reconstruction (b). The radii of the atoms decrease for layers further from the surface.

The displacements of the atoms in the slab in going from the relaxed unreconstructed surface to the reconstructed surfaces are given in Table I. In both cases the reconstruction causes the atoms of the first layer to contract toward the bulk and move along the (110) direction in the plane. For the p(2x1) reconstruction the atoms in layers 2, 5, 6, 9 and 10 move along the (110) direction, with no in-plane displacements observed for layers 3, 4, 7 and 8. The z displacements (the direction perpendicular to the surface) are observed to decrease in magnitude for atoms farther from the surface. The displacements of atoms for the c(2x2) reconstruction decreases more rapidly for atoms farther from the surface and no z-displacement is found past the fourth layer of atoms. Atoms in layers 1 and 5 move along the (110) direction while those in layers 3 and 7 move along the (110) direction for the c(2x2) reconstruction.

TABLE I

The relative displacement of Si atoms (in reduced units) from the unreconstructed layers for the p(2x1) and c(2x2) reconstructions.

	p(2x1)		c(2x2)	
layer	$\Delta x, \Delta y$	Δz	$\Delta x, \Delta y$	Δz
1	±.242	-.044	±.238	-.052
2	±.039	.009	.0	.010
3	.0	±.053	±.019	.0
4	.0	±.035	.0	.017
5	±.012	.0	±.006	.0
6	±.005	.0	.0	.0
7	.0	±.006	±.001	.0
8	.0	±.004	.0	.0
9	±.002	.0	.0	.0
10	±.004	.0	.0	.0

ADATOM ADDITION

An adatom was added to the unreconstructed and reconstructed surfaces at a particular x and y location, and its z position as well as positions of the other atoms of the surface, were allowed to relax by solving the steepest-descent equations to find the potential energy minimum. The bottom layer of atoms was held fixed so that the adatom did not pull the slab over to find the absolute minimum of the surface. In this way the minimum energy channels for motion of an atom at low temperature along the surface were mapped out. Figure 2 shows contour plots of energy of an adatom on the surface for the p(2x1) and c(2x2) reconstructions as a function of adatom position.

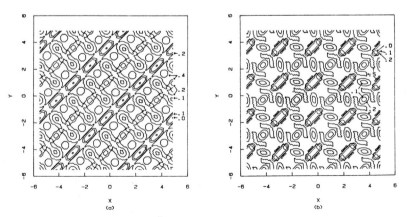

Figure 2. a) Contour plots of the energy as a function of adatom position for the p(2x1) reconstruction. The energies are relative to the lowest energy contour of -589.8 (reduced units). b) Contour plots of the energy as a function of adatom position for the c(2x2) reconstruction. The energies are relative to the lowest energy contour of -588.8 (reduced units).

166

For the p(2x1) reconstruction the lowest energy path is found to be between the dimer pairs with the absolute minimum along this path located directly above the Si atoms in layer 4. The adatom is also found to be at the lowest elevation at these locations. A barrier of 0.76 eV is encountered in going from the position of one Si atom in layer 4 along the channel to the next. The highest energies are found for adatoms placed nearest the surface dimers. Such adatoms are also at highest elevation above the surface.

For the c(2x2) reconstruction zig-zag paths parallel to the x and y axes are found to be the lowest energy channels for adatom motion. Low energy minima are found near the atoms in layer 3 with barriers found along the path between the layer 2 and layer 3 Si atoms.

CONCLUSIONS

The potential used to model the melting of bulk Si [1] is found to produce both stable p(2x1) and c(2x2) reconstructions. To model the out-of-plane tilting of dimers observed in the c(4x2) reconstruction of the (100) surface, modifications would need to be made in the potential function, possibly adding 4-body and higher-order interactions. The p(2x1) and c(2x2) reconstructions are adequately modeled by the potential because these surface reconstructions are attempts by the system to recover the tetrahedral bonds lost by the introduction of the surface. The dimers of the p(2x1) reconstruction were tilted out of the plane to see if the c(4x2) reconstruction could be produced, but negligible energy differences were found between the tilted and non-tilted dimers with this potential.

Other parameterizations of a more general potential of the type of ref. [1] have also been used to study the surface reconstruction. These potentials were chosen so that the energies of several different lattices as a function of density were in close agreement with the Yin and Cohen Si calculations [16]. These different potentials do not qualitatively change the picture of surface reconstruction as reported here.

REFERENCES

[1] F. H. Stillinger and T. A. Weber, Phys. Rev. B **31**, 5262 (1985).
[2] P. N. Keating, Phys. Rev. **145**, 637 (1966).
[3] W. Weber, Phys. Rev. B **15**, 4789 (1977).
[4] D. A. Smith, Phys. Rev. Lett. **42**, 729 (1979).
[5] R. Biswas and D. R. Hamman, preprint.
[6] F. F. Abraham and I. P. Batra, Surf. Sci. Lett., in press.
[7] F. F. Abraham and J. Q. Broughton, preprint.
[8] I. NoorBatcha, L. M. Raff, and D. L. Thompson, J. Chem. Phys. **82**, 1543 (1985).
[9] I. NoorBatcha, L. M. Raff, and D. L. Thompson, J. Chem. Phys. **83**, 6009 (1985).
[10] R. E. Schlier and H. E. Farnsworth, J. Chem. Phys. **30**, 4 (1959).
[11] F. Jona, M. Deke, D. E. Johnson, S. J. White, D. P. Woodruff, J. Phys. C **10**, 1109 (1977).
[12] J. J. Lander and J. Morrison, J. Chem. Phys. **37**, 229 (1962).
[13] T. D. Poppendieck, T. C. Ngoc, and M. B. Webb, Surf. Sci. **25**, 287 (1978).
[14] M. J. Cardillo and G. E. Becker, Phys. Rev. B **21**, 1497 (1980).
[15] A. Sakai, M. J. Cardillo and D. R. Hamann, Phys. Rev. B, in press.
[16] M. T. Yin and M. L. Cohen, Phys. Rev. B **24**, 2303 (1981).

NEW SILICON (111) SURFACE (7X7) RECONSTRUCTION
BENZENE-LIKE RING MODEL

YOU GONG HAO AND LAURA M. ROTH
Physics Department of State University of New York at Albany, 1400
Washington Avenue, Albany, N. Y. 12222

ABSTRACT

A new silicon (111) surface (7X7) reconstruction model is proposed.
The model for the unit mesh consists of 13 benzene like rings formed by
lowering triples of top layer atoms and 12 raised top layer atoms, with
three of the rings combining to give depressed corners.   The construction
of the model is based on some considerations of the physical properties of
the surface and on cluster calculations using the semi-empirical MNDO-
geometrical optimization method.   A cluster calculations with 9 silicon
atoms gave the ring structure as an energy minimized result and also gives
evidence that the two halves of the unit mesh should be at different height
levels.   A 13 silicon atom "flower" cluster calculation gave the height of
the raised atoms to be .851A. The structure of the unit mesh resembles two
types of (2X2) structure which can be formed from the rings and raised atoms
("flowers"), one of which is readily transformed into pi-bonded chain or
molecular structures.   The model gives a good qualititative account of the
various hills and valleys seen in scanning tunneling microscopy, and since
lateral displacement is small, it agrees with low energy channeling
experiments.   In addition good agreement with Cardillo's helium atom
scattering experiment and Yang et al.'s LEED experiment is found as the
model has different heights for the two halves of the unit mesh and
characteristics of a mixture of two types of structure.

INTRODUCTION

The silicon (111)-(7X7) surface reconstruction problem has attracted a
great deal of attention because it has both practical and theoretical
importance.[1]   In recent years, several experiments have revealed new
information about the surface.[2,3]   Scanning tunnelling microscopy (STM)
has shown a pattern of hills and valleys which tends to support adatom
models of surface reconstruction.   However the adatom models need a source
of adatoms, and recently we have found that the most likely source,
migration from surface steps, is ruled out because the surface diffusion
coefficient is too small at the conversion temperature[4]. Low energy
channeling (LEC) has shown that the surface atoms are mainly displaced in a
direction perpendicular to the surface, with little or no lateral
displacement, which tends to support smooth models, but the smooth models
thus far proposed are not so good for explaining the STM [3] result.
Here we propose a new model which consists of structural units of
benzene-like silicon rings and raised atoms or "flowers".   Using this
model, we believe that we can explain both STM and LEC experiments very well
and also other experiments on the surface[10-12]. We have carried out a
locally energy minimized geometrical optimization calculation of ring and
flower clusters by using the semi-empirical MNDO method. The calculations
mainly support our geometry of the unit mesh.

MODEL CONSTRUCTION

Each surface atom of the cleaved silicon (111) surface has one dangling
bond, which is naturally unstable, and it is expected that surface
reconstruction will take place in such a way as to reduce the total number
of dangling bonds and so to reduce the total energy. We have assumed a
perpendicular motion of the surface atoms, and we think of them as combining
together into elementary units as Kanamori has suggested [5]. If a triangle
of 3 atoms in the top layer is pushed down to the plane of the second layer
a benzene like ring is formed (Fig.1a). In the model, neighboring surface
atoms are raised a little above the top layer, and form the center of
"flowers" (Fig.1b), and "corners" are formed by combining 3 rings (Fig.1c).
These structural units can be combined to build the structures shown in
Fig.1d-h:
    d).(2X2) structure with two-fold symmetry-(2X2-A)
    e).(2X2) structure with three-fold symmetry-(2X2-B)
    f-h).((2n+1)X(2n+1)) structures.
In this set of configurations, we found immediately that Fig. 1h
coincides with the STM experimental result (Fig.1 of [2]). The two-fold
symmetry structure of Fig.1d. is probably unstable both because of the
three-fold nature of the substrate and according to our calculations which
we will give in the next section. This structure can readily be changed to
a (2X1) structure, by the procedure shown in Fig.2.
    Since the bond length of an SP hybrid orbital is proportional to the
fraction of S orbital[6], the SP_2 bond should be shorter than the usual SP_3
bond of bulk silicon. This is needed for the ring structure when we push
the atoms down, but we believe that this SP_2 bond should be little longer
than the 2.216A bond length here, as in Pandey's pi-bonded chain case. This
gives a strain force along the bond direction on the surface, and so the
center atoms of the flowers are raised to relieve this strain. In other
words, the raising of the atoms in the flower center is a result of the
lowering of the atoms in the ring. We believe that this is a possible
mechanism in the silicon (111) surface reconstruction.

MNDO CALCULATION

The MNDO (Modified Neglect of Diatomic Overlap) method was developed by
Dewar et al. [7] for molecular cluster calculations, and is a semi-empirical
calculation of the Roothaan-Hall (RH) SCF-LCAO-MO method. The method has
enjoyed success in organic molecular cluster calculations[8] and has been
shown reliable in silicon crystal impurity calculations[9]. Using MNDO we
calculated three clusters: the ring (Fig.3a); the flower and the corner
(Fig.3b), which are used to build our (7X7) unit mesh.
    In the ring cluster calculations we gave some atoms freedom to move in a
direction perpendicular to the ring plane defined by the three original
second layer atoms (1, 3, 5), and kept the other atoms fixed. By doing so,
we obtained a structure in which six silicon atoms form a benzene-like ring
with the three outer silicon atoms raised. The Pz bond orders of the ring
atoms are also shown in Fig. 3a. The bond orders show very clearly that
there is Pz pi-bonding along the ring. In order to compare with other bond
orders, we give the bond orders of the Pz orbital of atom # 5 with the
nearest two atoms # 4 and # 6 in Table 1. As we pointed out above the
three outer atoms being raised is a necessary condition of the three inner

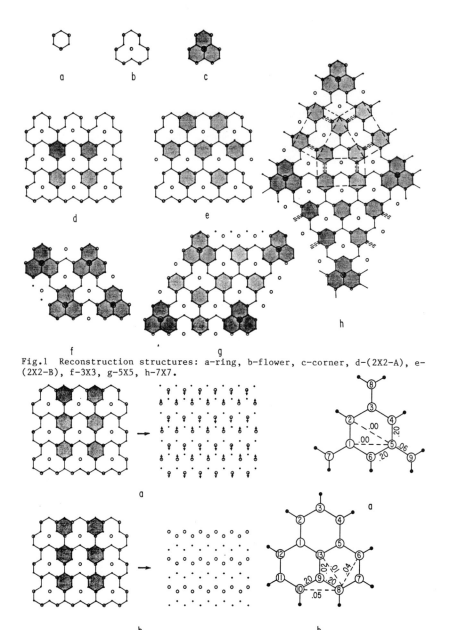

Fig.1 Reconstruction structures: a-ring, b-flower, c-corner, d-(2X2-A), e-(2X2-B), f-3X3, g-5X5, h-7X7.

Fig.2 Relation between (2X2-A) and (2X1) structures: a-with buckled (2X1), b-with pi-bonded chain (2X1).

Fig.3 Calculation clusters

atoms being lowered , which turns out to be very important in our total
(7X7) unit mesh. The calculation gives the result that the height of the
three raised atoms is 0.75A and the total energy of the system is 2ev lower
than the unreconstructed surface.

TABLE 1.

	4S	4Px	4Py	4Pz	6S	6Px	6Py	6Pz
5Pz	.00	.00	.01	.20	.00	.01	.00	.20

Before calculating the above ring structure, we calculated a cluster
with six silicon atoms in a ring with hydrogens simulating neighboring
silicons. The result shows that the ring formed and the 3 outside hydrogen
atoms were forced to rise. In this case the total energy increased when we
pushed even one of the hydrogen atoms down. In the 9 silicon atoms cluster
of Fig.3a which gives the same structure of a ring with raised atoms, when
we push one of the outside silicon atoms down, the total energy first
increases and then goes down. In other words there is a barrier. In our
model we assume that pushing one outside atom down raises the energy. The
MNDO calculation does not quite support this, but possibly a larger cluster
or a more accurate calculation would support it.
   In the flower cluster calculation we simply let the center atom of the
flower move vertically and let all other silicon atoms be on a same plane.
We calculated the height of the center atom to be 0.851A, a little higher
than in the ring calculation which might be caused by the presence of more
lowered silicon atoms around it, and the total energy is about 3ev lower
than the unreconstructed surface. Also we have the Pz pi-bonding around the
flower edge which is shown in Fig. 3b.
   The corner of our model is the same cluster as the flower except that
the center atom is lowered down to the plane. The main reason for choosing
this structure is that the lowering or removing of the central atom is
needed to reproduce the experimental result. We found that the total
energy increses continuously as the central atom is moved upward. Lowering
the atom with the atoms around it relaxed out 0.15A raises the total energy
less, about 1 eV. We believe that a more complete calculation would show
that this structure is stable based on our raising-lowering relation of
atoms since around it there are six raised atoms.

OUR (7X7) MODEL

   Following the above physical considerations and MNDO calculations we
propose our silicon (111)-(7X7) surface model as follows. The unit mesh
consists of 98 atoms, including first and second layers, which are combined
into rings and flowers as shown in Fig.1h. The two halves of the unit mesh
are at different height levels connected by the (dashed) bonds, the height
difference being about 0.3-0.8A. The lower half has one ring in the center
and two rings along each side. The upper half has three rings in the
center and two rings along each side. The six flower center atoms of the
lower half are at almost the same height, about 0.8A higher than the rings
of the lower plane. The six flower center atoms in the upper half are at
different heights, the center three being at a lower level which is 0.8A
higher than the upper half rings, and the outer three being higher than the

center three by about 0.25A. The height differences of the raised atoms are
thought to be a result of the different number of rings around them. In the
corner we have basicaly a flower except the center atom is at the ring
level.

In our model, the two halves of the unit mesh are at diffferent heights
which comes about because the bond connecting the 2 rings of the boundaries
should be tilted. The (7X7) unit mesh has a structure resembling the (2X2-
B) structure with boundaries of the (2X2-A) structure. We have not given a
structure of lower layers beneath the surface. We expect these to be
ripple-like as we have a rippled surface. The absolute heights of the
raised atoms as well as the height difference between the two halves could
be adjusted, and some rings could be tilted.

TOTAL ENERGY CALCULATION

We have estimated the total energy of several configurations of the
surface by assigning energies to the various structural units and to
interactions among them.   A minimal number of parameters would be energies
for the flower (F), ring (R), and corner (C), an interaction between two
rings (A_2 for $(2n+1)X(2n+1)$ structures and A'_2 for the (2X2-A) structure),
and an interaction among 3 rings (A_3). We might expect F and R to be
negative, C and A_3 to be positive, and we postulate because of our raising-
lowering argument that $A_2 < A'_2$. The energy per surface atom for the various
structures is then

$$4E((2n+1)X(2n+1))=e+6a/(2n+1)+4c/(2n+1)^2$$
$$4E(2X2-A)=e+a'$$
$$4E(2X2-B)=e$$

where $e=F+R+A_3$, $a=A_2-A_3$, $a'=A'_2-A_3$, and $c=C-(F+9R+18A_2-9A_3)/4$.   If we assume
that the (7X7) is the lowest energy $((2n+1)X(2n+1))$ structure then we must
have $c/a=-21/4$, and we find

$$4E(7X7)=e+(3/7)a$$

Assuming $a<0$, this structure has lower energy than the (2X2-B) structure.
The condition that it have lower energy than the (2X2-A) structure is that
$a<(7/3)a'$.

We can intepret the (7X7) strcture as consisting of islands of the (2X2-
B) separated by (2X2-A) boundaries. The energy A_2-A_3 being negative means
that it is energetically favorable for boundaries to form, and then the
corners being less favorable (c>0) there is an optimal size.   But the (2X2-
A) structure has many such boundaries, and we must argue that these
boundaries have higher energy than in the (7X7) case because the A'_2 bond is
flat while the A_2 bond is tilted, and that is ralated to the fact that the
upper half of the unit mesh is higher than the lower half.   Then relating
this to the pi-bonded chain model of the (2X1) surface reconstruction, the
energy for the pi-bonded chain might be lower than the (2X2) structures, but
not by as much as the (7X7) because the latter is the thermodynamically
stable structure. One must look for an activation energy to form this
structure and we propose that this must be the energy to go from the pi-
bonded chain structure to produce the islands of (2X2-B). It would be
interasting to obtain an estimate of this energy.

COMPARISION WITH EXPERIMENTS

1). Our (7X7) reconstruction structure coincides with the STM experimental result when one compares our Fig.1h with Fig.1 of [2].

2). Since our model does not have lateral displacements, the model is consistent with the LEC results.

3). The center atom in the corner has a totally empty dangling bond which is the best candidate to give a unique site for the 1.5% special behavior hydrogen chemisorption. [10]

4). Recently Yang et al. did an experiment which revealed the existence of (2X1) or (2X2) structure blocks in the (7X7) reconstruction [11]. Our model has (2X2-A) and (2X2-B) regions in it.

5). The (2X1) to (7X7) transition happens at a rather low temparature (600K-700K) [12], which indicates that there is not a drastic change of structure between the two structures. Our model seems to give a account of this.

REFERENCES

1. N. P. Lieske, J. Phys. Chem. Solids 45, 821 (1984).
2. G. Binning, H. Rohrer and E. Weibel, phys. Rev. Lett. 50, 120 (1983) .
3. R. J. Culbertson, L. C. Feldman, P. J. Silverman, Phys. Rev. Lett.45, 2043 (1980).
4. Y. G. Hao and L. Roth, Surf. Sci., 163, L776 (1985).
5. J. Kanamori, Solid State Comm. 50, 363 (1984) .
6. M. J. S. Dewar, in THE MOLECULAR ORBITAL THEORY OF ORGANIC CHEMISTRY (Mcgraw-Hill,New York, 1969) p. 149.
7. M. J. S. Dewar and W. Thiel, J. Am. Chem. Soc. 99, 4899 (1977).
8. D. E. Gabell, A. H. Cowley and M. J. S. Dewar, J. Am. Chem. Soc. 103, 3290 (1981).
9. G. D. Watkins, Phys. Rev. B 29, 3193 (1984).
10. B. Lindgren and D. E. Ellis, Phys. Rev. B 26, 636 (1982).
11. W. S. Yang and F. Jona, Solid State Comm. 48, 377 (1983) .
12. P. P. Auer and W. Monch, Japan J. Appl. Phys. Supp. 2, pt. 2 (1974).

Classical Two and Three-Body Interatomic Potentials for Silicon
Simulations

R. BISWAS AND D. R. HAMANN
AT&T Bell Laboratories, 600 Mountain Avenue, Murray Hill, NJ 07974

ABSTRACT

We develop two and three-body classical interatomic potentials that model structural energies for silicon. These potentials provide a global fit to a database of first-
principles calculations of the energy for bulk and surface silicon structures which spans
a wide range of atomic coordinations and bonding geometries. This is accomplished
using a new "separable" form for the 3-body potential that reduces the 3-body energy
to a product of 2-body sums and leads to computations of the energy and atomic forces
in n^2 steps as opposed to n^3 for a general 3-body potential. Simulated annealing is performed to find globally minimum energy states of Si-atom clusters with these potentials
using a Langevin molecular dynamics approach.

The computer-based microscopic description of the structure and properties of
materials, is an area of much current interest. First − principles quantum-mechanical
calculations, for example, have recently enjoyed great success in predicting the properties of simple structures [1]. However, the computational complexity of quantum-
mechanical energies and forces precludes their use for applications with more than
10-20 atoms per unit cell. Such applications include crystal growth and epitaxy, melting, laser annealing, defect motion and materials strength, amorphous structures, surface and interface reconstruction. These problems are of much current interest and
generally require molecular dynamics studies of 10^2-10^3 atoms evolving through 10^4-10^6
configurations. Simulated annealing techniques are very useful in determining low
energy states of such complex systems.

With a view towards studying some of the above dynamical problems for semiconductor materials, we explore in this paper the extent to which a classical potential
model *can* succeed in providing an accurate global description of its structural energetics. Pair potentials, for example, have been extensively used in simulations of simple
systems such as rare gas solids and liquids. However for covalent materials such as Si,
pair forces alone are inadequate, since the equilibrium diamond lattice is unstable relative to close-packed structures without 3-body forces [2]. In fact, the 3-body Keating
model [3] which is fitted to small distortions of the diamond structure has had much
success in describing local distortions and phonons, but has also been extrapolated to
compute energies of complex Si structures, sometimes far beyond the range of its validity. To investigate the feasibility of a global model of classical potentials, Si is the
material of choice because of the huge computational effort that has produced accurate
quantum-mechanical energies for simple Si structures spanning a wide range of atomic
coordinations, bond lengths, and bond angles [1,4]. While only a few of these structures have low enough energies to be experimentally accessible in extended form, their
features may occur as local distortions or dynamic intermediates associated with complex structures or processes. There are neither *a priori* arguments nor empirical tests to
suggest that a classical model can accomplish such a global fit without arbitrarily many
multi-atom potentials. The computational complexity of such a model would render it
impractical, so we have confined our explorations to 2 and 3-body potentials. We find
that it is possible to achieve excellent qualitative separation of structural groups, and
quantitative accuracy for individual structures which should be acceptable for many
purposes.

A significant innovation in the present work is a new "separable" form for the 3-body potential which permits the energy to be calculated in n^2 computational steps instead of the n^3 steps generally required, where n is the number of interacting atoms. This reduction in the computational complexity of the 3-body model is a considerable aid for simulation purposes. This is particularly true for the simulated annealing of Si-clusters, with the present 2 and 3-body potentials, discussed in this paper.

Our 2 and 3-body potential model is defined by the following expression for the structural energy,

$$E = \frac{1}{2}\sum_{1,2}{}' V_2(1,2) + \sum_{1,2,3}{}' V_3(1,2,3), \tag{1}$$

where primes indicate that all summation indices are distinct. There is no explicitly volume dependent term, since atomic volume is not a useful physical concept for inhomogenous structures. The assumption of neglecting 4-body and higher terms is used and tested in the present work. Any 3-body potential $V_3(1,2,3)$ may be expressed as a function of 2 lengths r_{12}, r_{13} and the included angle θ_1. This potential is symmetrized over the 3 particles in the sums in (1). Without losing generality we can expand the angular dependence of this potential in the complete set of Legendre polynomials. The coefficients in this expansion are functions F_ℓ of bond lengths multiplied by linear coefficients C_ℓ,

$$V_3(r_{12}, r_{13}, \theta_1) = \sum_\ell C_\ell F_\ell(r_{12}, r_{13}) P_\ell(\cos\theta_1). \tag{2}$$

Our key simplification is to assume that the functions F_ℓ are separable and symmetric products of functions ϕ_ℓ of each bond length. This leads to the symmetric separable form

$$V_3(r_{12}, r_{13}, \theta_1) = \sum_\ell C_\ell \phi_\ell(r_{12}) \phi_\ell(r_{13}) P_\ell(\cos\theta_1) . \tag{3}$$

Generally, separability is consistent with a local picture of the atomic bonding interactions. The addition theorem for spherical harmonics now reduces the 3-body energy to a rotationally invariant scalar product of vectors $\Phi_{\ell m}$ that are simple 2-body sums i.e.:

$$\sum_{2,3} V_3(r_{12}, r_{13}, \theta_1) = \sum_\ell C_\ell \left(\frac{4\pi}{2\ell+1}\right) \sum_{m=-\ell}^{+\ell} \Phi_{\ell m}^{*1} \Phi_{\ell m}^1 , \tag{4}$$

where

$$\Phi_{\ell m}^j = \sum_2 \phi_\ell(r_{j2}) Y_{\ell m}(\hat{r}_{j2}) . \tag{5}$$

The $\Phi_{\ell m}^j$ vectors are the moments of the structure around atom j, that describe its local environment. The calculation of the energy in (1) requires the sum in Eq. 4 to be corrected for the case when indices 2,3 are identical. This introduces a modification of the 2-body interaction,

$$\sum_{2,3}{}' V_3(r_{12}, r_{13}, \theta_1) = \sum_{2,3} V_3(r_{12}, r_{13}, \theta_1) - \sum_2 f_3(r_{12}) , \tag{6}$$

where

$$f_3(r) = \sum_\ell C_\ell \phi_\ell^2(r) \ . \tag{7}$$

We have examined a few short range, monotonic functional forms for the 3-body functions ϕ along with similar functions for the 2-body potential. Overall, our best results were obtained with the family of simple exponentials, $\phi_\ell = e^{-\alpha_\ell r}$. In addition, we used generalized Morse 2-body potentials,

$$V_2(r) = A_1 e^{-\lambda_1 r} + A_2 e^{-\lambda_2 r} \ . \tag{8}$$

Very general functional forms for ϕ_ℓ often led to unphysical solutions that provided good fits but performed poorly on test structures.

The nonlinear parameters of our model are the decays of the radial functions, whereas the coefficients C_ℓ, A_1, A_2 are linear variables. The parameters in the potential were least-squares fitted to the database of accurate quantum-mechanical energies of diamond, wurtzite, the high pressure β-tin and simple hexagonal structures, and simple hypothetical Si structures [1,4,5] as shown in Fig. 1(A), and a 4-layer slab (Fig. 2). The nonlinear parameters were varied with a simplex routine, and for each nonlinear parameter set the least-squares fit was performed to obtain the linear variables. Because of symmetry, only moments $\Phi_{\ell m}$ with certain values of the angular momentum ℓ are allowed for each structure, e.g. the $\ell = 0,3,4,6 \cdots$ moments are allowed in diamond. To account for bond breaking energies, we have added to the database our own Linear Augmented Plane Wave calculations for the energy of a 4-layer Si(111) slab as a function of the positions of the outermost atomic layers (Fig. 2). The (111) slab is the only structure in the above database that permits an $\ell = 1$ moment and is therefore essential to the fit. Fits to slabs have the important effect of requiring the cohesive energy to decrease monotonically from bulk structures to low-dimensional structures. For the fits, we chose a set of structures from Fig. 1 that, together with the (111) slab, contained all the moments up to $\ell = 6$.

The results of our fit compare very well with the quantum-mechanical energies for the Si(111) slab, as shown in Fig. 2. Our global fit to the crystal structures shown in Fig. 1(B), agrees with the quantum-mechanical results of Fig. 1(A), to within an rms error of 0.05 eV and displays the correct structural trends over a large range of atomic volumes. The absolute energies are plotted in Figs. 1 and 2. The first high-pressure phase is correctly predicted to be β-tin. In Fig. 1(B), bcc and si-hexagonal were test structures that were not fitted. Si-hexagonal is very close to β-tin as it should be [4]. Wurtzite is higher energy than diamond-Si. hcp is not as well fitted as other phases and s-cubic is somewhat lower in energy than the quantum-mechanical result. Our diamond Si-phase has an equilibrium bond length of 2.32Å (experimentally 2.351Å), but the nonlinear decay parameters could be uniformly scaled to produce the experimental bulk bond-length if desired.

Table 1 lists the parameters for our fitted potential. The coefficients in the angular expansion (C_ℓ) decay uniformly indicating the convergence of the solution and a well-defined angular decomposition. The solution is stable, and remains in the same region of parameter space when the structures in the fitting database are altered or weighted differently, or when the number of 3-body nonlinear parameters is altered by 1. The range of the potentials is comparable to that of the atomic valence wavefunction overlaps, indicating a physical reasonableness in the overall fit. A real space cutoff of 10Å only causes an error of order 0.01 eV in the total energies.

Other classical models, beyond the restricted Keating model [3] include that of Stillinger and Weber (SW) [6], and Pearson, Takai, Halicioglu and Tiller (PTHT) [7], and Tersoff [8]. SW used 3-body potentials that have a Keating [3] angular form ($\frac{3}{2} C_0 = C_1 = C_2$; $C_3, C_4 \dots = 0$), and are separable (although the separability was not utilized). The potentials were confined within a very small cutoff radius of 3.78 Å.

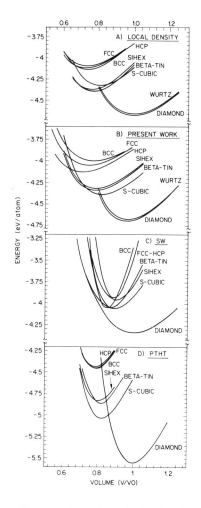

FIGURE 1. Energies for simple silicon structures as a function of atomic volume. SW potentials are from Ref. 6, and PTHT from Ref. 7. Wurtzite could not be distinguished from diamond in (C) and (D).

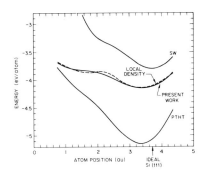

FIGURE 2. Energy of a 4 layer Si(111) slab. Starting from the ideal (111) geometry the two outermost layers are symmetrically displaced in the normal direction. The distance of either layer from the slab center is plotted. At 2.22 au, the slab reduces to two graphitic atomic planes.

Table 1

The values of the parameters for the present 2 and 3-body potentials.

ℓ,i	$\alpha_i,\lambda_\ell(\text{Å}^{-1})$	$A_i,C_\ell(\text{eV})$
	2–body	
1	3.946668	0.2682936×10^5
2	1.191187	-0.4259863×10^2
	3–body	
0	1.246156	0.9139775×10^2
1	1.901049	0.1644013×10^5
2	1.786959	0.9580299×10^4
3	1.786959	0.6663147×10^4
4	1.786959	0.3987710×10^4
5	1.786959	0.2046722×10^4
6	1.786959	0.7018867×10^3

Their 9 parameters were adjusted to fit the bond length, cohesive energy and melting temperature of bulk diamond Si, and satisfy other qualitative criteria [6]. The structural energies for the SW potential are plotted in Fig. 1(C) and Fig. 2. We were unable to obtain satisfactory fits with functions as short-ranged as the SW potential. Alternatively, PTHT [7] used the nonseparable Axilrod-Teller 3-body potential [9] which is based on the generalization of Van der Waals fluctuating-dipole forces for 3 particles. These potentials were long ranged with algebraic decays ($V_3(r) \approx r^{-9}$). The 3-parameter PTHT potential was obtained through an average fit to the bond lengths and cohesive energies of bulk diamond-Si and molecular Si_2, and the structural energies for this potential are shown in Fig. 1(D) and Fig. 2.

Clearly from Figs. 1 and 2, the present potentials compare with the quantum-mechanical results *much* better than previous work. The grouping of structures into three energy classes and the values of their equilibrium volumes are improved compared to SW, and the compressibilities and energy differences between the classes is improved compared to PTHT.

Recently, Tersoff [8] has developed a 2-body potential for silicon in which the strength of the attractive part is a complex function of the local bonding geometry, i.e., of the bond lengths, bond angles etc. This ansatz effectively couples in 3-body as well as higher-body correlations. The model is confined to nearest neighbors and its 7 free parameters are fitted to diamond-Si properties and cohesive energies for Si_2, s-cubic and fcc silicon. Energies of models for the reconstructed Si(111) surface have been computed with this potential.

The present model has a weak 2-body potential that has a strength of -1.09 eV at the equilibrium position of 2.77Å. Clearly this pair potential is inappropriate for the multiply-bonded Si_2 dimer. These properties of our 2-body potential are consistent with work of Carlsson et al. [10]. Our attempts to constrain our 2-body potential to fit Si_2 did not yield satisfactory fits. We believe that a model which simultaneously fits small clusters and extended systems must include N-body potentials beyond $N = 3$. Alternatively, the 3-body potential, exhibits (for $r_{12} = r_{13} = 2.35$Å) a very weak angular dependence for $90° \leqslant \theta \leqslant 180°$, with a very shallow minimum at $\theta \cong 110\text{-}115°$. Bond angles of $\theta \leqslant 70°$ generate strong repulsions. It is important that the $\ell = 0$ part of the 3-body is purely repulsive (see Table 1). This provides increased repulsions with more nearest neighbors, and counteracts the attractive 2-body energy.

In tests of the transferability of the potential to low symmetry situations, we obtain symmetric dimers for a relaxed Si(100) surface, with bond length $\cong 2.50$Å, and an energy gain of 1.22 eV/dimer relative to the ideal surface. The potentials yield an outward relaxation of atoms at a vacancy, with a vacancy formation energy of 4.82 eV in comparison to the quantum-mechanical result of ≈ 4.5 eV [11]. The model is too stiff for small distortions around the diamond structure, with some phonon frequencies 25 to 50% too high.

The atomic forces, for the present model can be analytically derived by differentiating (4) and the expressions have been previously published by us [12]. The forces can also be obtained through n^2 computational steps and requires a prior computation of the $\Phi_{\ell m}$ vectors. Since the atomic forces can be calculated with a computational speed comparable to that of energy, silicon systems are more efficiently studied with molecular dynamics where the position of *all* the particles is updated at each time step, rather than with Monte-Carlo methods. Monte Carlo methods require an updating of *one* of the particle positions in each iteration followed by a expensive recalculation of the $\Phi_{\ell m}$ vectors and the energy.

We demonstrate the usefulness of the present potentials by performing simulated annealing calculations within a molecular dynamics framework. Simulated annealing is a powerful technique for finding globally minimum energy configurations [13], and is most useful for complex, low symmetry physical systems such as atomic clusters or surface reconstructions, where the ground state is unknown. As a prototypical system we have considered the problem of finding globally minimum energy states of an isolated

178

cluster of Si-atoms. The molecular dynamics is modelled by a Langevin equation where the cluster is connected to a heat bath that provides a stochastic force and a viscous friction [14]. The stochastic forces drive the system into a thermal distribution characterized by the heat bath temperature alone. The equations of motion are integrated with a technique developed by Helfand [15].

Fig. 3 displays potential and kinetic energies for a 32-atom cluster during a typical annealing run. Instances of hill climbing out of metastable minima, are seen in Fig. 3, together with a gradual lowering of the cluster energy to a globally minimum value. The kinetic energy shows large fluctuations around the average value of $\frac{3}{2} k_B T$.

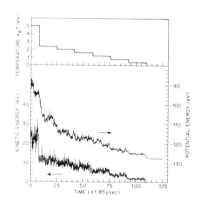

FIGURE 3. Potential and kinetic energies for a 32-atom cluster during a typical anneal. The linear temperature profile of the cooling schedule is shown in the top panel. The temperature is zero in the final step.

A high initial temperature is necessary to randomize the initial configuration. The final annealed configuration has a much higher degree of coordination than the initial diamond-like cluster, with an average of 6-8 nearest neighbors. The annealed cluster represents a completely different configurational minimum from the starting structure. Such a minimum can not be obtained with steepest descent methods.

We have studied systematics of the annealing process [14]. Generally, a minimum anneal time is required, beyond which anneals are not significantly better, although annealed configurations are appreciably worse for shorter times and resemble the effect of a quench. The well annealed configurations are independent of the initial geometry, and insensitive to the details of the cooling schedule. There is a statistical distribution of annealed clusters, but all annealed configurations show similar, reproducible nearest neighbor peaks in the radial distribution function and similar distributions of bond angles [14].

In conclusion, we have described a new "separable" form for the 3-body potential that leads to efficient energy and force calculations. The present interatomic potentials for Si interpolate among first-principles energy calculations for both semiconducting and metallic structures, and improve upon previous classical Si-models. Simulated annealing, such as that described here, can be performed on completely general systems. Classical potentials such as the present one may be extended to compound semiconductors, and have a potentially wide range of applications in simulations of materials properties.

We wish to thank W. A. Tiller for sending us a pre-publication copy of Ref. 7.

REFERENCES

[1] M. T. Yin and M. L. Cohen, Phys. Rev. B 26, 5668 (1982); Phys. Rev. Lett. 45, 1004 (1980).

[2] For a model in which the entire cohesive energy is expressed as a sum of 2, 3, ... body potentials, with no implicit or explicit "volume" term, we believe that only single-minimum, smooth 2-body potentials are physically plausible, and exclude the type of oscillatory long-ranged potentials that could be mathematically constructed to counter this assertion.

[3] P. N. Keating, Phys. Rev. 145, 637 (1966).

[4] R. Needs and R. M. Martin, Phys. Rev. B 230, 5372 (1984); K. J Chang and M. L. Cohen, Phys. Rev. B 30, 5376 (1984).

[5] M. T. Yin private communication, (equations of state for Si structures and Si-hexagonal energies).

[6] F. Stillinger and T. Weber, Phys. Rev. B 31, 5262 (1985).

[7] E. Pearson, T. Takai, T. Halicioglu and W. A. Tiller, J. Cryst. Growth 70, 33 (1984); T. Takai, T. Halicioglu, and W. A. Tiller, Scripta Metal. 19, 709 (1985).

[8] J. Tersoff, (to be published).

[9] B. M. Axilrod and E. Teller, J. Chem. Phys. 11, 299 (1943).

[10] A. E. Carlsson, C. D. Gelatt and H. Ehrenreich, Phil. Mag. A41, 241 (1980).

[11] G. A. Baraff and M. Schluter, Phys. Rev. B 30, 3460 (1984).

[12] R. Biswas and D. R. Hamann, Phys. Rev. Lett. 55, 2001 (1985).

[13] S. Kirkpatrick, C. D. Gelatt and M. P. Vecchi, Science 20, 671 (1983).

[14] R. Biswas and D. R. Hamann, (to be published).

[15] E. Helfand, Bell Sys. Tech. J. 58, 2289 (1979).

SPECIAL PURPOSE PROCESSORS
FOR COMPUTING MATERIALS PROPERTIES.

A. F. BAKKER
AT&T Bell Laboratories, 600 Mountain Ave., Murray Hill,NJ 07974

ABSTRACT

The need for computational power in the modeling of physical systems is rapidly increasing. Realistic simulations of materials often require complex interactions and large numbers of particles. For most scientists, full-time access to supercomputers is not possible, and even this might not be sufficient to solve their problems. As most of the calculations involved are straightforward and repetitive in nature, a possible solution is to design and build a processor for a specific application with a low cost/performance ratio. This approach is to be contrasted with the use of a general purpose computer, which is designed to treat a large class of problems and includes many expensive features (e.g. software) that are not utilized in the simulations. The architecture of a special purpose computer can be tailored to the problem; e.g., parallel and pipelined operations can be incorporated to obtain efficient computational throughput, and memory organization and instruction sets can be optimized for this purpose. A few of such algorithm oriented processors have been built in the last decade and have been utilized for certain specific jobs; for example: molecular dynamics simulations of systems of Lennard-Jones particles and Monte Carlo calculations of Ising models. An overview of some existing algorithm oriented processors and expectations for the future will be presented.

I. INTRODUCTION

The need for sophisticated models of materials has increased dramatically in the last few years. Recent advances in atomic-scale microscopy have greatly improved our knowledge of the structure of materials. For example, the scanning tunnel microscope has been used to map out the topography of several crystal surfaces, including the complex silicon 7X7 reconstruction of the (111) face. Standing wave X-ray scattering studies and ion scattering techniques have given precise locations of atoms at the surfaces of semiconductors. By combining these results a very detailed picture of the surface structures can be obtained. However, in most cases the implications of this information for material properties and atomic interactions are not at all clear. Atomic-scale models can provide the necessary link between the interatomic potential and the structure and properties of the material. But their usefulness is restricted by the large amount of computation required. The molecular dynamics technique, for example, involves the frequent evaluation of the forces between all ·interacting particles. Furthermore, there are often several thermodynamic variables and external fields that influence the properties of the system. The evaluation of the properties throughout this multi-dimensional parameter space can consume vast quantities of computer time. The development of fast processors and special computers designed for efficient modeling can greatly facilitate work in this field. They are, in a sense, the "laboratory apparatus" of the theorists working on these problems, and are almost as essential for their work as are vacuum systems for experimental surface scientists.

Molecular dynamics and Monte Carlo studies of systems of interacting particles have already provided important insight into certain physical processes such as phase transitions. In the past such models have often been used to test the validity of theories. Models employed for this purpose require relatively little computation. Generally they involve simple interatomic potentials and the properties of the system need to be evaluated only at a small number of points where the assumptions of the theory are most suspect. Approximate values of the important parameters and correlation functions can be obtained from theory, and thus the initial simulation runs can be programmed to give the necessary information.

A second application of these models is exploratory in nature and involves a study of the influence of processing conditions on the structure and properties of materials. In this case it is almost essential to have some form of computer graphics available to represent the system. A qualitative understanding of the structures generated by the simulations can be obtained from simple drawings of the atomic positions. These can give critical information needed to decide which parameters and correlation functions should be measured in order to best characterize the system. Two-phase coexistence and other types of inhomogeneities can be readily detected from even the most rudimentary graphics.

Finally, atomistic computer models have potential applications for extrapolating the properties of materials from regions of the phase diagram in which they have been measured into regions that involve extreme conditions of temperature or pressure, for example. In some cases modeling may be the most economical method for exploring the effect of various conditions on the chemical and physical properties of materials. Optimum conditions to produce certain desired properties could thus be determined without the need for experiments. Rapid increases in the cost of experimental apparatus and the decreasing cost of computation will enhance the feasibility of this approach in the future.

The tools required for a computer simulation are an efficient algorithm, a programming language and a fast computer. Reductions in the time it takes to do the simulation may often be obtained by writing a better algorithm, by using the assembler language, or dedicating a computer to the exclusive function of performing the simulation. Most algorithm improvements are developed by specialists in the field of numerical analysis. Their impacts on the efficiency of the computation are usually of the same order of magnitude as those resulting from technological improvements. Compilers on general purpose computers are currently of such quality that assembly code hardly improves the computational speed. The computer, however, will always have hardware limits, and ultimately these will determine the time required for producing the results of the simulation.

II. COMPUTATIONAL TOOLS.

The easiest way to start computer simulations is to use the computer that is immediately available at the university or research laboratory. This is likely to be a general purpose computer (e.g., VAX 780) with a Fortran77 compiler and a speed of about a million floating point operations per second (1 MFLOP). When simulating an atomic system, it will be found that the size of the system must be limited to a few thousand particles and the duration of the simulation to a few thousand time steps. These limitations are due to the nature of general purpose computers; their architectures were designed to solve a wide range of problems and are in general not efficient for a specific algorithm. Furthermore, most general purpose computers are shared with other users, hence only a fraction of the computer time is at your disposal.

To speed up the calculations, one can attach a "function oriented processor" to the general purpose computer. An array or vector processor may speed up the floating point calculations dramatically when the problem can be vectorized; factors of 10-50 are possible in some cases. The remaining scalar operations and the data transfer bandwidth between the host and array processor will be the limiting factors in the total simulation time, and in some cases this type of processor will not improve the total speed at all.

The supercomputers available today (e.g., a Cray 1) will permit simulations of large systems (up to 100,000 particles). They have the same limits, however, as the local general purpose computer: the algorithm does not map efficiently onto the architecture, the computer is shared with others For most institutes, the expense makes it difficult to purchase sufficient computer time to solve the problem, or to buy the computer itself. It is rather difficult to program such a "datacruncher" to deal efficiently with the calculations involved. (Is the nature of the problem such that it is easily vectorizable or suitable to parallel processing?) Although the Cray 1 has a theoretical speed of 200 MFLOPS in the vector mode, it is often difficult to utilize more that 25% of that speed on a specific problem. In addition, vector computers have memory limitations that would require frequent disk access to treat large systems. A limitation of parallel computers is the intercommunication bandwidth restrictions; unless the problem is "local", in the sense that each processor can be assigned a portion of the calculation to be performed with relatively little data transfer with other processors. Even so, considerable programming effort is required to take advantage of this property.

The conclusion is that commercially available computers have not been designed to compute a specific problem efficiently and are therefore relatively expensive and often not affordable.

III. ALGORITHM ORIENTED PROCESSORS.

If the problem is important enough and it would take years of supercomputer time to obtain a satisfactory solution, an alternative approach is to design and build a low-cost high-performance computer. The architecture is then tailored to suit your algorithm, i.e., an algorithm oriented processor (AOP) or special purpose computer (SPC). Parallel and pipelined operations can be implemented in hardware wherever the algorithm permits. Analysis calculations, often a major portion of the computational load, can be executed in parallel with the simulation part. Simulation models of atomic systems often can utilize parallel architecture quite effectively, since the motion of atoms in locations separated by distances larger than the range of the potential can be obtained simultaneously. The calculation of interatomic distances involves differences in the three Cartesian coordinates which can also be obtained independently. The word length can be chosen to give the exact precision required for the problem, thus eliminating the overhead of processing unnecessary bits. The memory size, type and organization can be selected appropriately, with fast access memory utilized at locations where speed is critical. The instruction set can be chosen to perform the basic routines required for the algorithm, thus simplifying the job of software development. The newest microprocessors and memory chips can be incorporated to provide fast floating point operations and short access times. Special purpose computers tend to be very reliable, since the number of chips required is generally much smaller than that for a general purpose computer. Furthermore, they can be designed to run faster than commercially available supercomputers when applied to the specific problems for which they are optimized. This kind of power can be devastating when it is applied to the simulation of one system, 24 hours a day!

The main factor limiting the development and use of AOP's is the time and skills required for their construction. In addition, the maintenance of the AOP can be difficult, since the architecture is unique and maintenance contracts are not available. For this reason, it is essential that accurate and complete documentation is generated at the time the computer is designed and constructed.

The building of a special purpose computer today is facilitated by a number of recent technological developments. Computer-aided design (CAD) software is available on general-purpose computers. Some of these programs can even run on the faster personal computers. Computerized wire-wrap machines make it possible to wire complicated circuit boards with a minimum number of errors. Advanced test equipment is available to test the logic of the chips and the connectivity of the circuits. Low-cost minicomputers are available to serve as the user interface and to store and analyze the results generated by the special purpose computer. Thus, initial configurations can be set up on the minicomputer using high-level languages, and then loaded into the special purpose computer. Reliable VLSI function blocks provide powerful operations in a single unit. However, the integration of a knowledge of physics, algorithms, hardware and software is required in order to build such a computer and to generate useful data.

IV. ALGORITHM ORIENTED PROCESSOR EXAMPLES.

Special purpose processors have been utilized a long time in the application fields of signal processing. The first AOP for research purposes was a Monte Carlo Ising processor, built in Delft in 1968. Its success led to the design and construction of a crystal growth processor (Delft, 1970), a second Monte Carlo Ising processor (Delft 1974) and another crystal growth processor (Delft 1977). These processors could simulate systems of modest size, due to the technological limitations at the time, but their cost/performance ratios were about a factor of 100 lower than those of commercially available computers. In the 70's, advances in technology resulted in the commercial availability of more sophisticated digital components, so that larger and more complex AOP's were feasible. The "know how" and improved infrastructure resulted in two low-cost AOP's, with Cray 1 speed: the first molecular dynamics processor (Delft 1980) and a large Monte Carlo Ising processor (Delft 1981). At that same time AOP's from two other research institutes became operational, a Monte Carlo Ising processor was built by Pearson (Santa Barbara, 1982), and a Monte Carlo Ising processor with random interactions was constructed by Ogielski and Condon (Murray Hill, 1983). The new results from these processors stimulated many other research institutes and universities to design and construct AOP's in many different research fields. In order to provide some understanding of the relationship between the AOP architecture and computational efficiency, the designs of the three new Monte Carlo Ising processors will be analyzed and then the molecular dynamics processor will be discussed.

V. MONTE CARLO ISING PROCESSORS.

An exact calculation of average quantities in an N-spin system requires the consideration of all 2^N possible configurations. This is not feasible for $N > 40$, because of the required computational time. However, a Monte Carlo method can be used to obtain an estimate of these quantities [1]. This can be done efficiently by means of "importance sampling" in configuration space. This involves the generation of a Markov chain of configurations, in which the probability of finding a particular state is proportional to the Boltzmann factor for that state. The procedure is to choose, in a random way, a (central) spin $S(k)$, the neighboring spins of which determine the difference in energy δE of the two possible states of $S(k)$. Then a random number is compared with a weighting function of δE, $1/(1+\exp\delta E)$, the outcome of which determines the new state of $S(k)$. This procedure is repeated for many or all spins (randomly or sequentially selected), and a new configuration is realized. The accuracy of the estimate depends upon the number of configurations generated in this way. This algorithm is very suitable for implementation in special purpose hardware, due to its repetitive nature, simple operations, and single bit storage. The three existing Ising processors differ in the way the spin values are stored in and fetched from the spin memory, whereas their spin update parts are similar. Figure 1 shows this spin update part schematically. The decoding hardware generates the address for the probability look-up table from the spin values of the relevant neighbors. This look up table is either downloadable from the host, to which the processor is attached, or permanently stored in a read-only memory. The efficiency of the spin update processor can be increased by pipelining (up to 25×10^6 spin updates per second has been achieved). Special arrangements for fetching the neighbor spin values are required to make full profit of the speed of the spin update part.

In the Pearson Ising processor [2], the spin memory is organized in the form of a large ring shift register, through which all spin values are shifted at a 1.5 MHz. rate. Using skewed boundary conditions, the six nearest neighbor values of a simple cubic spin lattice can be tapped from the ring at six fixed locations in parallel and fed to the spin update section. The result is stored in the ring at the location of the central

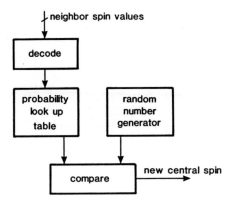

Fig. 1. Schematic diagram of the spin update section of a typical Monte Carlo Ising processor.

spin (see fig. 2). The speed of the update hardware allows the data tapping to be implemented in sixteen-fold at non-overlapping parts of the ring, and multiplexing of the data at the input of the pipelined update section. In parallel with the spin update, the magnetization and the total energy of the spin system is updated at a 25 MHz rate. Additional calculations of quantities are carried out in the host, to which the whole spin memory must be transferred in that case. The maximum spin memory is 2 Mbit and is implemented in 16 bit wide random access memory circuits. The processor is 10 times faster than the Cray 1, and the cost was $2000 for components only; but it is restricted in its applications because of a relatively inflexible design.

In the Delft Ising Spin Processor (DISP), built by Hoogland [3,4], the spin memory is organized into 64 independently addressable one bit wide memory banks. Figure 3 shows an application to a two-dimensional lattice, the numbers represent the memory

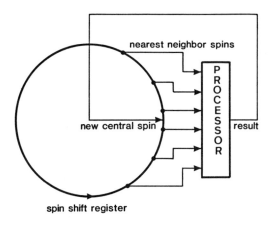

Fig. 2. Spin memory scheme in the Pearson Ising processor.

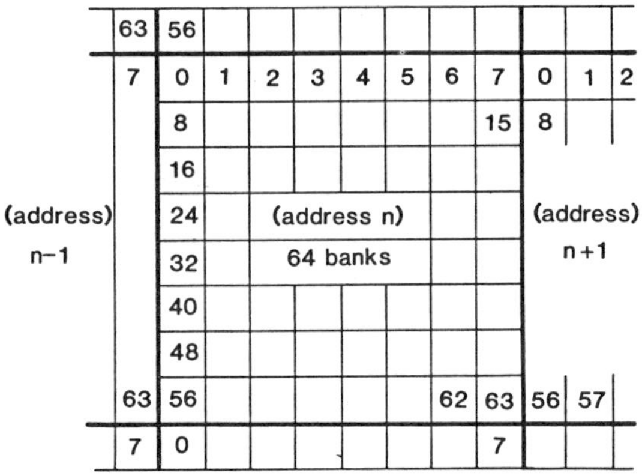

	63	56										
	7	0	1	2	3	4	5	6	7	0	1	2
		8							15	8		
		16										
(address)		24			(address n)				(address)			
n−1		32			64 banks				n+1			
		40										
		48										
	63	56						62	63	56	57	
	7	0							7			

Fig. 3.   Spin memory scheme of the Delft Ising processor.

bank numbers in an address plane. The spin lattice is mapped onto the memory in such a way that the value of the central spin and all surrounding neighbors are located in different banks. In this way all neighbor spin values can be fetched in parallel when a central spin is selected in either a sequential or random fashion. Although the number of independent memory banks would seem to imply that lattices with up to 63 neighbors could be handled, the hardware configuration limits this number to 31. The processor has a spin update rate of 1.5 MHz (independent of the number of neighbors involved), which is greater than Cray 1 speed. The lattice structure, the dimensionality of the space (1, 2, 3 or 4 dimensions), and the number of spins (up to 2^{22}) can be downloaded from the host into address modifier sections for every memory bank. Also periodic boundary conditions are automatically taken care of in this way. The processor is very flexible and many multi-spin interactions can be

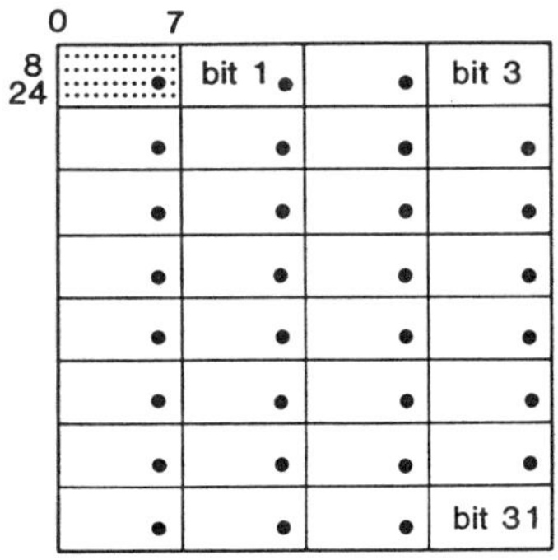

0	7		
8 24	bit 1		bit 3
			bit 31

Fig. 4.   Spin memory arrangement of the Ogielski-Condon Ising processor.

chosen and additional hardware is present for front-end processing as may be required for renormalization-group applications, for example. The cost of the components of this machine was approximately $10,000.

The Ogielski-Condon Ising processor [5-7] employs commercially available 32 bit wide memory boards for the VME bus. Both the dedicated spin update part and a general purpose microcomputer use the same VME bus to access that memory. In contrast with the Delft processor, which fetches all neighbors of one central spin in parallel, this processor selects 32 central spins in parallel, and fetches one neighbor for every central spin. These 32 neighbor spin values are packed in one 32 bit word of the memory and thus the number of neighbors involved dictates the number of memory accesses. Hence the spin area is divided into 32 subsections, each of which keeps their spin values in one fixed bit of the spin memory (see fig. 4.). The 32 neighbor sets are stored in shift registers, and they are fed sequentially into a high-speed pipelined spin update section. The resulting central spins are stored from the output register into the spin memory as a 32 bit word again. Special addressing hardware assures periodic boundary conditions. The spin update rate of 25 MHz makes this processor ten times faster than the Cray 1, at component costs of $10,000. This processor contains most of the features of the Delft processor, but it can also treat random interactions, an essential element for the simulation of spin glasses. The 16 Megabyte spin memory allows simulations of very large systems, and up to 16 neighbors can be involved in 1, 2, 3 or 4 dimensions. The general purpose microcomputer can directly access the spin memory, and is used to pre-process the results of the simulation (data reduction). Furthermore the total system is attached to a host. It took only eight months to design, construct and test the hardware, and after seven additional months several major results were produced. This work would have taken about ten years on a dedicated Cray 1!

VI. MOLECULAR DYNAMICS PROCESSOR.

In a molecular dynamics simulation Newtonian equations of motion of a number of particles are solved, and extensive information on the technique is available in the literature [8]. For each particle the position and momentum is specified at a certain time. The temporal evolution of the system is obtained by applying repeatedly the discretized equations of motion; in this case the leapfrog scheme was employed. The calculation of the force on a particle due to interactions with all other particles is the most time-consuming part. For short range pair interactions (e.g. truncated Lennard-Jones potentials) the N^2 interaction calculations can be reduced a number proportional to N. This requires a pre-sort algorithm such as the linked list method, which assures that only the neighboring particles are involved in the calculation (only particles in neighboring linked-list cells can interact). Still the work required for these calculations is dominant: calculating the distances between every pair of particles, looking up the forces and calculating the resulting changes in momentum at every time step. In the Delft molecular dynamics processor, built by Bakker [9-14], these calculations are implemented in parallel for the three Cartesian coordinates using pipelined hardware, which can treat 4,000,000 particle pairs per second (see fig. 5). All of the attributes of a particle are contained in one addressable word of the particle memory (188 bits), which is designed for a maximum of 64,000 particles. In parallel, the momenta and positions in the x-, y- and z-directions for the particles in two neighboring linked list cells i and j are moved to twelve high-speed memories X_i and X_j (positions) and P_i and P_j (momenta). Dynamically a vector of all particle pairs is generated by microcoded hardware, and the positions are pushed through the calculation pipe. Here the coordinates are subtracted and squared in three parallel hardware subtracters and multipliers (one set for every dimension, illustrated by the three dashed planes in fig. 5). Summation of the three squared coordinate differences gives the squared distance of the particle pairs and is used as an address of a force look-up table, which is downloaded from the host computer at the start of the simulation.

188

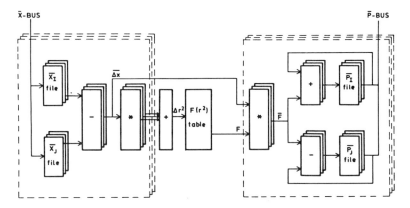

Fig. 5.   Calculational pipe of the Delft molecular dynamics processor.

The value obtained from the force table is multiplied, in parallel, by the x, y and z coordinate differences to obtain the three components of the force. These results are proportional to the change in momentum and an add or subtract operation is performed with the previous momenta for both particles in the three directions in parallel. After all particle pairs have been treated in this way the resulting momenta are moved back to the particle memory. Then the next set of linked list cells is processed, until all relevant linked list cell combinations are treated. The hardware is implemented in nine pipe stages and runs at 4 MHz. Other hardware is included to set up and use the linked list, to download microprograms of the various microprocessors involved and to calculate standard quantities of the system in parallel with the simulation. The processor is microprogrammable and this guarantees the flexibility to modify the control of the processor or to add new hardware for front-end processing of the results. The processor has Cray 1 speed and has been producing results since 1980. The cost of the components was $30,000.

VII. CONCLUSIONS.

Algorithm oriented processors have already had a substantial influence on the field of computational physics. The success of the pioneering work described here has encouraged many researchers to begin designing processors to solve their particular problems. The ability to tailor the architecture to the specific algorithm employed, and to utilize the latest technology in high-speed processors makes this a cost-effective approach for many research groups. As a note of caution, however, we would like to emphasize that the projects described above were selected precisely because they were successful, and it is not known how many have failed to come to a point where the processors could produce useful results. Some of the common elements of the successful projects can be enumerated. First, in each case the research group included researchers from both the computer science and physical science areas. In no case did a single person have sufficient knowledge and experience to design, construct and use the computer to solve an important physical problem. Second, in every case the processor was designed to solve one specific problem, with little attempt to make the hardware sufficiently general to solve a range of problems with common elements, although the more successful AOP's were those that were built to solve the more interesting problems. In view of the complexity of computer architecture, it seems to be self-defeating to attempt to design AOP's with sufficient flexibility to solve several different problems. Third, in every case discussed above, there was a need for large amounts of computer power in order to adequately define the solution to the problem.

Many designs are proposed or under construction, and it is to be expected that numerous research results will be published in the next few years, which would have been impossible without special purpose processors. The upgrading and development of parallel algorithms should not be rejected, since most new AOP's will have a parallel architecture in order to reach the computational power required. Next to parallel general purpose computers, highly parallel AOP's will be realized by multiprocessor systems interconnected by a network. Both the processor nodes and the network will be dedicated to the algorithm involved, and this will make the AOP more powerful and much less expensive than the commercially available parallel computers.

Acknowledgments

The author is grateful to G. H. Gilmer and M. H. Grabow for help with the preparation of the manuscript, and to J. H. Condon for numerous discussions.

References.

1. J. P. Valleau and S. G. Whittington, in: *Statistical Mechanics, Equilibrium Techniques*, B. J. Berne, Ed. (Plenum, New York, 1977), p. 137.

2. R. B. Pearson, J. Richardson and D. Tousssaint, J. Comp. Phys. 51 (2), 241 (1983).

3. A. Hoogland, J. Spa, B. Selman, and A. Compagner, J. Comp. Phys. 51 (2), 250-260 (1983).

4. A. Hoogland, A. Compagner and H. W. J. Blote, Physica 132A ,593-596 (1985).

5. J. H. Condon and A. T. Ogielski, Rev. Sci. Instrum. 56 (9), 1691-1696 (1985).

6. A. T. Ogielski and I. Morgenstern, Phys. Rev. Lett. 54 (9), 928-931 (1985).

7. A. T. Ogielski and I. Morgenstern, J. Appl. Phys. 57 (1), 3382-3385 (1985).

8. R. W. Hockney and J. W. Eastwood, *Computer Simulations using Particles*, (McGraw-Hill, New York, 1981).

9. A. F. Bakker, Ph. D. Thesis, Delft University of Technology.

10. A. F. Bakker, C. Bruin, F. van Dieren and H. J. Hilhorst, Phys. Lett. 93A (2), 67-69 (1982).

11. A. F. Bakker, C. Bruin and H. J. Hilhorst, Phys. Rev. Lett. 52, 449 (1984).

12. C. Bruin, A. F. Bakker and M. Bishop, J. Chem. Phys. 80, 5859 (1984).

13. J. H. Sikkenk, H. J. Hilhorst and A. F. Bakker, Physica 131A, 587-598 (1985).

14. J. C. van Rijs, I. M. de Schepper, C. Bruin, D. A. van Delft and A. F. Bakker, Phys. Lett. 111A, 58-59 (1985).

MATERIALS BY DESIGN —
A HIERARCHICAL APPROACH TO THE DESIGN OF NEW MATERIALS

JAMES J. EBERHARDT,* P. JEFFREY HAY,** AND JOSEPH A. CARPENTER, JR.***
* U.S. Department of Energy, Energy Conversion and Utilization
Technologies (ECUT), Washington, DC 20585
** Los Alamos National Laboratory, P.O. Box 1663, Los Alamos, NM 87545
*** Oak Ridge National Laboratory, P.O. Box X, Oak Ridge, TN 37831

ABSTRACT

Major developments in materials characterization instrumentation over
the past decade have helped significantly to elucidate complex processes and
phenomena connected with the microstructure of materials and interfacial
interactions. Equally remarkable advances in theoretical models and computer
technology also have been taking place during this period. These latter now
permit, for example, in selected cases the computation of material structures
and bonding and the prediction of some material properties. Two assessments
of the state of the art of instrumental techniques and theoretical methods
for the study of material structures and properties have recently been conducted. This paper will discuss aspects from these assessments of computational theoretical methods applied to materials. In addition, an approach
will be presented which uses advanced instrumentation and complementary
theoretical computational techniques in tandem in an effort to construct and
verify hierarchies of models to translate engineering materials performance
requirements into microscopic-level and atomic-level materials specifications
(composition, structure, and bonding). Areas of practical interest include
catalysis, tribology (contacting surfaces in relative motion), protective
coatings, and metallurgical grain boundaries. A first attempt involving
modeling of grain boundary adhesion in Ni_3Al with and without boron additions
will be discussed.

INTRODUCTION

The concept of using a hierarchy of structural models to design materials is based upon the presumption that the properties exhibited by a material
ultimately can be casually related to and based upon the composition, structure (including defects), and bonding in the assembly of atoms of which a
material is composed. It is of interest, therefore, to determine how such
relationships may be established to permit *a priori* design of materials.
This paper outlines an approach and gives a synopsis of two recent studies[1,2]
which address various aspects of the state of knowledge in constructing
models to establish these linkages among composition, structure, bonding, and
materials properties at the most fundamental levels (electronic, atomic/
molecular) of the hierarchy. An outline of some research needed to achieve
the objective of materials by design through such hierarchies also is
presented.

MATERIALS BY DESIGN — DEFINITION AND HIERARCHY OF MODELS

The materials by design (MBD) approach can be defined as the use of
scientific principles to predict and thereby to eventually improve materials
properties and performance characteristics. In this age of computers,
scientific principles, more often than not, are expressed in terms of computational models having predictive capabilities. Accordingly, for the purpose
of designing *a priori* new materials for specific functions, it is useful to
conceive of a hierarchy of models which can relate the performance characteristics of a material to its structural features. Figure 1 shows such a

192

ORNL-DWG 85C-18462

DEFINITION: USE OF SCIENTIFIC PRINCIPLES TO PREDICT MATERIALS PROPERTIES AND PERFORMANCE CHARACTERISTICS

HIERARCHY OF MODELS - "EXPERT SYSTEM"

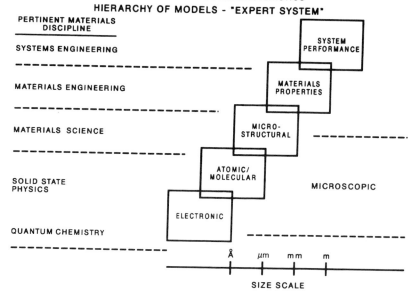

Fig. 1. Hierarchy of models for materials by design.

conceptual hierarchy; this also could be called an "expert system" because the models embody the knowledge and judgement of many kinds of experts across the many fields of knowledge needed to form the hierarchy. Some materials-related disciplines whose knowledge is required for MBD are indicated along the ordinate in Figure 1. It should be noted also that the size of features modeled declines as one proceeds down the hierarchy from right to left. Microscopic-level models include microstructural, atomic/molecular, and electronic.

Examples of systems performance models are computer-aided design and manufacturing (CAD/CAM) models for performance desired, say, from an automobile engine. This desired engine performance implies materials requirements and properties such as high-temperature capability. The desired material performance capabilities imply certain types of materials (e.g. ceramics, metal alloys, polymers) and other properties (ductility, strength, corrosion resistance) determined by microstructure. Ultimately, the description of the material properties can be reduced to the two lowest levels of the hierarchy, viz atomic/molecular-level simulations and electronic structure. The state of the art of the models at these latter two levels of the hierarchy has been the subject of the two recent assessments[1,2], which are summarized here.

The concept of designing materials is not new. Zener[3] refers to the design of a "new high-temperature alloy for steam turbines" as early as 1956, and von Hippel[4] wrote eloquently of the quest to design new materials in 1962. However, four major developments over the past fifteen years are resulting in *the historic convergence* of theory and experiment at these two lowest levels in the hierarchy. These developments are: (1) advances in electronic-level and atomic/molecular-level theoretical computational methods (and associated software); (2) computer architectures specifically suited to mathematical formulation of physical phenomena (e.g. approximating solutions to second-order nonlinear partial differential equations via the matrix-eigenvalue mathematical formulation); (3) developments in solid-state electronic devices; and (4) proliferation of new, more powerful materials characterization instrumentation. It is this historical convergence which is new, and these developments taken together give us confidence that at long last we are on the verge of being able to use the most fundamental scientific principles to design new materials to meet the future needs of our civilization.

Theoretical models can provide: (1) complementary information or insights unobtainable by experiment (as, for example, information on short-lived species); (2) scientific as opposed to heuristic foundation for design efforts; (3) framework to guide design; and (4) methods for prescreening and testing of design prior to experimental verification. Here it is well to reemphasize the *complementarity* of theoretical techniques to experimental methods as opposed to the displacement of experiment by theory.

SOME TECHNOLOGIES OF INTEREST

Three technological areas were addressed by one study[2] as being of particular interest for treatment by microscopic computational structural models: heterogeneous catalysis, tribological phenomena (lubrication and wear), and material interfaces (viz bonding of solid coatings to solid substrates and interactions involved in grain boundary adhesion). The fundamental mechanisms describing phenomena of interest in all three of these areas are: defect stability/mobility, physisorption, chemisorption, surface reactivity, interfacial bonding (adhesion), intermolecular bonding, and electron transfer at surfaces.

These fundamental mechanisms, in turn, can be described by the electronic states and geometrical configurations of the atoms and molecules during the various stages of the process of interest and, as such, these mechanisms are amenable to simulations by models at the electronic- and atomic/molecular-levels of the hierarchy.

ELECTRONIC-LEVEL AND ATOMIC/MOLECULAR-LEVEL SIMULATION MODELS

Table I shows an outline of the two lowest levels of microscopic computer simulation techniques (electronic-level calculations and atomic/molecular-level simulations), the scientific disciplines generally involved in their development, and the relationship of the electronic-level to the atomic/molecular-level models.

Electronic-level calculations are computations in which the "electron locations" (more correctly, electron probability distributions) are computed in the potential field of a fixed nuclear framework (the so-called Born-Oppenheimer Approximation). These in turn are divided into techniques developed with either chemical (quantum chemistry) or solid state (solid state physics) features in mind. There is some overlap between the two schools, of course, but in general the solid state techniques were pursued as alternatives to constructing approximate solutions to the many-body problem, which is the approach of the quantum chemists. This was necessary because the computational difficulty of *ab initio* quantum chemistry techniques scales by at least a factor of N^4, where N is the number of particles (electrons) in the physical system of interest. Because many more atoms of high atomic number (particularly metal alloys) are of interest in solid state systems, the development of alternate methods to those being pursued by chemists was desirable. Table II lists some of the more common quantum chemistry techniques, related acronyms, and a short description of each. Table III does the same thing for the solid state physics methods.

Table I. FUNDAMENTAL MECHANISMS CAN BE DESCRIBED AT THE LOWEST
LEVELS OF THE HIERARCHY

I. Electronic-Level Simulations

- Quantum Chemistry
  1. ab initio
  2. semiempirical

- Solid State Physics
  1. first principles approaches
  2. semiempirical

II. Atomic/Molecular-Level Simulations

- Molecular Dynamics
- Monte Carlo

Table II. ELECTRONIC-LEVEL SIMULATIONS

Quantum Chemistry Techniques

Name	Description
Ab Initio Methods	
Hartree-Fock (HF)	Single particle description in terms of orbitals in average field of electrons
	Electron correlation effects not included
	Predicts molecular structures well but chemical energies only qualitatively well
Correlated wave functions	Treat electron correlation effects and proper dissociation of molecule (lattice)
• Configuration interaction (CI)	
• Generalized valence bond (GVB)	Accurate description of potential energy surfaces and chemical energies
• Multiconfiguration self-consistent field (MC-SCF)	
• Green's function	
• Moeller-Plesset Perturbation Theory (MPPT)	
Local density methods	Non-local exchange-correlation interaction replaced by local potential
• Rigorous approaches	
- Discrete variational method (DVM)	Good description of one-electron properties
- Gaussian type orbitals (GTO)	
- Localized muffin-tin orbital (LMTO)	Total energies reliable only in non-muffin-tin calculations
• Scattered wave (muffin-tin) approaches (X_α-SW)	
<u>Semiempirical Methods</u>	Matrix elements determined from experiment or rigorous calculations
• Extended Huckel	
• (HMO, INDO, Fenske-Hall)	Permits global correlation with experiment and of structure-reactivity relations

Table III. ELECTRONIC-LEVEL SIMULATIONS

Solid State Physics Methods

Name	Description
First Principles Approaches	
Local Density Functional (LDF) Approaches	Local potential describes exchange and correlation effects
- Korringa-Kohn-Rostoker (KKR) - Augmented plane wave (APW) - Coherent potential approximation (CPA) - Linear muffin-tin orbitals (LMTO) - Augmented spherical wave (ASW)	Predicts lattice spacings, densities of states, modulus, elastic constants Linearization of energy minimization search and muffin-tin approximation for potential result in order of magnitude savings in computations
Semiempirical Approaches	
- Tight binding method (TBM) - Empirical pseudopotential method (EPM) - Embedded atom method (EAM)	Matrix elements determined from experiment or from rigorous calculations Global studies of structure-properties relationships

Atomic/molecular-level simulations are computations in which atomic locations are predicted. These computations require a description of the interactions (forces) among the atoms in the system of interest. These many-body interaction potentials can, in principle, be derived from the electronic-level computations but generally are not available because of the enormous amount of computing required (e.g. the electronic wave function would have to be recomputed for a wide range of interatomic distances). For this reason, atomic/molecular-level simulations usually are carried out by using assumed analytical potentials of a simple form such as the Lennard-Jones or Morse Potential. This lack of accurate, transferable many-body potentials is hampering the realistic simulation of many of the fundamental mechanisms involving defect structures in materials. Table IV summarizes the two predominant atomic/molecular-level simulation methods and gives a brief description of the approach and limitations of each.

Phenomenological Models

To describe the macroscopic properties of materials involving defects, grain boundaries, dislocations, and other microstructures, phenomenological theories need to be employed to address issues such as fracture, ductility, and brittleness in metals. In this paper we shall not attempt to address theories at this level of the hierarchy except to state that the goal of an MBD approach is to provide more quantitative information on the parameters used in these macroscopic approaches on the basis of the microscopic simulations discussed above.

Table IV. ATOMIC/MOLECULAR-LEVEL SIMULATION TECHNIQUES

Name	Description
Molecular dynamics	Integrates classical equations of motion of particles
	Determines dynamical (e.g. transport coefficients, rate constants) and equilibrium properties of systems
Monte Carlo approaches	Determine stable equilibrium structures by Boltzmann-weighted probabilities
	Determine equilibrium (not dynamical) properties of system
Both approaches	Require description of interaction potential between particles
	Higher than pairwise interactions necessary for accurate properties

Calculations for systems of 100 to 1000 particles and time
evolutions on order of nanoseconds are presently possible.
Simulations of microsecond (infrequent) phenomena must wait
for increased computational power.

CURRENT AND FUTURE CAPABILITY OF ELECTRONIC- AND ATOMIC/MOLECULAR-LEVEL
SIMULATION MODELS FOR PREDICTION OF MATERIALS STRUCTURES AND PROPERTIES

Electronic-Level Simulations

The electronic properties of a particular system can be determined in
principle by solving the Schroedinger equation for the electronic wavefunction, from which one can derive such information as charge distribution and
densities of electronic states for the material. These calculations are
carried out with fixed nuclear coordinates, because the Born-Oppenheimer
Approximation allows separation of the problem into electronic and nuclear
motion. By carrying out calculations as a function of nuclear geometries
and by computing the total energy of the system at each point, one can compute the equilibrium structure, the bulk modulus, and other measures of the
interaction potentials among the constituent atoms.

For periodic systems traditional solid state approaches employ local
density techniques wherein the problem is reduced to the solution within a
given unit cell. In these approaches, which have become practical for a
variety of metallic, ionic, and covalent solids, the electron-electron
interactions are described by *local* potentials for the classical coulomb and
nonclassical exchange potentials. In addition, electron correlation effects
have been incorporated into the "exchange-correlation" potentials. Within
this local density formalism (including methods with acronyms such as KKR,
CPA, APW, LMTO, and ASW), systems ranging from elemental solids with one
atom per unit cell to alloys with 20 or more atoms per unit cell typically
can be treated.

Ab initio quantum chemical approaches differ from the solid state
approaches in the two important respects that (1) the "infinite" cluster of
atoms in a solid state material is represented by a finite cluster of atoms
to treat the solid or surface and (2) the exchange interactions are described

by a nonlocal potential in the Hartree-Fock Approximation. These approaches have proven useful in describing properties that are *not* dependent on the long-range interactions of the solid, such as the description of chemisorption and dangling bonds at surfaces, defects in semiconductors, and mechanisms of catalytic reactions. Techniques are now available to treat aggregates of 40 to 50 atoms reliably within the *ab initio* (Hartree-Fock) framework.

Finally, semiempirical techniques can be used to understand a broad range of materials without having to explicitly perform first principles calculations on each to understand the overall trends. Such approaches as the tight binding method (TBM), the empirical pseudopotential method (EPM), and the embedded atom method (EAM), which parametrize the matrix elements needed in the theory from experimental or theoretical information, have proved very valuable for global searches of materials and in correlating structural properties with electronic properties.

Atomic/Molecular-Level Simulations

Once the interactions among the atoms are known, either from electronic structure calculations or from experimental information, simulations can be performed to determine thermodynamic and kinetic properties of the material. These properties can include the equilibrium structures and energies of grain boundaries in metals and ceramics and the diffusion rates of species on metal surfaces. The two major approaches can be divided into *molecular dynamics* and *Monte Carlo* techniques. In the former, the classical equations of motion are solved for all the atoms or molecules in the system (or a unit cell of the system), and the results can be used to obtain information on processes occurring in real time intervals of picoseconds to nanoseconds. In the latter, one finds the thermodynamically stable configurations of the system at a given temperature and pressure by varying the positions of the atoms according to Boltzmann weighted probabilities. Typically 100 to 1000 atoms can be treated in such simulation studies.

Table V is a synopsis of the types of materials, structural features, and properties which are commonly within reach of current atomic/molecular-level and electronic-level simulation techniques. These generally fall in the categories of materials having strongly ionic bonding, materials (especially molecules with covalent bonds) composed primarily of first and second row atoms, and alkali metals. The computation of the electronic structure of clusters of transition metal atoms presently constitutes a very active research area. Simulation of materials interfaces and point defect dependent phenomena, such as impurities in noncrystalline solids, interaction of point defects with surfaces, some types of diffusion, and high-temperature behavior of point defects, may be feasible over the next five to ten years. Simulation of extended defects (dislocations, microstructures, rapid transients, vibronic and diffuse electronic states) also fall into this category of simulations which are generally five years or more away. Problems which require revolutionary methods or advances in computing power include the simulation of the electronic structure of a dissociated edge or a dissociated screw dislocation (the electronic structure near an extended defect is needed to address the issue of whether or not an atom will segregate at that location), point defect/extended defect interactions, and materials processing (sintering, forging, machining).[1]

DESIGN PROBLEM OF "REAL WORLD" INTEREST

The potential for MBD using simulations based on electronic and atomic structures may perhaps best be seen in the case of ordered metallic alloys. It has been known that many long-range ordered (LRO) alloys and ordered

Table V. **MATERIALS AND PROPERTIES WITHIN REACH OF CURRENT
THEORETICAL TECHNIQUES**

Material types	References
Ionic materials (halides, binary oxides)	
• Accurate potentials based on shell model	• Sangster and Dixon (1976); Stoneham (1981)
• Applications to point defects, surfaces, dislocations, and grain boundaries	• Catlow and Mackrodt (1977)
• Self-trapped holes in KCl	• Tasker and Stoneham (1977)
• Quantum mechanical models for localized regions of lattice	• Vail (1984); Baraff et al. (1983)
Related ionic and semi-ionic materials	
• Bulk properties and defects in sulfides	• Pandey and Harding (1984)
• Charge transfer and interstitials in selenides	• Newmark (1980)
Complex oxides	
• Electrical and magnetic properties and properties at high temperature and pressure using techniques similar to those for ionic materials (above)	• Lewis and Catlow (1983)
Minerals and silicates	
• Structural properties	• Tossell (1977); Parker et al. (1984)
• Binding sites in zeolites	• Sanders et al. (1984)
Semiconductors	
• Structural properties and defects using valence-force potentials	• Stoneham (1985)
Metals	
• Vacancy formation enthalpies	• Taylor (1982)
• Monovalencies in Al and Cu	• Chakraborty et al. (1981)
• Positron annihilation characteristics	• Chakraborty and Siegel (1983)
• Electric field gradients for hyperfine probes and nuclear quadruple resonance	• Wichert (1982); Minier et al. (1978)

Table V. (continued)

References

M. J. L. Sangster and M. Dixon, *Adv. Phys.* **25**, 247 (1976).
A. M. Stoneham, *Handbook of Interaction Potentials*, AERE-R10205 (1981).
C. R. A. Catlow and W. C. Mackrodt, *Computer Simulation of Solids: Lecture Notes in Physics*, vol. 166 (Springer, Berlin, 1977).
P. W. Tasker and A. M. Stoneham, *J. Phys. Chem. Solids* **38**, 1185 (1977).
J. M. Vail, A. H. Harker, J. H. Harding, and P. Saul, *J. Phys. C.* **17**, 3401 (1984).
G. A. Baraff, M. Schulter, and G. Allan, *Phys. Rev. B* **27**, 1010 (1983).
R. Pandey and J. H. Harding, *Phil. Mag. B.* **49**, 135 (1984).
G. T. Newmark, *J. Appl. Phys.* **51**, 3383 (1980).
G. V. Lewis and C. R. A. Catlow, *Radiation Eff.* **73**, 307 (1983).
J. A. Tossell, *Am. Mineral.* **62**, 136 (1977).
S. C. Parker, C. R. A. Catlow, and A. N. Cormack, *Acta Cryst. B* **40**, 200 (1984).
M. J. Sanders, C. R. A. Catlow, and J. V. Smith, *J. Phys. Chem.* **88**, 2796 (1984).
A. M. Stoneham, to be published (1985).
R. Taylor, "Point Defect Calculations in Metals," in *Computer Simulations of Solids*, ed. C. R. A. Catlow and W. C. Mackrodt (Springer, Berlin, 1982), p. 195.
B. Chakraborty, R. W. Siegel, and W. E. Pickett, *Phys. Rev. B* **24**, 5445 (1981).
B. Chakraborty and R. W. Siegel, *Phys. Rev. B* **27**, 4535 (1983).
T. Wichert, *Point Defects and Defect Interactions in Metals*, (U. of Tokyo, Tokyo, 1982), p. 19.
M. Minier, P. Andreani, and C. Minier, *Phys. Rev. B* **18**, 102 (1978).

intermetallic compounds have strength and temperature capabilities exceeding those of normal disordered solid solution alloys and even approaching those of ceramics; however, in its search for new high-temperature structural materials, the general technical community more or less bypassed this general class of materials in favor of ceramics because of a perceived lack of any, or very little, ductility without the very-high-temperature capability of the ceramics. However, recent work[5] at the Oak Ridge National Laboratory and elsewhere has shown the feasibility of producing very ductile (30–50% elongation at room temperature) LRO alloys based on the $(Fe,Ni)_3V$ system and intermetallic alloys based on the the Ni_3Al system. Such alloys should eventually find use in many high-temperature applications such as heat engines and high-temperature industrial reactors and heat exchangers.

The $(Fe,Ni)_3V$ alloys were "designed" by using a standard metallurgical guide, namely, the electron-to-atom ratio, in order to achieve ternary alloys with cubic crystal structures. Obtaining a cubic crystal structure now seems to be the first and foremost criterion for achieving ductility in an ordered alloy because such high symmetry structure provides a greater number of slip systems than other crystal structures. The logic behind the "design" of the $(Fe,Ni)_3V$ alloys has been described in detail elsewhere;[5] suffice it to say that the design worked in that case because of a base of experimental data and certain empirical correlations which happened to hold for these specific atoms. If more sophisticated models were available to predict that certain combinations of atoms might have cubic ordered crystal structures and therefore be amenable to suitable macro- or microalloying to form a useful structural material, what other alloys might be identified? The development and use of models to predict the existence of heretofore unknown compounds with ordered crystal structures has already been reported.[6,7]

In the case of the Ni_3Al alloys the "design" was based on knowledge[8] that Ni_3Al, which naturally has a cubic crystal structure, is quite ductile in the single crystal form, even at room temperature, but is very brittle in the normal, polycrystalline form, as shown in the left of Fig. 2, usually failing by intergranular fracture. Following the logic that there must be something "wrong" with polycrystalline Ni_3Al grain boundaries, several dopants known to act as grain boundary "strengtheners" for other alloy systems were added systematically and one, boron, was indeed found to dramatically alter the ductility properties.[8,9] The "design" of the ductile Ni_3Al alloys has also been described elsewhere.[9] Had models been employed to predict the properties of grain boundaries in ordered alloys, perhaps the empirical search used in the Ni_3Al work could have been shortened. The use of these kinds of models has now begun to be applied to an extension of the solution of the Ni_3Al grain boundary embrittlement problem to other possible A_3B alloy systems. This ongoing work is described below.

There is still another opportunity for modeling in the case of Ni_3Al and perhaps other ordered alloys. Ni_3Al alloys lose much of their strengths and ductilities at temperatures above about 800°C (Fig. 3) despite the fact that they remain ordered all the way up to the melting point of Ni_3Al at around 1400°C. The reasons for these reductions in strengths and ductilities are not known. Perhaps models of the dislocation structures and slip systems in ordered alloys could help to explain the reasons and indicate approaches for extending the use temperatures of such alloys even higher.

CURRENT THEORETICAL APPROACHES TO THE Ni_3Al PROBLEM

Electronic-Level Approaches

Currently several theoretical approaches are being employed to understand the relationship between the electronic properties of Ni_3Al in

Fig. 2. Effect of boron in improving the ductility of Ni_3Al.

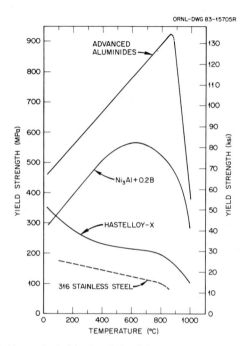

Fig. 3. Variation of yield strength with temperature for Ni$_3$Al alloys and selected high-temperature alloys.

the bulk and at grain boundaries and the influence of solute atoms in the bulk and at grain boundaries. Table VI outlines the types of models being employed and the kinds of information expected from them. *Local density* methods, such as the Linear Muffin-Tin Orbital (LMTO) method, are being applied to pure Ni$_3$Al and to supercells of 20 to 40 atoms representing the interface region of a high-angle grain boundary. From these results one should obtain information on the bulk cohesive energy, the cohesive energy of the grain boundaries with and without solute atoms such as boron, and correlations of such energies with densities of states (DOS) and charge distributions in the solid. The *coherent potential approximation* (CPA) method for random alloys is also being explored to study similar properties and to investigate the role of vacancies in the lattice (as in Ni$_3$Al$_{1-5}$) and of B atoms in the lattice.

The *tight binding method* (TBM) is being used to make global searches of binary alloys which are expected to have cubic ordered crystal structures and, hence, to have high ductility.

Atomic/Molecular-Level Simulations

Monte Carlo atomistic simulations are being carried out using empirically derived potentials and theoretically based variants of the Embedded Atom Method (EAM) to study the equilibrium geometries and energies of grain boundaries in Ni$_3$Al. Improved potentials for use in molecular dynamics calculations also are being investigated.

Table VI. BORON EFFECT IN Ni_3Al AND EXTENSION TO NEW MATERIALS

Current Theoretical Activities

Methods and investigators	Goals
Electronic-Level Simulations	
• Supercell LMTO — F. Mueller, G. Daalderop [Los Alamos National Laboratory, New Mexico (LANL)]	• Properties of high-angle Ni_3Al grain boundaries with and without B atoms; extraction of many-body potentials
• CPA alloy studies — A. M. Boring (LANL)	• Properties of lattices with defects and with B atoms
• Semiempirical TMB studies — D. Pettifor (University of London, England)	• Global search for cubic ordered intermetallics
Atomic/Molecular-Level Simulations	
• Monte Carlo studies — D. Srolovitz, S. Chen, A. Voter (LANL)	• Relaxed geometries and energies of grain boundaries. Thermodynamic properties of segregation and disorder
• Embedded atom studies — M. I. Baskes, M. S. Daw (Sandia National Laboratory, Livermore, California)	• Bulk defect energies and structures, geometry and strength of grain boundaries, segregation and diffusion of impurities, dislocation structure
• Hard sphere geometric models of grain boundaries — D. Farkas (Virginia Polytechnic Institute and State University)	• Diffusion and segregation to grain boundaries

EXPERIMENTAL TECHNIQUES APPLIED TO THE Ni_3Al SYSTEM

Currently, several experimental approaches are being applied to explain the observed boron effect on Ni_3Al by either obtaining direct experimental evidence or by experimentally verifying predictions coming out of the simulation calculations. The experimental studies, the investigators, and the goals are summarized in Table VII. In all likelihood, no one experimental technique will be able to provide a definitive understanding of the boron effect any more than any one type of simulation calculation will. Through the use of theory and experiment in tandem the probability of gaining the desired understanding is increased. It then may be possible to extend this understanding through the simulation models to the effects of dopants and impurities on metallic grain boundaries in general.

RESEARCH NEEDS

While the present state of the art of microscopic simulation techniques provides hope that computational techniques can significantly aid our quest

Table VII. BORON EFFECT IN Ni₃Al AND EXTENSION TO NEW MATERIALS

Current Experimental Activities

Activities	Goals
Transmission Electron Microscopy	
• Structure of grain boundaries in B-doped polycrystalline Ni$_3$Al — J. A. Horton, Jr., et al. [Oak Ridge National Laboratory, Oak Ridge, Tennessee (ORNL)]	• Elucidate grain boundary structural features in "real" Ni$_3$Al in order to provide conceptual guidance as to where B may prefer to reside
• Effects of B on dislocation structures of grain boundaries of Ni$_3$Al — S. Sass et al. (Cornell University, Ithaca, New York)	• Understand effects of B on structures of Ni$_3$Al grain boundaries
• Effects of deformation on Ni$_3$Al — H. Birnbaum et al. (University of Illinois, Urbana, Illinois)	• Understand phenomena associated with in-situ deformation and fractures in vicinity of Ni$_3$Al grain boundaries
Atom Probe Field Ion Microscopy	
• Structure and compositions of Ni$_3$Al containing B — M. K. Miller (ORNL) and S. S. Brenner (University of Pittsburgh, Pittsburgh, Pennsylvania)	• Determine distribution of boron and substitutional element atoms in Ni$_3$Al
Diffusion and Segregation	
• Electron microprobe studies of diffusion in polycrystalline and bicrystal Ni$_3$Al — Y. T. Chou et al. (Lehigh University, Bethlehem, Pennsylvania)	• Measure the rates of diffusion of Ni, Al, B, and other species both within the grains and in the grain boundaries and determine possible effects of various structural defects on those rates
• Auger electron spectroscopy studies of segregations of B to grain boundaries and free surfaces of Ni$_3$Al — C. L. White et al. (ORNL)	• Determine the segregation of boron to free surfaces and grain boundaries as functions of boron concentration and alloy stoichiometry
• Radioactive tracer studies of diffusion in Ni$_3$Al — N. L. Peterson (University of Illinois, Urbana, Illinois)	• Elucidate the effect of boron on grain boundary diffusion and determine the effects of non-stoichiometry on transport of boron to Ni$_3$Al grain boundaries

Table VII. (continued)

Activities	Goals
Positron Annihilation	
• Positron annihilation studies of atomic-scale defects (e.g. vacancies) in B-doped Ni_3Al — A. DasGupta et al. (ORNL) and K. Lynn [Brookhaven National Laboratory, Upton, New York (BNL)]	• Elucidate numbers and types of atomic-scale defects on Ni and Al sublattices of Ni_3Al stoichiometries in order to explain possibilities of influences on B effect
Photoelectron Spectroscopy	
• XPS and UPS investigations of bonding of B in rapidly solidified and annealed B-doped Ni_3Al — K. Lynn (BNL)	• Determine possible shifts in bonding state of B as it segregates to Ni_3Al boundaries
Microscopic Deformation Processes	
• Studies of the effects of grain size on deformation of Ni_3Al — E. M. Schulson et al. (Dartmouth College, Hanover, New Hampshire)	• Establish trend of strength and ductility as functions of grain size in order to provide possible support to grain boundary dislocation hypothesis of effect of B in Ni_3Al
Ion Beam Channeling	
• Ion channeling studies of B-doped Ni_3Al — J. M. Williams (ORNL)	• Determine locations of B and other alloying elements in bulk lattice sites and grain boundary sites in Ni_3Al alloys
Small-Angle Neutron Scattering	
• SANS studies of atomic-scale defects in B-doped Ni_3Al — J. Epperson et al. (ORNL and Argonne National Laboratory, Argonne, Illinois)	• Characterize the nature of atomic-scale defects observed in positron annihilation studies and elucidate the effects of B and alloy stoichiometry on ductility of Ni_3Al
Small-Angle X-ray Scattering	
• SAXS studies of atomic-scale defects in B-doped Ni_3Al — A. DasGupta et al. (ORNL)	• Characterize the nature of atomic scale defects observed in positron annihilation studies and elucidate the effects of B and alloy stoichiometry on ductility of Ni_3Al

toward the objective of MBD, a number of areas for additional research are evident from this analysis. Better methods of displaying simulation results are needed to permit extraction of qualitatively meaningful features from the computations. Reliable techniques are needed to obtain rigorously transferable and predictive many-body potentials for atomic/molecular-level simulations, and simple analytical expressions (a la Lennard-Jones) are needed for these many-body potentials. Models relating parameters in models at one level in the hierarchy to models at another level in the hierarchy (e.g. quantum statistical mechanics) are needed. Algorithms for rapid, accurate estimates of the total energy of any given geometrical configuration of atoms are needed. New, more sophisticated numerical procedures are required to significantly reduce computation time. "Smart" systems should be developed to assist researchers to select appropriate computational techniques for a given problem of interest. Efficient algorithms which take maximum advantage of computer architectural features such as parallelism and vectorization must be developed. Finally, combinations of several simulation and modeling techniques must be pursued to treat extended defect phenomena such as brittle fracture in ceramics, yield phenomena, and creep/fatigue in alloys. This list is by no means exhaustive.

SUMMARY

To summarize, a hierarchy of models has been outlined to structure our thinking on the kinds of models necessary in order to be able to design *a priori* materials with desired performance specifications. The state of the art is such that microscopic computational methods at the electronic- and atomic/molecular-level can contribute significantly to gaining insights into the mechanisms functioning in, and in predicting properties of, several types of materials. However, presently there is no one computational technique which can best simulate all kinds of materials features. The most urgent need appears to be for the development of many-body potentials to permit the simulation of dynamic phenomena such as the movement of atoms or vacancies in material structures. Several theoretical and experimental approaches are currently being used to understand the influences of solute atoms on the structure and properties of Ni_3Al.

REFERENCES

1. "Theory and Computer Simulation of Materials and Imperfections," Division of Materials Sciences, U.S. Department of Energy (August 1984).

2. "Assessment of Theoretical and Experimental Tools for Applied Research and Exploratory Development in Certain Energy Technologies," Energy Conversion and Utilization Technologies (ECUT) Program, U.S. Department of Energy (in progress).

3. C. Zener, *Westinghouse Engineer*, p. 148, September 1956.

4. A. von Hippel, *Science* **138**, 91 (1962).

5. C. T. Liu and J. O. Stiegler, "Ductile Ordered Intermetallic Alloys," *Science* **226**, 636 (1984).

6. A. Zunger, "Ternary Semiconductors and Ordered Pseudobinary Alloys: Electronic Structure and Predictions of New Materials," paper presented at 1985 Sanibel Symposium, March 23, 1985, Marineland, Fla. (To be published in *International Journal of Quantum Chemistry, 1985 Symposium Issue.*)

7. D. G. Pettifor, "The Structural Stability of Binary Compounds," in *High Temperature Alloys: Theory and Design*, J. O. Stiegler, ed., The Metallurgical Society of AIME, Warrendale, Pa., 1984, p. 61.

8. K. Aoke and O. Izumi, *Nippon Kinzoku Takkaishi* **43**, 1190 (1979).

9. C. T. Liu and C. L. White, "Design of Polycrystalline Ni_3Al Alloys," in *High Temperature Ordered Intermetallic Alloys*, C. C. Koch, C. T. Liu, and N. S. Stoloff, eds., proceedings of Materials Research Society symposium held November 26–28, 1984, Boston, Mass., Materials Research Society, Pittsburgh, Pa., 1985, p. 365.

Acknowledgments

One of us (JJE) would like to thank Dr. Thomas Kitchens, U.S. Department of Energy, Office of Basic Energy Sciences, Division of Materials Sciences, for providing a draft copy of reference 1. The other two of us (PJH and JAC, Jr.) would like to acknowledge support for this work by the U.S. Department of Energy, Office of Conservation, Division of Energy Conversion and Utilization Technologies (ECUT), under contract DE-AC05-840R21400 with Martin Marietta Energy Systems, Inc.

PROGRESS IN SPIN GLASSES AND RANDOM
FIELDS — RESEARCH WITH A SPECIAL PURPOSE COMPUTER

A. T. OGIELSKI
AT&T Bell Laboratories, 600 Mountain Av., Murray Hill, N.J. 07974

ABSTRACT

Extensive numerical simulations of random magnetic materials have been recently performed at AT&T Bell Laboratories with a fast specially designed computer. I will discuss certain issues concerning the use of specialized computers in research, and I will review some major results obtained in simulations of a three-dimensional spin glass and an antiferromagnet with random fields.

In this report I will talk about some very large numerical "experiments" on models of disordered, irregular magnetic materials such as spin glasses, dilute magnets and systems with local random fields. However, since similar Hamiltonians appear also in the description of other phenomena, one expects that these studies contribute to our understanding of the physics of disordered systems in a wider sense.

So far we have been rather unsuccessful in developing efficient analytic methods for computation of physical properties of such materials to the extent that one could calculate numbers that would directly explain the experiments. The role of the theory, accordingly, has been to provide insights into the nature and origin of observed phenomena rather than to lead to numerical predictions. In such situations, where theory usually supplies a model Hamiltonian and a plausible framework for description, computer calculations can be essential for i) verification that theoretical assumptions are actually correct, and ii) computing physical quantities that can be compared to experimental data.

It turns out, unfortunately, that for most disordered systems one needs an enormous amount of computation to obtain trustworthy numerical solutions, and we have seen already many times that insufficient calculations can only contribute to confusion.

I will concentrate on the statistical-mechanical description of a class of systems whose Hamiltonian can be written in the general form

$$H = - \sum_{xy} J_{xy} S_x S_y - \sum_x h_x S_x \qquad (1)$$

The fundamental degrees of freedom are classical spin variables S_x, distributed either randomly or regularly on sites of a d-dimensional lattice. The interactions J_{xy} and local fields h_x can in general be static random variables drawn from some probability distributions.

There are several features of such disordered systems which significantly complicate numerical calculations. One is the spatial inhomogeneity, which implies the need to perform calculations on very large systems, and to average the results over many distinct configurations of disorder variables. This effectively leaves only the Monte Carlo (MC) technique [1] as a viable computational scheme at non-zero temperatures. But the MC method is intrinsically dynamical in nature (time corresponds to the number of MC steps), and one must confront very slow relaxation which is ubiquitous among random "glassy" systems. Obviously, one must follow the configurations generated by the MC process over times much longer than some characteristic time scale to exploit ergodicity and obtain meaningful thermodynamic averages; in addition stochastic dynamics is often appropriate for modelling of real physical processes and long-time behavior is of great physical interest.

In short, good numerical work on random systems requires a truly enormous amount of computation, often well beyond the limits of available supercomputers (such as cost, restricted allocation of time, or insufficient speed).

A strategy which, we think, has proved its worth is to employ specialized computers adapted for very efficient execution of the algorithm. Such "hardwired subroutine" optimizes the integration of database, algorithm and hardware in order to achieve maximal processing speed. It turns out that beyond simply making it possible to solve numerically some hard problems, the major advantages of specialized processors are low cost of computation and concentration of resources on particular problems. The design and construction time and cost — which are essential for any special computer project — have been substantially reduced by recent advances in logic integration and computer-aided design, so that combined time of hardware and software development now can be acceptably short.

The special purpose computer optimized for very fast Monte Carlo calculations for a class of systems where the Hamiltonian can be written in form Eq. (1) has been constructed by J. H. Condon and myself at AT&T Bell Laboratories two years ago. Design, hardware and operating software are described in detail in our paper [2]. Here I will mention some aspects of the design. Our system has been constructed with two main objectives: i) speed and ii) flexibility. Necessary flexibility has been secured by choosing a bus-oriented system, with a dedicated custom-designed Monte Carlo processor executing the MC "heat bath" algorithm, and a general purpose computer (CPU) sharing the common memory.

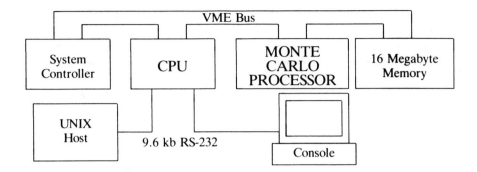

Fig. 1.   Architecture of the special purpose computer

A block diagram of the system is shown in Fig. 1. The Monte Carlo processor contains a number of interdependent modules executing various subtasks is parallel. Each of these modules is different and is optimized for very fast execution of its job. With this architecture the calculations that usually consume most time — recurrent loop updating all spins according to the MC algorithm — are delegated to fast special hardware, while analysis of configurations and evaluations of physical averages, where a lot of flexibility is needed, is left to programs written in a high level C language executing on the CPU. This is a high performance system, permitting the updating speed of about 25 million spin updates per second, and allowing for maximum lattice sizes of up to 8192^2, 512^3, $64^4, \cdots$, Ising spins. With currently installed commercial memory cards the updating speed of 17 million spin/sec can be maintained. This performance compares very favorably with that of available supercomputers, with a gain of speed by a factor of 5-10 over Cray-1 as a reference.

The most interesting calculations performed to date on the processor are: a very thorough study of three-dimensional Ising spin glass with short-range interactions, and of a three-dimensional dilute cubic antiferromagnet in magnetic field, which is a typical random field system realized in experiments. This is a largely completed research, other work which is still in progress is not reviewed here.

We have examined the behavior of the spin glass model with discrete nearest-neighbor random interactions $J_{xy} = \pm J$ with probability 1/2 for each value. While this model does not describe exactly the structure of experimentally realizable spin glass materials, it is thought to contain essential ingredients — randomness and frustration — and can well be in the same universality class as some anisotropic insulating materials. The spin glass problem has occupied many theorists and experimentalists for more than a decade now [3], and has a very rich history full of controversies. Among the main debated issues were: The very existence of a phase transition to a new type of state, nature of the unusually slow dynamics, relation of surprisingly rich mean field theory to real materials, etc. Additional thrust to spin glass research has been given by recent realization of its relation to other areas of study such as neural networks [4], combinatorial optimization problems [5] and others.

In our studies we concentrated first on static thermodynamic properties of the spin glass [6], and a detailed study of its dynamics followed [7].

We used a range of lattice sizes, from 8^3 to 64^3, and very long simulation times reaching 10^8 updates per spin even for 64^3 lattices per temperature step. We found that the spin glass phenomena indeed can be explained by an equilibrium phase transition. The definitions of critical exponents can be found in references cited above, currently our best estimates are: $\nu = 1.3 \pm .1$, $\gamma = 2.9 \pm .3$, $\eta = -.22 \pm .05$ and from scaling relations $\beta \cong .5$. To be sure, these are effective exponents just as available experimental data can be interpreted only in terms of effective exponents; it is reassuring, however, that most experimental estimates for insulating spin glasses find $\Delta = \beta + \gamma$ in the range 3-4, and $\gamma \approx 3$. Detailed study of equilibrium dynamic correlation functions, especially the Edwards-Anderson order parameter $q(t) = \overline{\langle S_x(0)S_x(t)\rangle}$ which in this system is directly related to the magnetic structure factor revealed in addition that very slow spin glass dynamics can be well described by conventional critical slowing down, albeit with a large dynamic exponent $z\nu = 7.2 \pm 1$ for the average correlation time (experimental values for several insulating materials are in the range 6-8 as well). In addition, we have determined the empirical formula describing the shape of dynamic correlation functions.

These calculations took well over a year of computing time (equivalent to several years of Cray-1 time!) and quite convincingly established the critical behavior of short-range Ising spin glasses, and were able to give estimates of critical exponents in reasonable agreement with (still not too precise) experimental data. The work of other groups which also numerically investigated the spin glass behavior can be found in references to cited papers. We think that recent large simulations finally give us correct quantitative description of the spin glass transition, but there are still some interesting problems which require more work.

Another interesting problem that we have been recently investigating concerns the behavior of magnetic systems in presence of random fields. Although apparently simpler than spin glass, this problem also has a ten-year-plus history marked by swings in theoretical predictions and controversies among experimentalists. It appears that just recently a correct scaling hypothesis has emerged [8], and it was a pleasure to see it verified by numerical simulations and at the same time to provide estimates of critical exponents. In addition, novel dynamic behavior has been observed in simulations.

We have performed [9] simulations of a dilute antiferromagnetic Ising model in uniform magnetic field h on a cubic lattice, which is a random field system closely modelling magnetic systems studied experimentally. We find that experimentally observed rapid onset of nonequilibrium behavior at $T > T_c$ can be attributed to predicted (D. S. Fisher, ref. 8) very rapid growth of correlation time τ

$$T \ln \tau \sim \xi^\theta$$

where ξ is the magnetic correlation length and a new independent exponent $\theta \approx 1.5$ characterizes the zero-temperature fixed point governing the transition. By extensive calculations along various paths in the h-T plane we could demonstrate that previously reported critical exponent estimates, which violated certain exact bounds [10] were strongly influenced by crossover effects. Our estimates of critical exponents $2 - \eta = 1.5 \pm .1$ and $2 - \bar{\eta} = 3.0 \pm .3$ for connected and disconnected order parameter susceptibility, respectively are found in agreement with these inequalities. The demonstration of very strong crossover effects and of new type of dynamic scaling should have an impact on analysis of experimental data and planning of new experiments.

In conclusion, I think that for certain classes of very large computational problems in theoretical physics the construction of specialized computing machines can significantly advance our understanding of physics. In our case, we have been certainly able to make progress by performing simulations a few orders of magnitude larger than related work, and thus to extend the simulations to larger systems and very long time scales with good precision which was necessary to obtain the correct interpretation of results.

4. J. J. Hopfield, Proc. Natl. Acad. Sci. USA, *79*, 2554 (1982).

5. S. Kirkpatrick, C. D. Gelatt, Jr., and M. P. Vecchi, Science *220*, 671 (1983).

6. Andrew T. Ogielski and Ingo Morgenstern, Phys. Rev. Lett. *54*, 928 (1985); J. Appl. Phys. *57*, 3382 (1985).

7. Andrew T. Ogielski, Phys. Rev. *B32*, 7384 (1985).

8. D. S. Fisher Phys. Rev. Lett. (in press) 1985; J. Villain, J. de Physique (in press); A. J. Bray and M. A. Moore, J. Phys. *C18*, L927 (1985).

9. Andrew T. Ogielski and David A. Huse, preprint (1985).

10. M. Schwartz and A. Soffer, Phys. Rev. Lett. *52*, 2499 (1985).

THE STONY BROOK MULTIPROCESSING MATERIALS SIMULATOR

HERBERT R. CARLETON
College of Engineering and Applied Sciences, State University
of New York, Stony Brook; New York   11794

ABSTRACT

     A parallel processing system with substantial computa-
tional power for the simulation of materials has been
designed and constructed at Stony Brook. Although primarily
designed to implement the molecular dynamics algorithm, the
system is highly efficient in carrying out simulations that
exhibit high degrees of local interaction. Our studies show
that this system is capable of performance, for this application, which is comparable to the Cray-1 at a small fraction
of the cost.

MOLECULAR DYNAMIC SIMULATION

     Computer simulation of classical gasses and fluids has
been widely utilized to study materials properties [1].
Since a direct simulation of an assembly of N particles
requires a calculation of the order of N(N-1) steps at each
simulated timestep, techniques have been developed to
systematize the particle behavior [2] and to reduce the order
of the calculation by taking advantage of the short-range
order of the interaction potential [3].
     Even with these simplifications, the time required to
complete a single timestep for a 1000 particle system on a
Cray-1 supercomputer is .1 seconds [4]. Simulations of a
system with a significant number of particles over a desirable time period is very expensive.
     The advent of inexpensive microcomputers on a chip with
substantial computational power make it feasible to construct
a dedicated computer for the purposes of materials simulation. Various dedicated computer innovations have been
developed which show promise for the simulation of materials
properties [5], [6].

The Algorithm

     The computational overhead in carrying out a molecular
dynamic calculation can be reduced by truncating the range
of the interaction potential [3], which has the effect of
reducing the number of binary interations from N to n, where
n is the number of particles within "range" of the potential.
Analysis shows that at ordinary temperatures and pressures of
interest, the number of particles within range of the
Lennard-Jones potential is around 60 - 100. We have employed
a "leapfrog" algorithm [5], in which the position of all
particles in the system are assumed fixed while the incremental velocities of each particle are computed from the
interaction force. The particle positions are then updated
from the current velocities which are held constant.

The interaction force is provided from a lookup table
and interpolation technique.

HARDWARE IMPLEMENTATION

The SBMMS was conceived as a topological mapping of the
molecular dynamic algorithm into the hardware space of the
simulator in such a way that maximum parallelism of the
individual processing elements could be achieved. For this
purpose, the assumed physical volume occupied by the particles was divided into cubic subcells each containing about 16
particles at the beginning of simulation. This division is a
software decision and is based on performance optimization.
The system will handle up to 16,000 particles or 256,000
degrees of freedom in its basic form.
The simulated volume is then a cube of 4 subcells on a
side. Best performance would be achieved with a processor
for each subcell, for a total of 64. We adopted a more
practical approach for a prototype and designed for 16 processors arranged in a 4x4 array. Each processor is then
assigned 4 subcells which we assumed for convenience to
correspond to the z direction.
Following a calculational philosophy that the best way
to look at a simulation problem is to concentrate on the
mapping of the degrees of freedom of the particles in the
data memory space of the computer, the SBMMS provides an
"indpendent" memory for each subcolumn of subcells which
"belong" to a slave processor. The position vectors of each
particle then determine the subcell location in which it will
reside in computer memory. Since the number of subcells
which are envisioned, even in multiply-connected SBMMS units,
is about 512, only three of the high-order bits of each of
the position vectors are required to determine subcell
location.

SYSTEM DESCRIPTION

A photograph of the SBMMS is shown in Fig. 1. Major
system elements consist of a host computer, which serves as
controller of the SBMMS and as a program development station
and file server, a host-SBMMS interface, processor units
consisting of four processor sub-units (PSU) each, a GATEWAY
to provide access to data memory for both host and slave
processors, and data memory units consisting of four 16K-word
memory subunits (MSU) each. An arbitrary number of processor
and memory units can be included in any system, but the
modular nature of the gateway dictates inclusion of four
units, i.e. 16 processors per basic system unit. We used a
Digital Equipment Corporation PDP-11/23 with a DMA interface
as a host computer.

Fig. 1 Photograph of the SBMMS simulator with PDP-11/23 host.

The key to the SBMMS architecture is the articulation of the processing subunits to the memory subcells, which contain only particle information. This is accomplished by a synchronous network, which we call a GATEWAY illustrated in Fig. 2. This arrangement makes it possible for each cell, via its associated processor, to access each of the 26 adjacent cells. Access conflicts are avoided by the host, which configures the GATEWAY so that only one memory cell is articulated to a processor at any one time. The gateway is capable of supporting sustained data transfer at a rate of 5 megabytes/sec at each node.

At the start of a computation run, the host downloads all software to the instruction memory of each processor. In a SIMD configuration, which is appropriate to the molecular dynamics calculation, all processors are downloaded simultaneously. The host-SBMMS interface provides the capability to access processors individually or in groups of four at a time for other operational modes. The host also downloads all particle initial conditions to the data memory, which is mapped into the linear address space of the host via the host-SBMMS interface.

At a commmand from the host, all processing associated with the particles in a single cell is initiated. Dedicated hardware determines when this process has been completed by all processors, which then interupts the host. The host then re-configures the gateway (in one microsecond), if required, and initiates a second phase of processing on a data cell which is adjacent to the initial cell, carrying intercell information with it. This process continues until all cells which are adjacent to the first cell have been processed. In our initial algorithm this requires five phases to exhaust all possibilities.

Processing Elements

Following the concept that all other computer functions
are to be separated from the data memory, the processing
elements of the SBMMS are independent, self-contained units
with their own instruction memory, register set, and communication links to both the host computer and to the switching
network which provides access to the data memory, the
GATEWAY.

These processor nodes are "medium grain" in the parlance
of computer architecture in that they have a substantial
instruction set. Our prototype system uses the Texas Instru-
ments TMS32010 processor for this purpose which is a 16-bit,
5 Mips processor with internal fixed-point multiplier. Since
the accumulator is 32 bits wide, all mathematical algorithms
in our system are written for double-word precision of 31
bits.

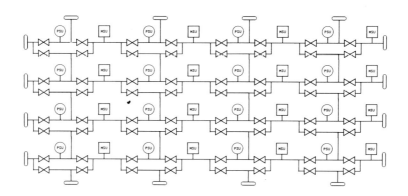

Fig. 2 Diagram of processor-memory architecture. The diamond
shaped elements are bi-directional bus drivers under control
of the host.

Memory Unit

Since a large number of independent memory subunits are
envisioned, static memory chips were selected, rather than
more inexpensive dynamic rams, in order to eliminate complicated refresh circuitry and to eliminate possible problems
with memory access/refresh conflicts. Each memory subsystem
consists of a 16Kx16-bit array of static ram with associated
address generator and control circuitry. Larger subunit
memory size is inconsistent with the philosophy of highest
speed since this would imply more data per processor. Hence,
large systems must be built using equal numbers of processors
and memory elements. This is made possible with the SBMMS by
articulating any number of 16-processor basic system units.

Gateway

The connection between each processor and a corresponding data cell is accomplished by a high speed multiport
bidirectional network which is capable of supporting data
transfers at the full speed of the slave processor. Each
memory cell is accessed by four slave processors as determined by a configuration register which is only under the
control of the host comptuer. Switching is accomplished in
such a way that no access conflicts can occur. This feature
of the architecture places the burden of conflict avoidance
on efficient host programming rather than on run-time avoidance. The host posts a single configuration word for the
whole system each time a new computational phase is
initiated.

Software

A macro library of routines in the machine language of
the slave processor have been written. These routines are
very compact and have been optimized for speed and written in
in-line code where necessary. The routines are primarily
double-precision math routines and zero-address data transfer
operations. Additional software is provided for system test
and maintenance and for continuous status monitoring during
runtime so that results of long runs will not be lost in the
event of a system crash.

PERFORMANCE

Although a full-scale simualtion has not been run on the
SBMMS as yet, our measurements and analysis on a four
processor system verify the initial design concepts. The
current measurements have been limited to 32-bit fixed-point
data structures since the TMS32010 is faster in this mode,
with a 32-bit multiply of 3 microseconds. Total time for
updating all parameters of a single particle is 3.8 milli-
seconds per processor. This implies that the timestep of a
system of 1024 particles (16 processors, 64 cells) will be
accomplished in .24 seconds.
A comparison of the timestep to that of the Cray-1 for
other system sizes is provided in Table I. In this comparison, we consider three system sizes consisting of one, two,
and four Basic System Units of the SBMMS as indicated. The
Cray-times are those obtained by J. Broughton on a thousand-
particle system. Times for larger numbers of particles are
extrapolated from this result.
This comparison shows that Cray performance levels are
achievable with a 32 processor system, which would cost under
$100,000. A ten-fold improvement in speed can be obtained by
incorporation of single-chip floating-point arithmetic units
capable of 10 MHz operation which are now available.

Table I. Performance comparison of SBMMS to that of the Cray-1 running the molecular dynamics algorithm for various number of particles N. Timestep in seconds.

cell configuration	number proc.	number cells	number particles	timestep SBMMS	Cray-1
4 X 4 X 4	16	64	1024	.24	.1
4 X 8 X 4	32	128	2048	.24	.2
8 X 8 X 8	64	512	8192	.48	.8

CONCLUSIONS

The SBMMS is very promising as a dedicated high-speed computer for the simulation of materials properties. Performance will be achieved comparable to a supercomputer for the molecular dynamics problem. Since the cost of a system module is so low, about $2,000 per processor for a complete system, large-scale simulations which require dedication of the computer system for long periods of time will become feasible.

ACKNOWLEDGEMENTS

This project was initiated with J. Broughton who has continued to provide help and inspiration.
We are grateful to the New York State Science and Technology Foundation and Stony Brook University for their financial support of this project to H. Austerlitz and to the many graduate students who have contributed to this effort.

REFERENCES

1.   R.W. Hockney and J.W. Eastwood, Computer Simulations using Particles, (McGraw-Hill, New York 1981).

2.   L. Verlet, Phys. Rev. 159, 98 (1967).

3.   O. Buneman, J. Comput. Phys. 1, 517 (1967).

4.   J. Broughton (private communication).

5.   A.F. Bakker, C. Bruin, F. van Dieren and H.J. Hilhorst, Phys. Let 93A (2), 67-69 (1982).

6.   K.C. Bowler and G.S. Pawley, Proc. I.E.E.E. 72, 42 (1984).

A HIGHLY PARALLEL COMPUTER FOR MOLECULAR
DYNAMICS SIMULATIONS

D. J. Auerbach, A. F. Bakker[†], T. C. Chen[‡], A. A. Munshi, and W. J. Paul
IBM Almaden Research Center, San Jose, California 95120 6099

ABSTRACT

We are constructing a highly parallel computer array well suited to the requirements of molecular dynamics simulations of the behavior of large systems of interacting particles. A tightly coupled message passing structure is used based on a specially designed processor with 12 MFLOPS peak performance. High speed synchronous communication over a full permutation network and hardware support for operations important in molecular dynamics codes are also provided. An analysis of this configuration indicates that efficient use can be made of 1000 processors working simultaneously.

INTRODUCTION

The use of computer simulation of the motion of a large number of interacting particles is now a well established practice in many branches of science and engineering[1]. Examples include the study of the evolution of the spiral structure of galaxies, the dynamics of large biological molecules, the stability of plasmas, the nature of phase transitions, and a wide variety of problems in materials science and statistical surface physics[2]. While these studies have been very fruitful, there are important questions which can not be addressed due to the limitations of available computers. Essentially, these limitations result in restrictions of the number of particles under study and/or the maximum time of the simulation. Such limitations become particularly severe for problems with multiple length or time scales. The extension of such calculations to the size and time scale of actual laboratory experiments represents a significant scientific opportunity. However the computational resources required are extensive: a *dedicated* computer 10-1000 times faster than those currently available.

We describe here a massively parallel computer, currently under construction in our laboratory, to provide this resource. This computer will be applicable to a broad class of scientific and engineering problems, but we limit the discussion here to molecular dynamics methods. We begin with an analysis of the algorithms used for molecular dynamics. This analysis shows that the application of a parallel computer to molecular dynamics is both simple and natural. Next we present a brief description of the overall parallel architecture of the machine under construction and illustrate the application to molecular dynamics. An analysis of parallel molecular dynamics algorithms for this structure indicates that a small fraction of global (non-parallel) computation limits the number of processors which can be efficiently used to of order 1000. Thus to achieve the desired overall performance, a very high performance node computer is required. The final section of the paper presents a brief description of the special processor we have designed and constructed for this purpose. Hardware support is provided for operations found to be important in our analysis of the molecular dynamics algorithms.

---

† Permanent address, Delft Technical University, Delft, The Netherlands.
‡ Permanent address, United College, The Chinese University of Hong Kong, Shatin, Hong Kong.

MOLECULAR DYNAMICS ALGORITHMS

Consider a system of a large number of particles with positions and velocities $\vec{r}_i$ and $\vec{v}_i$ for $i = 1, 2, ... N$, where N is the number of particles. For simplicity we take the case of N identical atoms moving under the influence of a potential which is the additive contribution of pairwise spherically symmetric potentials. The total potential is then

$$V(\vec{r}_1, \vec{r}_2, ..., \vec{r}_N) = \frac{1}{2} \sum_{i \neq j} \Phi(\mid \vec{r}_i - \vec{r}_j \mid) \tag{1}$$

The motion of the particles is found by solving Newton's equations of motion.

$$m \frac{d\vec{v}_i}{dt} = -\frac{\partial V}{\partial \vec{r}_i} = \vec{F}_i = \sum_{j \neq i} \vec{f}_{ij} \ , \ i = 1,..,N \tag{2}$$

The key to an efficient algorithm is an efficient technique for handling the sums over particles which appear in eqn. 2. If we had to carry out these summations over all N interacting particles for each of the N particles, we would have a problem with execution time scaling as N^2, a disaster for large systems. By using the fact that we have short range forces, this may be reduced to a problem with execution time proportional only to N. To do so, we restrict the summation in eqn. 2 to those particles which satisfy $\mid \vec{r}_i - \vec{r}_j \mid \ < r_{max}$.

Two techniques are used to efficiently find the pairs which satisfy this inequality: the use of coarse-grain tables and the use of Verlet (neighbor) tables with infrequent update[1,2]. The computational space is divided up into coarse-grained cells which are chosen so that in searching for the neighbors of a given atom, only the coarse-grained cell in which it is located and the nearest and next nearest neighbors need to be considered. Since placing the particles in the proper cells can be done in linear time, we have reduced the problem from one which scales as N^2 to one which scales with N. In two dimensions we will have to consider 9 n_{cell} potentially relevant neighbors for a given particle, where n_{cell} is the number of particles in a course grain cell. By building explicit tables of the neighbors of a given atom (and updating infrequently) we can gain another factor of about 6.

PARALLEL ARCHITECTURE

The use of coarse grain cells immediately suggests a natural mapping of algorithms described above onto a parallel computer. We assign the responsibility for computing the forces and updating the positions of the particles in a given cell to a processor of the parallel computer. To perform this task, the processor will first have to learn the positions of neighboring particles. Thus we are led to divide the parallel algorithm into a communication phase followed by a computation phase. Efficiency demands that the communication bandwidth grow linearly with the number of processors. This can be achieved by connecting the processors with a network that allows the communication to be done in parallel also. Each processor sends coordinates of its particles to its neighbors say to the north and simultaneously receives coordinates from its neighbor to the south. For a 2-D problem, 8 such permutations are required, while for 3-D 26 permutations are required.

Figure 1 illustrates the overall architecture of a computer built along these lines. The processors used are SPARKs, a custom processor described in the following section. They are interconnected by a network which allows for communication following any permutation.

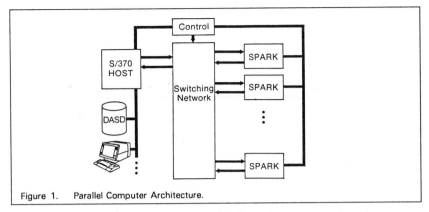

Figure 1.   Parallel Computer Architecture.

The use of a full permutation network adds reliability and generality to the design and permits an efficient simulation of other parallel machine architectures.[3] Control of the array is done by an enhanced SPARK which controls the network, synchronizes communication and computation phases, and communicates with a host computer for I/O.

The permutation network for an arbitrarily large system can in principle be built up from elementary switch elements via a recursive construction[4]. This is illustrated in Figure 2 for binary switch elements. The number of switches required scales as $N \log_k N$ where k is the number of states of the elementary switch. To reduce the required number of circuits we will use k=16, basing the elementary switch on a custom integrated circuit.

Efficiency Considerations

A perfectly efficient parallel computer would exhibit linear speedup: the computation rate of N processors would be N times that of a single processor. Several factors are important in achieving efficient parallel computation:

● Load Balancing. The work must be shared equally among all the processors. Load balancing is naturally achieved here since each processor is responsible for an equal volume of space. Of course there will be a limit to the number of processors we can use since to avoid imbalance from fluctuations, each processor should have 5-10 particles in it's domain.

● Small Communication Overhead. The time spent communicating must be a small fraction of the computation time. The large amount of computation which must be done on each data item and the large bandwidth of the permutation network together meet this requirement.

● Small fraction of serial code. If the problem under consideration has any portion which must be done serially, this will ultimately limit the number or processors which can be efficiently used.

To gain a more quantitative understanding of these factors we have analyzed one of the FORTRAN molecular dynamics codes used in our laboratory. The analysis was done by counting (approximately) the number of "simple" assembler language instructions, floating point additions and floating point multiplications required for table set up, force accumu-

222

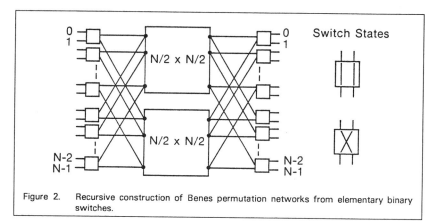

Figure 2. Recursive construction of Benes permutation networks from elementary binary switches.

lation, integration, etc. Formulae for the run time developed and checked by comparison with measured run time on an IBM 3081. Agreement to better than 10% was achieved. A corresponding analysis for the parallel algorithm was also performed, and formulae developed for run time as a function of communication rate and number of processors. The code studied performed a "temperature renormalization" which required computation of the total energy of the system at each time step. This global computation of course presents a serial bottleneck. Analysis indicates that of order 1000 processors can nonetheless be used with 70-90% efficiency (depending on the communication speed/computation speed assumed).

This analysis teaches us a very important lesson: the maximum number of processors which can be effectively used is limited. To achieve 10-1000 times the performance of current mainframes requires a very fast processor at each node of the computer array.

THE SPARK PROCESSOR

We are using a custom designed processor as the node for the parallel machine. The processor, called SPARK (Scientific Processor ARray Kernel), is a compact, high performance floating point processor. The design is based on VLSI building blocks and the extensive use of pipelining to achieve both high performance and a favorable cost/performance ratio. A sketch of SPARK is shown in Figure 3. The machine consists of four tightly coupled execution units, the floating point unit, FAU, the fixed point unit, XAU, the branch control unit, BAU, and the main memory, MM. Concurrent operation of these units allows for overlapping of address calculation and floating point operations, pipelined vector processing and hardware support for index table driven algorithms (scatter-gather).

Major Units

The scientific computing power comes from the floating point unit, FAU, on the right of Figure 3. In the initial design, 32-bit IEEE floating point is supported, but this will be upgraded to 64-bit precision. The operands and results of calculations are serviced by a 3 port register file implemented with VLSI static RAM to allow a large (4K word) register space and a flexible indexing scheme. This scheme, described further below, allows the registers to

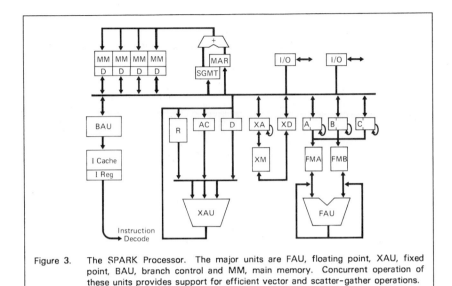

Figure 3.   The SPARK Processor. The major units are FAU, floating point, XAU, fixed point, BAU, branch control and MM, main memory. Concurrent operation of these units provides support for efficient vector and scatter-gather operations.

be viewed as in various ways: as 16 vector registers of length 256, or 8 vector registers of length 128, ..., as 1 vector register of length 4096, or as sets of 16 scalar registers.

The fixed point unit, XAU, to the left of the FAU is mainly devoted to handle the bookkeeping chores required in floating point computations, including the preparation of subscript addresses, generation of memory addresses, and handling of tables of indices (neighbor tables in molecular dynamics). It is also equipped with a 4K word register file which can be accessed as a vector register.

Leftmost in Figure 3 is the Branch Unit (BAU), responsible for branch and loop handling based on condition codes generated by the FAU and XAU. The BAU is equipped with an instruction cache (I-Cache) of 4K 64 bit words.

Above these units is the Main Memory Unit (MM), responsible for holding all code and data. The MM is 64 bits wide with a 16M word segmented address space. A block move operation allows data and code to be moved to and from the MM and various other memories and registers at a rate of 48 MBytes/sec.

Address Indexing Scheme

To keeping the floating point processor running at full speed requires supplying three addresses to the FAU in one machine cycle. If a completely general calculation were required for each of these three addresses, the hardware burden for address calculation would be very heavy. An analysis of the inner loops of molecular dynamics and other scientific codes indicated that in most cases *at most one non-trivial* address calculation was required where in trivial address calculations we include incrementing and selecting from a (small number) of vector registers. Based on this analysis we implemented a flexible address indexing scheme which sustains the full pipeline rate in most cases.

The address of a floating point operand or result is derived from three sources: (1) an index register which optionally can be automatically incremented under control of an instruction bit, (2) 4 bits of the instruction word which are logically combined with the 4 high order register bits (denoted A, B, C in Fig. 3), and (3) an optional overwrite of one or more of the registers by the fixed point unit. Vector operations are performed by initializing the index registers to the base of the vectors, enabling the optional autoincrement, and maintaining a count in the BAU. Indexing off a table (scatter-gather) can be done at the full pipeline rate by placing the table of indices in XM, using the XAU to add a base address and ship the result to an FAU index register. By setting appropriate bits of A, B, or C to zero, the floating registers can be "reshaped" as 16 vector registers of length 256, 8 vector registers of length 512, and so on down to 1 vector register of length 4096.

Performance

The first version of SPARK has an instruction cycle time of 168 nsec. giving a pipeline rate of 6×10^6 operations/sec. The peak floating point performance is thus 6 MFLOPS. For chained floating point operations, the peak floating point performance is 12 MFLOPS, but this feature is of little use in molecular dynamics simulations and has not been implemented on the prototype SPARK. Since 3 indexing, 1 fixed point, 1 branch and 1 floating point operation can be programmed in each machine cycle, the peak instruction rate is 36 MIPS.

Sustained performance is of course of more interest and also much harder to quantify. Benchmarks based on kernels from Lawrence Livermore Laboratory indicate performance ranging from 1.1 to 6.0 MFLOPS for a SPARK without using chained floating point operations. These results are based on an analysis of hand coded programs where however the hand coding was done according to the algorithms of a FORTRAN compiler under development. On the same benchmarks, IBM 3081 ranged from .6 to 3.5 MFLOPS and the CRAY-1 from 2.8 to 82 MFLOPS. We are currently working on coding a molecular dynamics application, but no performance benchmarks are available yet.

ACKNOWLEDGEMENTS

It is a pleasure to acknowledge the many contributions made by Farid Abraham especially during the formative stages of this project. Outstanding work on debugging the SPARK prototype by Christopher Lutz, Joe Schlaegel and Behnam Tabrizi is also gratefully acknowledged.

REFERENCES

1. R. W. Hockney and J. W. Eastwood, *Computer Simulation Using Particles*, McGraw-Hill, (1981); K. Binder, *Monte Carlo Methods in Statistical Physics*, Springer, New York (1979).

2. F. F. Abraham, *Rep. Prog. Phys.*, **45**, 1113 (1982).

3. Z. Galil and W. J. Paul, *J. ACM* **30**, 360 (1983).

4. V. Benes, *Mathematical Theory of Connecting Networks and Telephone Traffic*, Academic Press, New York (1965).

MICROSTRUCTURAL DYNAMIC STUDY OF GRAIN GROWTH

M. P. ANDERSON*, G. S. GREST* AND D. J. SROLOVITZ**
*Exxon Research and Engineering Company, Annandale, NJ 08801
**Los Alamos National Laboratory, Los Alamos, NM 87545

I. INTRODUCTION

The complete prediction of microstructural development in polycrystalline solids as a function of time and temperature is a major objective in materials science, but has not yet been possible primarily due to the complexity of the grain interactions. The evolution of the polycrystalline structure depends upon the precise specification of the coordinates of the grain boundary network, the crystallographic orientations of the grains, and the postulated microscopic mechanisms by which elements of the boundaries are assumed to move. Therefore, a general analytical solution to this multivariate problem has not yet been developed. Recently, we have been able to successfully incorporate these aspects of the grain interactions, and have developed a computer model which predicts the main features of the microstructure from first principles [1,2]. The polycrystal is mapped onto a discrete lattice by dividing the material into small area (2d) or volume (3d) elements, and placing the centers of these elements on lattice points. Interactions and dynamics are then defined for the individual elements which are analagous to those postulated in continuous systems. This discrete model preserves the topological features of real materials, and can be studied by computer simulation using Monte Carlo techniques. In this paper we report the application of the Monte Carlo method to the metallurgical phenomenon of grain growth with isothermal annealing. Extension of the model to treat primary recrystallization is presented elsewhere [3,4].

Grain growth is the term used to describe the increase in average grain size which occurs upon annealing a polycrystalline aggregate after primary recrystallization is complete. Normal grain growth is a steady state kinetic process characterized by time invariance of the normalized grain size distribution function, $F(R/\bar{R})$, and the topological distribution function, $P(N_e)$, where R is the grain radius and N_e is the number of grain edges as measured from polished sections [5]. In the long time limit the mean grain size $\bar{R}$ increases with time as [6]

$$\bar{R} = \alpha \, t^n \qquad\qquad (1)$$

where α is a temperature dependent constant and n is the growth exponent. While scatter in the value for n exists, n has a maximum value of 0.5 and a mean value of approximately 0.4 [6].

Most of the existing grain growth theories [5,7-9] implicitly assume that grains can be described as spherical, and that growth occurs in an average environment. This, however, ignores the fact that adjacent grains share common boundaries, resulting in a microstructure that is topologically connected. These theories [5,7-9] predict long time growth kinetics of the form of Eq. (1) with n = 1/2. Since experimental studies of grain growth yield an exponent, n, less than 1/2, the discrepancy may be a consequence of neglecting topological constraints, and the detailed environment.

To examine this possibility, Monte Carlo studies have been made of grain growth in both two dimensions (2d) and three dimensions (3d).

Mat. Res. Soc. Symp. Proc. Vol. 63. ©1985 Materials Research Society

II. SIMULATION PROCEDURE

In order to incorporate the complexity of grain boundary topology, the microstructure is mapped onto a discrete lattice (Fig. 1). Each lattice site is assigned a number between 1 and Q corresponding to the orientation of the grain in which it is embedded. Q is chosen large enough so that grains of like orientation impinge infrequently, typically Q = 48 or 64. Since the speed of the computer algorithm decreases with increasing Q, larger values of Q are not routinely employed. However, for $Q \gtrsim 36$, the results are insensitive to the magnitude of Q. In the present model, a grain boundary segment is defined to lie between two sites of unlike orientation.

The grain boundary energy is specified by defining an interaction between lattice sites within a given distance (usually nearest neighbor in two-dimensions (2d) and up to the third nearest neighbors in three-dimensions (3d)),

$$E_i = -J \sum_j (\delta_{S_i S_j} - 1). \qquad (2)$$

Here S_i is the orientation of site i ($1 \leq S_i \leq Q$), δ_{AB} is the Kronecker delta and the sum is taken over all sites within a specified distance. J is a position constant which sets the scale of the grain boundary energy.

The kinetics of boundary motion are simulated via a Monte Carlo technique in which a site is selected at random and re-oriented to a randomly chosen orientation between 1 and Q. If the change in energy associated with the re-orientation, ΔE, is less than or equal to zero the

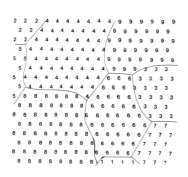

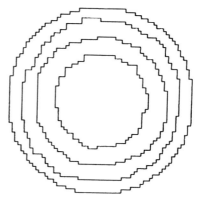

Figure 1 - Sample microstructure mapped onto a triangular lattice. The integers denote orientations and the lines represent grain boundaries.

Figure 2 - Evolution of an initially spherical grain of initial radius R_o = 30 (in units of lattice sites) in an infinite matrix on a simple cubic lattice for t = 0, 200, 400 and 600 MCS.

re-orientation is accepted. However, if the re-orientation attempt results in $\Delta E > 0$, the re-orientation is accepted with probability $\exp(-\Delta E/k_B T)$, where $k_B T$ is the thermal energy. A unit segment of grain boundary, therefore, moves with a velocity, $v_i = C[1 - \exp(-\Delta E_i/k_B T)]$, where C is a constant proportional to the boundary mobility. This description of the local boundary velocity is formally equivalent to that derived from classical reaction rate theory. Time, in these simulations, is proportional to the number of re-orientation attempts. N re-orientation attempts is used as the unit of time and is referred to as 1 Monte Carlo Step (MCS), where N is the number of lattice sites. The conversion from MCS to real time has an implicit activation energy factor, $\exp(-W/k_BT)$, which corresponds to the atomic jump frequency. Since the quoted times are normalized by the jump frequency, the only effect of choosing $T \cong 0$ (as in these simulations) is to restrict the accepted re-orientation attempts to those with $\Delta E < 0$. Re-orientation of a site at a grain boundary corresponds to boundary migration.

III. RESULTS

Extensive simulations have been performed in both 2d and 3d. Most studies in 2d were carried out on a triangular lattice of size 200 x 200. The results in 3d were obtained employing a simple cubic lattice of size 60 x 60 x 60 or 100 x 100 x 100.

A. Morphology

As a demonstration of the isotropy of the model, the shape of a circular grain (2d) or spherical grain (3d) was monitored as a function of time. The position of the grain boundary is plotted in approximately 200 MCS increments in Fig. 2 for the great circle of a shrinking spherical grain. In both 2d and 3d, the embedded grain remains nearly circular or spherical as it shrinks. This demonstrates that the product of the driving force and the boundary mobility is isotropic.

For the simulations of polycrystalline grain growth, the microstructure is initialized by randomly assigning an orientation between 1 and 0 to each lattice site. The temporal evolution of a polycrystalline microstructure is shown in Fig. 3 for growth on a 2d triangular lattice and in Fig. 4 for a (100) planar section on the 3d simple cubic lattice with first, second and third nearest neighbor interactions. Steps are clearly visible in the micrographs in Fig. 4 due to the small size of the simulated volume (100^3 lattice sites). These steps are less visible in Fig. 3 for the 200^2 system. In spite of these steps, good correspondence to planar grain morphologies in isotropic polycrystalline metals and ceramics is obtained. In both 2d and 3d, grain boundaries tend to intersect at 120° when examined on a scale larger than a unit step. Grains which are larger than the mean size (2d) or the mean size in cross-sections (3d) tend to have more than six sides and grow. Grains which are smaller than the mean size tend to have less than six sides and shrink.

B. Kinetics

Averaging the simulation data for the shrinking circular grain (2d) and shrinking spherical grain (3d) over 20 simulations shows that in both cases the radius decreases as the square root of time, $n = 1/2$ in Eq.(1). The kinetics obtained are in agreement with analytical treatments which are rigorous for these simple geometries. This demonstrates that the kinetic simulation technique employed is in accordance with the rate theory model.

The mean grain size was monitored as a function of time for the polycrystalline microstructures. In Fig. 5 the mean grain area for 2d and

228

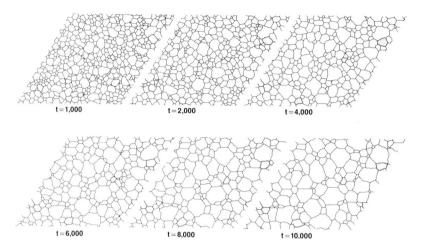

Figure 3 - Temporal evolution of a two dimensional polycrystalline microstructure on a 200^2 triangular lattice for a pure, isotropic material. Grains with six edges have been shaded.

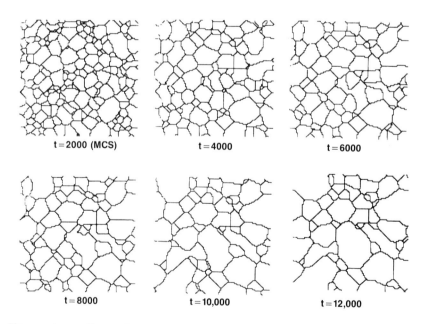

Figure 4 - Temporal evolution of the cross-section from the 3d microstructure for $Q = 48$ on a 100^3 simple cubic lattice.

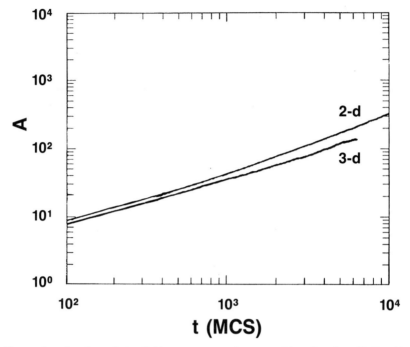

Figure 5 - Log-log plot of the mean area A versus time for Q = 48 for the two dimensional grain growth simulations and for the cross-sectional area of the three dimensional simulations.

mean cross-sectional area for 3d is presented as a function of time. Averaging the simulation data over more than 40 simulations for each dimension show that the growth kinetics are n = 0.41 ± 0.03 (2d) and n = 0.37 ± 0.02 (3d) in Eq. (1). It is concluded from these results that the grain growth kinetics are independent of dimensionality (for d ≥ 2). For comparison, the average grain growth exponent n derived from averaging the results for n from the literature for different metals and ceramics is 0.39 ± 0.07 [6].

C. Grain Size Distribution

The grain radius distribution was examined by plotting the frequency of occurrence versus the logarithm of the normalized grain radius determined from area in 2d and from area in cross-sections in 3d. These distributions were found to be time invariant when normalized by their respective means. Time averaged data are shown in Fig. 6. The size distribution functions are nearly identical between 2d and 3d. Comparison of the 3d model data with experimental data measured in our laboratory for pure Fe is presented in Fig. 7. Excellent agreement is obtained.

230

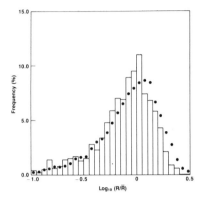

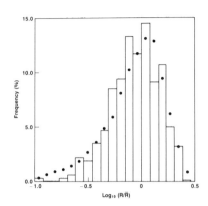

Figure 6 - Grain radius distribution as determined from cross-sections of the 3d lattice model (filled circles) and from the 2d lattice model (histogram).

Figure 7 - Grain radius distribution as determined from a cross-sectional area analysis of pure Fe (histogram) and from cross-sections of the three dimensional lattice model (filled circles).

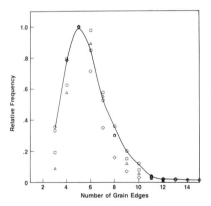

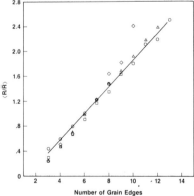

Figure 8 - Distribution of the number of grain edges measured from cross-sections. The circles represent data from the 3d simulations. The triangles correspond to MgO (10), the squares to Al (11), and the diamonds to Sn (5). The curve is drawn through the simulation results.

Fig. 9 - Mean grain radius for each topological class versus number of edges N_e. The circles represent data from cross-sections of the 3d simulation. The triangles correspond to MgO (10) and the diamonds to Al (11). The squares represent the results obtained from the 2d simulation. The line is drawn through the 3d simulation results.

D. Topology

The grain topology was also examined by plotting frequency of occurrence versus the number of grain edges, N_e. Data in 3d were taken from cross-sections. It was found that these distributions are also time invariant. The topological distributions are nearly identical between 2d and 3d, similar to the grain size distribution functions. Comparison of the 3d topological data with experimental data for Al, Sn, and MgO is given in Fig. 8. The relationship between grain size and topology for the 3d model microstructure and the above materials is given in Fig. 9, where the mean grain size for each topological class (grains with the same number of edges) is plotted versus the number of grain edges. As before, the agreement between the simulation results and experiment is excellent.

IV. CONCLUSIONS

A simulation procedure for grain growth has been developed which accounts for local structure and topology. The simulated microstructures shown excellent agreement with those found in isotropic polycrystalline metals and ceramics. The grain size distribution, topological distribution, and the topology-grain relations are all in excellent agreement with experiment. Similarly, the kinetics found also reproduce the experimental results. It is suggested that the growth exponent, n < 1/2, is inherent to the grain growth process and is not due to impurities, preferred orientation, etc. as had been previously thought.

References

1. M. P. Anderson, D. J. Srolovitz, G. S. Grest and P. S. Sahni, Acta Metall. 32, 783 (1984).
2. D. J. Srolovitz, M. P. Anderson, P. S. Sahni and G. S. Grest, Acta Metall. 32, 793 (1984).
3. M. P. Anderson, G. S. Grest and D. J. Srolovitz, in Computer Simulation of Microstructure (edited by D. J. Srolovitz), Metallurgical Society of AIME, Warrendale, PA, in press.
4. D. J. Srolovitz, G. S. Grest and M. P. Anderson, Acta Metall., in press.
5. P. Felthan, Acta Metall. 5, 97 (1957).
6. F. Haessner, Recrystallization of Metallic Materials (edited by F. Haessner), p. 63, Riederer Verlag, Stuttgart (1978).
7. M. Hillert, Acta Metall. 13, 227 (1965).
8. N. P. Louat, Acta Metall. 22, 721 (1974).
9. S. K. Kurtz and F. M. A. Carpay, J. Appl. Phys. 51, 5725 (1981).
10. D. A. Aboav and T. G. Langdon, Metallography 1, 333 (1969).
11. P. A. Beck, Phil. Mag., Suppl. 3, 245 (1954).

MODELLING OF THIN FILM GRAIN STRUCTURES AND GRAIN GROWTH

H.J. FROST* and C.V. THOMPSON**
*Thayer School of Engineering, Dartmouth College, Hanover, NH 03755
**Dept. of Materials Science and Engineering, M.I.T., Cambridge, MA 02139

ABSTRACT

We have generated two-dimensional grain structures by considering constant growth of regions which nucleate under various conditions. We have considered continuous nucleation at a constant rate, nucleation site saturation, continuous nucleation with nucleation-exclusion zones surrounding the growing regions, and decaying nucleation rates. The microstructures resulting from these varying conditions differ markedly in their topological and geometric properties.

We have also developed a program which models grain boundary motion in structures resulting from nucleation and growth. Results from these models will be compared to observed microstructures in thin metallic and semiconductor films and should provide insight into the mechanisms of microstructural evolution and control.

INTRODUCTION

A polycrystalline thin film results from nucleation and growth to impingement of individual crystals with differing orientations. In both vapor phase deposition and crystallization of amorphous films, nucleation often occurs on a surface and resulting grain structures are columnar in that grain boundaries intersect both surfaces of the resulting continuous films. The topology and geometry of thin film columnar grain structures should reveal the conditions for crystal nucleation and growth during film formation. In this paper we review the results of modelling various conditions for nucleation and constant growth. Detailed discussions of these results are given elsewhere [1] [2]. Observed thin film microstructures are often affected by grain boundary motion during or after formation. We also give here a preliminary report on the affects of capillarity driven grain boundary motion in computer generated microstructures.

We treat two extreme cases of the nucleation rate as a function of time. The first is continuous nucleation, in which the nucleation rate, per unit of remaining uncrystallized area, remains constant throughout the process of film formation. We assume that nucleation sites are randomly distributed in the plane and that the growth rate is constant for all crystals. The resulting structure was first described by Johnson and Mehl [3] and Evans [4]. The second case is site saturation, in which nucleation occurs simultaneously at a limited number of randomly distributed sites. Between these two extremes, we have studied intermediate cases in which the nucleation rate declines with time. In addition we have studied the structures produced when there is a nucleation-exclusion zone around each growing crystal. In all cases we assume a constant and isotropic crystal growth rate. These models are the two-dimensional analog of the three-dimensional models of nucleation and growth to impingement reported by Mahin, Hanson and Morris [5]. As an illustration of the grain growth model, we show the early stages of the evolution of the site saturation structure.

234

CALCULATIONS

Details of the methods used in the calculations are given elsewhere [1] [2]. Two neighboring grains which nucleate at the same time and grow at the same rate will impinge along the perpendicular bisector of the line joining the two nucleation sites. Triple points where three grains meet are formed where three perpendicular bisectors meet. For any array of points, simultaneous nucleation and constant growth rate will result in the well-studied Voronoi polygon structure. The Voronoi polygon for a particular nucleation point is that region of the plane that is closer to that point than to any other nucleation point. Studies of this structure are reviewed by Getis and Boots [6]. An example is shown in Figure 1, in which nucleation sites were chosen randomly.

If two neighboring grains nucleate at different times, but grow at the same rate, they impinge along a boundary which forms a hyperbola. Three grains that nucleate at different times produce hyperbolic boundaries which may meet at zero, one or two triple points, depending on nucleation times and site positions. The possibility of forming two triple points leads to the possibility of forming lenses defined by two hyperbolic boundaries. This is an important distinction between simultaneous and sequential nucleation.

The structure that results from continuous nucleation, with constant growth rate, has also been studied, with various properties derived by Meijering [7] and Gilbert [8]. An example of this structure is shown in Figure 1. It is similar, but not identical, to the planar sections through the three-dimensional continuous nucleation model of Mahin et al. [5].

The cases of declining nucleation rate are treated the same as continuous nucleation except that the nucleation rate is taken to decrease exponentially. For the models with a nucleation-exclusion zone, the continuous nucleation calculation is amended to exclude any nucleation event that is within a distance δ of the edge of any previously nucleated grain. We measure δ in units of the square root of the average area in the continuous nucleation model. The structures resulting from $\delta = 0.2$ and $\delta = 0.6$ are shown in Figure 1.

In order to model the effects of grain boundary migration driven by capillarity forces, we have chosen to follow the boundary itself as a linear array of moving points. In each time increment the points are moved, perpendicular to the boundary line, by a distance proportional to the local curvature, κ. The boundary velocity, v, is therefore $\mu\kappa$, where μ is a constant which includes the boundary mobility and the boundary energy. The program increases or decreases the number of points as required to keep the spacing between points along the line within certain bounds. It also provides for the elimination of small grains and the switching of grain neighbors. The program alternates between moving the boundary segments and moving the triple points to positions for which local angles approach 120°.

Our program differs in technique from other recent models such as the soap film model of Weaire and Kermode [9] [10] and the grain growth model of Anderson et al. [11] and Srolovitz et al, [12]. The Anderson, Srolovitz et al. model is based on a large array of incremental areas, each of which belongs to one grain or another. In that model, the structure evolves as each incremental area is repeatedly allowed to change from one grain to another, based on random chance, weighted according to the grain membership of the neighboring incremental areas. Although their model results in grain growth, with boundary migration rates which are, on average, proportional to curvature, the boundaries remain ragged on the scale of the area increment. The boundary migration therefore might not be exactly proportional to curvature on a very local scale.

235

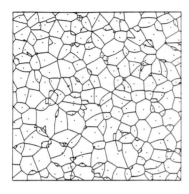

a. Continuous Nucleation
    (Johnson-Mehl Model)

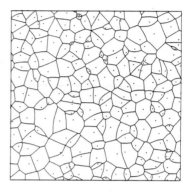

d. Nucleation-Exclusion Zone
    of width δ = 0.2

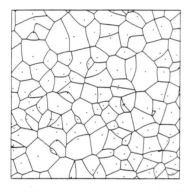

b. Declining Nucleation Rate (α=2)

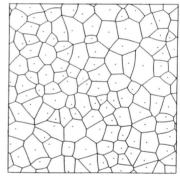

e. Nucleation-Exclusion Zone δ=0.6

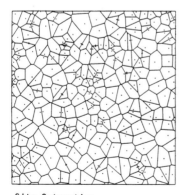

c. Site Saturation
    (Voronoi polygons)

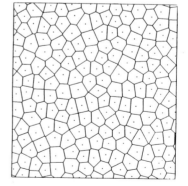

f. Sequential Packing of Non-
    overlapping Disks (δ → ∞)

Figure 1   Structures that result from different conditions of nucleation
and growth to impingement.

236

RESULTS

The various nucleation conditions produce structures which differ more
or less in almost all their topological and geometric properties. Actual
microstructures of thin films differ from any of these structures; qualitatively they most closely resemble the nucleation-exclusion zone models. If
grain boundaries are mobile during crystallization or film formation (after
grain impingement) altered microstructures will result.

Initial Structures After Growth to Impingement

The distributions of number of sides are shown in Figure ?. If no
more than three grains meet at any point, Euler's relation specifies that
the average number of sides per grain must be six. For the site saturation
case, no grain has fewer than three sides; there are few triangles, and few
grains with more than eight sides. For the continuous nucleation case
there are a few lenses (which have only two hyperbolic sides) and many
triangles and grains with more than eight sides. Including a nucleation-
exclusion zone sharply reduces the number of lenses and grains with more
than ten sides. The site saturation case is from Crain [13].
The difference between the various cases is shown more dramatically by
the grain area distributions (Figures 3). Once again, the site saturation
case is from Crain [13]. The continuous nucleation case produces far more
small grains (mostly lenses and three-sided grains) and more large grains.
The first grains to nucleate grow larger before impingement and those which
nucleate last have little free area remaining to grow into. The distribution is similar to an exponential decay and similar to that for a planar
section through a three-dimensional continuous nucleation model [5]. The
intercepts at zero area for continuous nucleation and declining nucleation
rate are clearly non-zero, while for the site saturation and nucleation-
exclusion zone cases they appear to be zero. The intermediate cases of
declining nucleation rate and of nucleation-exclusion zone show a logical
progression between the extreme cases.
We have measured several additional geometric and topological distributions, which are reported elsewhere [1] [2]. These include the
distribution of boundary segment lengths; the distribution of angles at
triple points; and the average number of sides on the grains that neighbor
grains with a particular number of sides.

Grain Boundary Migration

We here present preliminary results on the evolution of the site
saturation structure. Initially the boundaries are straight and the triple
point angles deviate markedly from local equilibrium. Figure 4a shows the
starting configuration. The square array has edge length 10 times the
square root of the average area. It is convenient to measure time in
dimensionless units of $\tau = t \mu/\bar{A}$, where t is the time, μ is the constant
relating boundary velocity to curvature, and $\bar{A}$ is the average area. For
comparison, an isolated circular grain of area $\bar{A}$ will collapse to annihila-
tion in dimensionless time $\tau = 1/(2\pi)$. As the structure evolves, those
grains with more than six sides grow in area and those with fewer than six
shrink. At time $\tau = 0.05$ (Figure 4b) the boundaries are noticeably curved
and the triple point angles have approached 120°. At time $\tau = 0.1$ (Figure
4c) the smallest grain in the array is about to vanish. Since the number
of grains is unchanged, the evolution has so far occurred without grain
growth. In the final picture, Figure 4d, at time $\tau = 0.15$, several grains
are about to disappear, and the structure appears qualitatively similar to
experimental microstructures.

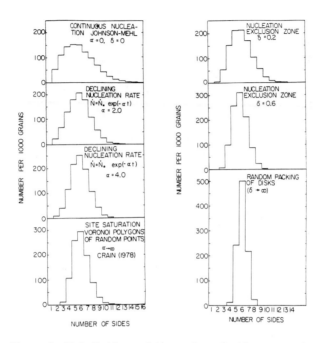

Figure 2 Distributions of the number of sides per grain.

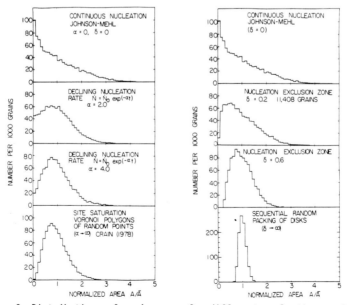

Figure 3 Distributions of grain areas for different nucleation conditions.

238

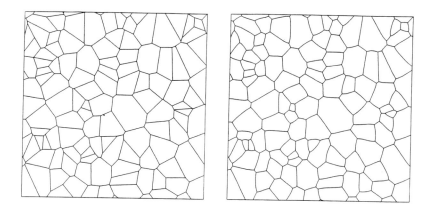

(A) Initial structure   (B) At dimensionless time τ = 0.05

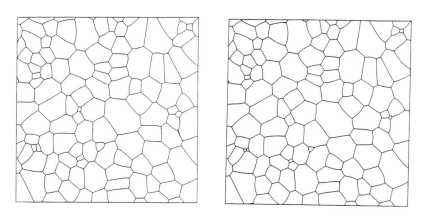

(C) At dimensionless time τ = 0.1   (D) At dimensionless time τ = 0.15

Figure 4   Evolution of the site saturation structure as grain boundary
            migration takes place.   The field of view is 100 times the
            average grain area.

SUMMARY AND CONCLUSIONS

The structures resulting from continuous nucleation at a constant rate were first described by Johnson and Mehl [3] and Evans [4]. We have reported more complete characterizations on larger numbers of grains. These results are similar to, though not identical to, results for planar sections through three dimensional Johnson-Mehl models reported by Mahin et al. [5]. Results reviewed here for the cases of decreasing nucleation rates and nucleation exclusion zones represent new analyses for variations in continuous nucleation conditions. The limit of rapidly decreasing nucleation rate is the site saturation case which results in the well characterized Voronoi polygon structure for random points. All of the nucleation conditions lead to microstructures with topological and geometrical characteristics which are readily distinguishable.

It is known that in real thin films, grain boundary motion may occur even during low temperature deposition (e.g. as low as 20% of the melting temperature in Au films deposited on SiO2) [14]. Grain boundary motion will, of course, have a profound affect on the geometry and topology of microstructures. We have developed a computer program which models capillarity driven grain boundary motion in computer generated microstructures. These models will be compared to real thin film microstructures. Surface-energy-driven secondary or abnormal grain growth can also affect thin film microstructures [15] [16]. The programs discussed here will be modified so as to allow modelling of this process as well.

ACKNOWLEDGEMENTS

The authors wish to thank Junho Whang of Dartmouth College for programming assistance, and Joyce Palmer of M.I.T. for discussions of her experimental work. C.V.T. acknowledges the support of the Joint Services Electronics Program, contract number DAAG-29-83-K003.

REFERENCES

1. H.J. Frost and C.V. Thompson (Symposium on Computer Simulation of Microstructural Evolution, TMS-AIME Fall Meeting, Toronto, Can., 1985).
2. H.J. Frost and C.V. Thompson, submitted to Acta Met., (1985).
3. W.A. Johnson and R.F. Mehl, Trans. AIME 135, (1939) 416-458.
4. U.R. Evans, Trans. Faraday Society 41, (1945) 365-374.
5. K.W. Mahin, K. Hanson, and J.W. Morris, Jr., Acta Met., 28 (1980), 443-453.
6. A. Getis and B. Boots, Models of Spatial Processes, (Cambridge University Press, 1979).
7. J.L. Meijering, Philips Res. Rep. 8, (1953) 270-290.
8. E.N. Gilbert, Annals of Mathematical Statistics, 33, (1962) 958-972.
9. D. Weaire and J.P. Kermode, Phil. Mag. B, 43, 245-259 (1983); Phil. Mag. B, 47,(1983) L29-L31.
10. D. Weaire and N. Rivier, Contemporary Physics, 25, (1984) 59-99.
11. M.P. Anderson, D.J. Srolovitz, G.S. Grest, and P.S. Sahni, Acta Met., 32, (1984) 783-791.
12. D.J. Srolovitz, M.P. Anderson, P.S. Sahni, and G.S. Grest, Acta Met. 32, (1984) 793-802.
13. I.K. Crain, Computers and Geosciences, 4, (1978) 131-141.
14. C.C. Wong, H.I. Smith, and C.V. Thompson, (Presented at the Spring Meeting of the Materials Research Society, San Francisco, 1985).
15. C.V. Thompson, J. Appl. Phys., 58, (1985) 763-772.
16. C.V. Thompson and H.I. Smith, Appl. Phys. Letters, 44, (1984) 603-605.

COOPERATIVE PREMELTING EFFECTS ON A (110) FCC SURFACE :
A MOLECULAR DYNAMICS STUDY

V. ROSATO, V. PONTIKIS[*] AND G. CICCOTTI[**]
[*] Centre d'Etudes Nucléaires de Saclay, Section de Recherches de Métallurgie
Physique, 91191 Gif sur Yvette, Cedex, France
[**] Dipartimento di Fisica, Universita degli Studi "La Sapienza", Ple Aldo Moro
5, 00185 Roma (Italy)

ABSTRACT

The thermodynamical and structural behavior of a (110) face of a (12-6)
Lennard-Jones fcc solid has been investigated by Molecular Dynamics computer
simulation on the solid-gas coexistence line up to a temperature $T = 0.94\ T_M$
(T_M : melting point). We have found evidence for cooperative defect production on free surfaces which leads to a structural transition above $T \approx 0.7\ T_M$.
This transition is studied using as an order parameter the excess energy for
surface layers due to missing bonds parallel to the surface with respect to
the bulk. Furthermore we report the values of the mean square displacement
for surface and bulk atoms as a function of temperature. Despite their high
values at the surface, surface layers are not molten but only highly disordered above the transition temperature.

INTRODUCTION

The absence of any superheating effect on most solids has suggested that
melting is a surface initiated process and, therefore, that the intrinsic
stability limit of solid surfaces is reached at temperatures $T < T_M$ [1].
There has been a tendency to identify the onset of surface instability with
the occurence of a local solid-liquid transition on surfaces before melting.
Such a point of view is justified by the possible validity of

$$\Gamma_{sv} > \Gamma_{sl} + \Gamma_{lv} \tag{1}$$

where $\Gamma_{\alpha\beta}$ are free energies of the solid-vapor (sv), solid-liquid (sl),
liquid-vapor (lv) interfaces for $T < T_M$ [2]. Several authors suggested the
possibility of a local melting on surfaces, at $T < T_M$ on considering the surface instability limit (i.e. its melting) as temperature leads to a breakdown of the self-consistent mean square displacement conditions [3]. Similar
conclusions were reached via Molecular Dynamics (MD) simulations [4] which
indicated successive discontinuities in the potential energy of surface layers
at certain temperatures $T < T_M$. Also the spontaneous temperature-induced production of surface dislocations was proposed as a possible mechanism for surface melting [5]. Although several experimental data are compatible with the
hypothesis of a surface solid-liquid transition before melting [6, 7], an alternative suggestion has been recently proposed by Rosato et al. [8, 9] according to which the onset of a surface instability is related to a local structural transition driven by cooperative point defect creation. Such cooperative phenomena on surfaces has been extensively studied by Burton et al. [10].
The MD results by Rosato et al. [8, 9] showed that, despite the high disorder
on a (110) face near the melting point, the surface retains its crystallinity
up to the higher temperature investigated ($T = 0.94\ T_M$). In this paper we
study by MD the bulk and surface atoms mean square displacements as a function of temperature. The values obtained at the surface are much greater than
those for the bulk at the melting point; this without any evidence of surface
melting. The surface structural transition is studied by using as an order

parameter the excess energy of surface layers, due to the increasing number of free bonds at the surface associated with the abundant point defect production. The calculations were carried out on an fcc (12-6) Lennard-Jones (σ = 3.405 Å, ε/K_B = 119.8 K for argon) system on the solid-gas coexistence line. The studied system is bounded by two (110) free surfaces in the x-direction, as periodic boundary conditions are applied along the y and z directions. The details of the calculations have been reported elsewhere [8].

RESULTS

1) Mean square displacement (MSD)

The quantity which is most directly related to the surface vibrations, and therefore to the stability condition used in lattice dynamics calculations, is the mean square amplitude of vibrations which is defined as

$$<u^2> = \frac{1}{3} \sum_{\alpha=x,y,z} <u_\alpha^2> = \frac{1}{3N} \sum_{\alpha=x,y,z} \sum_{i=1}^{N} \lim_{T\to\infty} \frac{1}{T} \int_o^T [u_\alpha^{(i)}(t) - <u_\alpha^{(i)}>]^2 \, dt \quad (2)$$

where N is the number of particles of the considered model, u_α the atomic displacement in the α (α = x,y,z) direction and brackets indicate a time average.

Table I and figure 1 summarize the results obtained for $<u_\alpha^2>$.

N	T^*	$<u^2>_{x\,b}$	$<u^2>_{y\,b}$	$<u^2>_{z\,b}$	$<u^2>_b$	$<u^2>_{x\,s}$	$<u^2>_{y\,s}$	$<u^2>_{z\,s}$	$<u^2>_s$
2688	0.358	2.44	1.97	2.42	2.13	6.21	4.42	7.50	6.04
672	0.465	3.92	3.21	3.57	3.56	11.25	11.49	16.07	13.35
2688	0.556	4.34	5.10	4.28	4.74	18.96	12.59	16.34	15.96
672	0.583	6.29	5.15	6.05	5.83	23.70	24.81	27.49	25.33
2688	0.614	5.88	4.80	6.07	5.58	34.70	30.00	34.17	32.96
2688	0.643	6.45	6.48	6.57	6.53	34.17	46.50	42.09	40.92
523(a)	0.334	2.08	1.84	2.01	1.97	7.37	4.02	10.38	7.26
864(b)	0.320				1.85				
864(b)	0.504				3.52				
864(b)	0.636				4.99				
864(b)	0.647				5.07				

TABLE I : MSD for surface and bulk atoms (the labels s and b stand for bulk and surface) for the crystallographic directions denoted as x, y and z (<110>, <1$\bar{1}$0> and <001> respectively). $<u^2>_a$ a = s,b represent the average. N is the number of particles of the system. References indicate relevant results obtained from other authors. Temperature and MSD are given in reduced units [8]. (a) ref. [13], (b) ref. [14].

For bulk regions, they show good agreement with the results obtained by Allen et al. [11-13] in the low temperature range (T < 0.5 T_M) and with those by Hansen and Klein [14] in the high temperature limit for a system without free surfaces. The quantitative differences with respect to [14] (see Table I) can be explained by the fact that the value of MSD increases by increasing

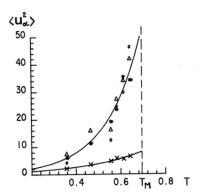

FIGURE 1 : Temperature dependence of surface (• , o , △) and bulk (X) MSD (x// <110> (•), y// <1$\bar{1}$0> (o), z// <001>(△); bulk data are averaged over the three directions). Temperature and MSD values are given in reduced units (ε/k_B for temperatures and $10^{-3} \sigma^2$ for MSD). Full lines are a guide for the eye.

the number the total number of particles N [14]. As expected, our results show an isotropic behavior of the MSD along the <110>, <1$\bar{1}$0> and <001> directions (Table I). Moreover, in the high temperature range, a marked departure from the linear behavior in temperature, which holds at T < 0.5 T_M has been found. This effect is due to the increasing importance of anharmonic contributions as temperature increases. The values of MSD on the surface, at each temperature, are bigger than those in the bulk. It must be noted that no significant anisotropy of MSD in the direction parallel to the surface (namely the <1$\bar{1}$0> and <001>) has been found, despite the anisotropy of the structure of the (110) face. To explain the same behavior experimentally detected on the (110) face of silver, it has been argued that changes in the force constants induced by density variations would tend to cancel out any apparent asymmetric effects [15]. As Allen et al. [11-13] pointed out, a decrease in density at the surface contributes to increase its MSD. In the high temperature range the change of surface local density with respect to the bulk one is of the order of 30% [8] as resulting from an high point defect concentration. Therefore the superposition of anharmonic effects and density decrease drives the surface MSD to values much bigger than those which can be predicted, on the basis of Lindemann's rule, indicating the stability limit of lattice vibrations. Rosato et al. [8] have shown that, even at high temperature close to the melting point, the surface retains its solid like character. This result suggests that Lindemann's criterion cannot be used as a local rule for the surface layers stability but that it must be rather considered as a free energy rule, for the whole system stability as pointed out in [16].

2) Order parameter

Previous work [8, 9] has shown that, on the fcc (110) face and for temperatures T > 0.7 T_M, the Arrhenius behavior of the surface concentration of vacancies breaks down so that defect production becomes nearly athermal in that temperature range. This can be interpreted as follows : by increasing

the temperature, the vacancy formation energy decreases because the probability of creating a defect in the vicinity of a pre-existing vacancy increases. This cooperative effect has been invoked by Burton et al. in studying
the high temperature structure of surfaces [10]. These authors, in their
nearest neighbours interaction model, introduced as an order parameter, a
quantity S which represents the number of free bonds parallel to the surface
resulting from defect creation.

In the present work we define an equivalent order parameter which
accounts for the continuous interatomic forces model we use. This order parameter is defined as follows : the total potential energy per particle of a N
point particles system can be expressed as

$$U = \frac{1}{2} \sum_{\lambda} U_{\lambda} \tag{3}$$

with

$$U_{\lambda} = \sum_{\substack{i \varepsilon \lambda \\ j=1}}^{N} V(r_{ij}) \tag{4}$$

where U_{λ} is the potential energy per particle of layer λ parallel to the
surface, $V(r_{ij})$ the value of the potential for particles i and j at a distance r_{ij} and $i\varepsilon\lambda$ means that particle i belongs to the layer λ.
U_{λ} can be expressed as the sum of two contributions

$$U_{\lambda} = \sum_{\substack{i \varepsilon \lambda \\ j \varepsilon \lambda}} V(r_{ij}) + \sum_{\substack{i \varepsilon \lambda \\ j \notin \lambda}} V(r_{ij}) = U_{\lambda}^S + U_{\lambda}^D \tag{5}$$

The first contribution to U_{λ}, U_{λ}^S, accounts for interactions within the considered layer and therefore will be affected mainly by the presence of defects
in the layer. The second term, includes in addition the effects of intralayer
relaxation resulting from the presence of free surfaces and thermal dilata-
tion. Our order parameter is then defined as

$$S_{\lambda} = \frac{U_B^S - U_{\lambda}^S}{U_B^S} \tag{6}$$

where U_{λ}^S is defined as indicated above and U_B^S represents the value of U_{λ}^S for
bulk atomic layers. At T = 0 K, S_{λ} = 0 and, for T ≠ 0 K S_{λ} is the relative
surface layers excess energy with respect to the bulk. Figure 2 shows the
variation of S_{λ} obtained for the system we studied. One can see that S increases rapidly above T > 0.7 T_M due to the increasing concentration of
point defects on the surface [9]. The slope dS/dT defines the surface excess
heat capacity whose rapid increase indicates the onset of a surface structural transition. However no any detailed indication can be reached on this
transition due to the small size of the system.

CONCLUSIONS

Present and previous results suggest the presence of cooperative
surface effects, on the atomic scale, in the high temperature range
(T > 0.7 T_M). In the particular case of the argon fcc (110) face, we have
found no evidence of solid-liquid transition on the surface below the bulk
melting point. The crystallinity of the surface at high temperature and the
saturation of the defect concentration in the first atomic layer of the surface suggest the onset of a surface structural transition due to the cooperative effects.

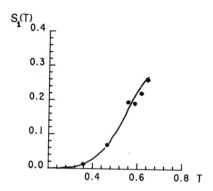

FIGURE 2 : Temperature dependence of the order parameter S_1. Temperature is given in reduced units.

Moreover we suggest that the high values of the surface MSD are due to the decrease of surface density by defect generation and to the strong anharmonicity. The high values of MSD cannot be interpreted as an indication of surface premelting.

ACKNOWLEDGEMENTS

We kindly acknowledge M. Guillopé for useful suggestions and M. Gillan for critical discussion and careful reading of the manuscript. Financial support from Euratom through the fellowship No JD/hp XII/355/82-IT to VR is greatfully acknowledged.

REFERENCES

[1] L.L. Boyer, Phase Transitions 5 (1985) 1
[2] L.D. Landau and E.M. Lifshitz, Course of Statistical Mechanics, ed. MIR Moscow (1965)
[3] L. Pietronero and E. Tosatti, Solid State Commun. 32 (1979) 255
     C.S. Jayanthi, E. Tosatti, L. Pietronero, Phys. Rev. B 31 (1985) 3456
[4] J.O. Broughton and L.V. Woodcock, J. Phys. C 11 (1979) 2743
[5] R.M. Cotterill, E.J. Jensen, W.D. Kristensen, in Anharmonic Lattices, Structural Transitions and Melting, edited by T. Rista (NATO Advanced Study Institutes Series, Noordhoff Leiden, 1974), p. 405
[6] G.H. Rhead, Surface Sci. 15 (1969) 353; 22 (1970) 223
[7] N.A. Gjostein, in Surfaces and Interfaces : I Chemical and Physical Characteristics, Eds J.J. Burke, N.L. Reed and N.V. Weiss (Syracuse University Press, 1975) p. 279
[8] V. Rosato, G. Ciccotti, V. Pontikis, in press on Phys. Rev. B
[9] V. Pontikis and V. Rosato, 7th European Conference on Surface Science ECOSS-7, 1-4 April 1985, Aix-en-Provence (France), Surface Sci. 162 (1985) 150
[10] W.K. Burton, N. Cabrera, F.C. Frank, Phil. Trans. Roy. Soc. London 243 A (1951) 299

[11] R.E. Allen, F.W. de Wette, Phys. Rev. 179 (1969) 873
[12] R.E. Allen, F.W. de Wette, A. Rahman, Phys. Rev. 179 (1969) 887
[13] R.E. Allen, F.W. de Wette, Phys. Rev. 188 (1969) 1320
[14] J.P. Hansen, M. Klein, Phys. Rev. B 13 (1976) 878
[15] J.M. Morabito, R.F. Steiger, G.A. Samorjai, Phys. Rev. 179 (1969) 638
[16] L.K. Moleko and H.R. Glyde, Phys. Rev. B 30 (1985) 4215

HARTREE-FOCK CLUSTER COMPUTATIONS
FOR IONIC CRYSTALS

J.M. VAIL AND R. PANDEY
Department of Physics, University of Manitoba,
Winnipeg MB R3T 2N2, Canada.

ABSTRACT

The ICECAP code is applied to charged and uncharged color centers in alkali halides and alkaline-earth oxides, to test the usefulness of complete-cation pseudopotentials for reproducing the cluster boundary conditions. The physical model includes consistency up to electrostatic octupole order between the Hartree-Fock cluster and the surrounding infinite shell-model lattice. The total energy of the system is determined variationally, including distortion and polarization of the cluster and lattice, and LCAO-MO gaussian-localized cluster wave functions. Electronic states with the lattice unrelaxed are also analysed, yielding color-center optical transition energies. Furthermore, consistency between quantum (cluster) and classical (shell-model) descriptions of the perfect lattice is tested.

1. INTRODUCTION

Recently an automated user-friendly computer program has been developed [1] to perform unrestricted Hartree-Fock self-consistent-field cluster computations in ionic crystals for which the embedding lattice is described by the shell model. The program, ICECAP (ionic crystal with electronic cluster, automated program), allows for electrostatic consistency up to octupole order between the Hartree-Fock cluster and the embedding lattice, and for the cluster includes pseudopotential ion cores as an option. It estimates the total energy of the infinite crystal variationally in the static-lattice approximation by incorporating a modified version of the HADES program [2], yielding ionic positions and polarizations, and from its incorporated Hartree-Fock cluster program [3] also yields detailed electronic structure information. Presently it deals only with an infinite, perfect (though distorted and polarized) embedding lattice. Plans are underway, however, to include surfaces, twins, grain boundaries, and dislocations in the lattice, and to provide for finite lattices.

An important problem in embedded-cluster calculations has to do with quantum-mechanical cluster boundary conditions, because the classical shell-model lattice does not provide any Pauli exclusion of the cluster wave function. In the present version of ICECAP, this effect must be provided for by using suitably localized cluster basis functions, or by choosing the quantum-mechanical region such that it is surrounded by cations which bear complete-ion pseudopotentials, or a combination of these features. Work is in progress [4] to incorporate a fully-consistent orthogonalization procedure [5] which is expected to have practical as well as theoretical advantages.

Since both the ICECAP code, and the procedure upon which it is based [6] are relatively new and not extensively tested, we have undertaken a systematic set of calculations to test the operation of the program and its options, and the physical accuracy of its method. We have begun to assess the physical accuracy in three aspects which are, of course, to some extent related. First, there is the question whether the Hartree-Fock description

248

of ions is compatible with the shell-model description. Second, there is
the question of how well complete-cation pseudopotentials represent the
ion-size effect as seen by electronic wave functions, and third, there is
the question of how well the composite model produces an accurate
representation of the detailed electronic properties. For the first
question, we have simply put a Hartree-Fock cation or anion in a
shell-model lattice. In the anion case, nearest-neighbor complete cation
pseudopotentials were also used, relating to the second question. The
principal applications of ICECAP will be to point-defect properties,
particularly those for which deviations from perfect-lattice electronic
structure are crucial. For the third question, we have therefore chosen to
examine color center properties, which involve not only perturbations of
ionic electronic structures, but also electronic states associated with
vacancies. These color center states may be considerably more diffuse than
ionic states, and, through their optical transitions provide a sensitive
test of the accuracy with which distortion and polarization of the
surrounding lattice are being treated. The optical transitions involve, in
their final states, unrelaxed lattice configurations, for which ICECAP
makes provision. In our color-center analysis, we have included
calculations with and without nearest-neighbor complete cation
pseudopotentials.

The results to be reported in this work relate to cluster calculations
for an oxide and a fluoride, namely MgO and NaF, and include perfect
lattice cations and anions, as well as one-electron F centers, considering
both optical absorption and emission for the latter. All of the ions are
therefore isoelectronic, and both crystals have the rocksalt structure, but
their different valencies mean that the F^+ center in MgO is charged, while
the F center in NaF is not, and therefore the optical transitions take
place in considerably different environments.

ICECAP was developed to allow convenient calculations in a
standardized model. Such calculations will only be easy if reliable sets
of model parameters are available. We have therefore used, as much as
possible, standard sets of data for our shell models, pseudopotentials, and
Hartree-Fock basis sets. Accordingly, our results contribute to an
assessment of the applicability of these data to perfect and defect lattice
problems.

In sec. 2, some specific details are given of our model elements and
parameters. In sec. 3, results of our calculations are given, principally
in terms of lattice distortion, electronic localization, and optical
transition energies in comparison with experiment. In sec. 4, conclusions
from these preliminary analyses are discussed.

2.   MODEL

The shell model, used for our embedding lattice, is a classical model
based on point-charge cores coupled harmonically to point-charge shells, to
express ionic polarizability, and with inter-ionic Coulomb and near-
neighbor short-range forces. For MgO, shell model parameters were taken
from Sangster and Stoneham [7], with Mg^{2+} ions unpolarizable and involving
only nearest-neighbor short-range interaction, but with O^{2-} ions
polarizable and including second-neighbor short-range interaction as well.
For NaF, the parameters were taken from table 2 of Catlow et al [8], with
both ion species polarizable and including up to second-neighbor
short-range interactions. In ref. 8, the ionic charge is given as 0.981e,
but we found that using 1.0e gave at least as good agreement between the
model and experiment for bulk lattice properties, so we used the latter
value.

For some of our calculations, nearest-neighbor cations include core pseudopotentials, which are designed to replace the effect that the many-electron density near the center of an atom has upon the more distant electronic density, by a relatively simple effective potential. These pseudopotentials are of two kinds, which we refer to as TOP, for Topiol [9], and BHS [10]. The TOP pseudopotentials are coreless Hartree-Fock effective potentials, and are adapted for use with gaussian basis sets, such as we use. The BHS pseudopotentials are of the norm-conserving localized-density-functional type, also adapted to gaussian bases.

The unrestricted Hartree-Fock self-consistent-field approximation, and its implementation for clusters in solids, is well described in ref. 5. In the present work we use the smallest optimal minimal basis sets of Huzinaga et al [11], except that for O^{2-} in MgO, additional p-type functions are added to the O basis set. For the vacancy-centered excess electron of the F center, s and p type bases are used for ground and excited states respectively, in both cases with the same range of values of the gaussian exponent α in exp $(-\alpha r^2)$. In Table I, the actual values of α are given, along with the corresponding ranges of the electronic densities, defined by $(2\alpha)^{-\frac{1}{2}}$. The F-center basis sets are chosen so that the ranges extend well beyond a nearest-neighbor distance, to allow diffuseness to occur in the variational solution.

ICECAP is written in such a way that data from references of the type given above can be easily transcribed from the original sources to an input file, with the exception that BHS pseudopotential coefficients first need to be transformed (Ref. 10, eq. (2.25)) using a subroutine, CONVERT [12].

3. RESULTS

In Table II, results are given from calculations in which a Hartree-Fock cation or anion is placed on a perfect lattice site, and the surrounding lattice is allowed to relax around it. The ion may stand alone in a shell-model lattice, denoted (SM), or in the case of an anion may have

Table I. Gaussian exponents α and ranges $(2\alpha)^{-\frac{1}{2}}$ used in basis functions of one-electron F centers in MgO and NaF. α unit, $(Bohr)^{-2}$; $(2\alpha)^{-\frac{1}{2}}$ unit, lattice nearest-neighbor distance.

MgO		NaF	
α	$(2\alpha)^{-\frac{1}{2}}$	α	$(2\alpha)^{-\frac{1}{2}}$
1.0	0.18	2.675	0.1
0.25	0.36	0.1063	0.5
0.06	0.73	0.02657	1.0
0.016	1.4	0.01181	1.5
0.004	2.8	0.006642	2.0
		0.002952	3.0
		0.001661	4.0

Table II. Calculated nearest-neighbor distances with given Hartree-Fock host ions. Units: latice nearest-neighbor distance.

	MgO		NaF	
	Mg^{2+}	O^{2-}	Na^+	F^-
SM	1.00	1.03	1.00	1.00
TOP	–	1.10	–	1.16
BHS	–	0.98	–	1.12

pseudopotential cations (TOP) or (BHS) as nearest neighbors. If the short range interactions of the pseudopotential clusters were the same as in the shell model, there would be no relaxation. In the case of a single ion in the shell-model lattice, it is a question of compatibility between the ion's internal configuration and the Coulomb potential due to the external lattice.

In Tables III and IV, results are given from calculations for one-electron F centers, either placed alone in a shell-model lattice (SM), or with nearest-neighbor pseudopotential cations (TOP) or (BHS). For each electronic optical transition energy, denoted ΔE_a for absorption and ΔE_e for emission, two calculations are required: one for the initial state, in which the lattice is relaxed to equilibrium with the cluster; and one for the final state, in which the ionic positions and polarizations are

Table III. Calculated F^+-center properties in MgO. Energy unit, eV; length unit, lattice nearest-neighbor distance.

SM	G	UNRES	ΔE_a	RES	UNRG	ΔE_e
energy	18.78	23.73	4.95	22.82	19.55	3.27
$d(d_{x_1},d_z)/2$	1.04			(1.10,1.00)		
$\langle r^2 \rangle^{1/2}$	0.82	1.03		1.04	0.83	
TOP						
energy	25.96	26.60	0.64	25.85	(26.90)	(-1.05)
$d(d_{x_1},d_z)/2$	1.07			(1.10,1.00)		
$\langle r^2 \rangle^{1/2}$	1.17	2.53		2.49	2.83	
BHS						
energy	21.57	26.64	5.07	25.54	24.52	1.02
$d(d_{x_1},d_z)/2$	1.05			(1.10,1.00)		
$\langle r^2 \rangle^{1/2}$	0.68	2.36		2.43	0.70	

Table IV. Calculated F-center properties in NaF. Energy unit, eV; length unit, lattice nearest-neighbor distance.

SM	G	UNRES	ΔE_a	RES	UNRG	ΔE_e
energy	2.51	6.14	3.63	5.55	3.63	1.92
$d(d_{x_1},d_z)/2$	0.95			(1.09,1.02)		
$\langle r^2 \rangle^{1/2}$	0.77	1.00		1.35	0.96	
TOP						
energy	6.89	7.16	0.27	7.15	(7.27)	(-0.12)
$d(d_{x_1},d_z)/2$	1.11			(1.10,1.05)		
$\langle r^2 \rangle^{1/2}$	2.72	2.90		2.70	2.69	
BHS						
energy	4.19	7.23	3.04	6.88	6.74	0.14
$d(d_{x_1},d_z)/2$	1.04			(1.11,1.04)		
$\langle r^2 \rangle^{1/2}$	0.70	2.34		2.38	1.84	

held fixed in the initial state configuration. Thus we have four states: ground (G), unrelaxed excited (UNRES), relaxed excited (RES), and unrelaxed ground (UNRG). In addition to these four calculated energy levels, nearest-neighbor distances and F-center root mean-square electron distances are given. The nearest-neighbor distances are the same for both states of a transition. For absorption they are determined by the ground state (G), and through symmetry are equal (d) for all six nearest neighbors. For emission they are determined by the RES, assumed to involve a p-type state oriented along the z-axis. In this case, the four nearest neighbors in the x-y plane are displaced equally, radially to (d_x), and the two axial neighbors likewise, to (d_z). The root mean-square electronic distances $\langle r^2 \rangle^{\frac{1}{2}}$ are different for each state. The calculated transition energies may be compared with the experimental values: for the MgO F$^+$ center, 4.96 eV [13] and 3.13 eV [14] for absorption and emission respectively; and for the NaF F center, 3.73 eV [15] and 1.66 eV [16] correspondingly.

4. DISCUSSION

Referring to Table II, we see that the Hartree-Fock ions distort the adjacent shell-model lattice (SM) negligibly. This is because the Huzinaga basis sets are highly localized, looking nearly like a point charge to the lattice, and the lattice Coulomb potential is weak compared to that of the nucleus. Note that for MgO we have added p-type functions of ranges 0.55, 1.0, and 1.5 to the Huzinaga oxygen-atom basis set. When pseudopotential ion-size effects are introduced, a rather large distortion (> 0.1) results in both crystals with TOP (obtained in the case of MgO by forcing convergence by density-matrix mixing), while with BHS the competition between attractive Madelung and repulsive Pauli effects produces weak inward relaxation in divalent MgO, but strong outward relaxation in monovalent NaF. We conclude that the combination of basis set, pseudopotential, and shell-model parameters that we have chosen is not very compatible for representing a perfect lattice cluster.

Referring to Tables III and IV, we see that a single quantum-mechanical electron in a shell-model lattice (SM) reproduces experimental absorption energies very well, and emission energies roughly, producing only small distortion, associated with reasonably well-localized wave functions in all four states. These results hold for both the MgO F$^+$ center and the NaF F center.

However, when pseudopotential ion-size effects are introduced, the picture changes. With BHS in both MgO and NaF, the ground state remains well-localized, with slight (~ 0.05) distortion, but the UNRES wave function spills well beyond the nearest-neighbor distance, because of Pauli exclusion, and agreement with the experimental absorption energy is poorer. The RES wave function is similarly diffuse, and although the UNRG wave function is much less so (in fact very well localized in MgO), the calculated emission energies are very wrong. We note how quadrupole consistency associates an oblate nearest-neighbor configuration with the prolate p-type RES wave function oriented along the z axis, in all cases. However, with TOP, the calculated results are completely unphysical, with extremely small absorption energies, with the RES p-type state below the s-type "ground" state (G) in MgO, and with the UNRG s-type state above the p-type RES in both crystals. It remains to be seen what the effect will be when the diffuse UNRES and RES p-type wave functions obtained with TOP and BHS pseudopotential nearest-neighbor cations are orthogonalized to anions on the z-axis. It is tempting to speculate that the superior performance of BHS at this stage is due to its norm-conserving feature. It appears that the ion-size effect of TOP is greater than that of BHS, judging by wave function diffuseness.

252

This work demonstrates the utility of the ICECAP program for systematic investigations that involve numerous different configurations of defect or perfect lattice clusters. Here we have used it to investigate the appropriateness of specified pseudopotentials as model elements. To obtain stronger conclusions than those stated above, further work with larger clusters, more flexible basis sets, and possibly improved pseudopotential and shell-model parameter sets, will be required. We anticipate that better quantitative results will be obtainable when more systematic orthogonalization [4], [5] to the embedding lattice is implemented.

ACKNOWLEDGEMENTS

We are grateful to A.H. Harker (AERE Harwell), and to A.B. Kunz, P.B. Keegstra, and C. Woodward (Michigan Tech.) for helpful communications. The work was supported in part by NSERC Canada and by the University of Manitoba Research Committee of Senate.

REFERENCES

1. J.H. Harding, A.H. Harker, P.B. Keegstra, R. Pandey, J.M. Vail, and C. Woodward, Physica 131 B + C, 151 (1985).
2. M.J. Norgett, Harwell Report No. AERE-R.7650 (1974).
3. A.B. Kunz, Michigan Technological University, unpublished.
4. P.B. Keegstra and A.B. Kunz, unpublished.
5. A.B. Kunz and D.L. Klein, Phys. Rev. B 17, 4614 (1978).
6. J.M. Vail, A.H. Harker, J.H. Harding, and P. Saul, J. Phys. C: Solid State Physics, 17, 3401 (1984).
7. M.J. Sangster and A.M. Stoneham, Phil. Mag. B 43, 597 (1981).
8. C.R.A. Catlow, K.M. Diller, and M.J. Norgett, J. Phys. C: Solid State Physics, 10, 1395 (1977).
9. S. Topiol, J.W. Moskowitz, C.F. Melius, M.D. Newton, and J. Jafri, Courant Institute of Mathematical Sciences Report COO-3077-105, New York University (1976).
10. G.B. Bachelet, D.R. Hamann, and M. Schlüter, Phys. Rev. B 26, 4199 (1982).
11. Gaussian Basis Sets for Molecular Calculations, edited by S. Huzinaga (Elsevier, New York, 1984).
12. C. Woodward and P. Keegstra, unpublished.
13. B. Henderson and R.D. King, Phil. Mag. 13, 1149 (1966).
14. Y. Chen, J.L. Kolopus, and W.A. Sibley, Phys. Rev. 186, 865 (1969).
15. A.E. Hughes, D. Pooley, and W.A. Runciman, Harwell Report No. AERE-R.5604 (1967).
16. P. Podini and G. Spinolo, Solid State Commun. 4, 263 (1966).

THEORETICAL INVESTIGATION OF THE JAHN-TELLER VIBRONIC ^{1}E ⊗ ε STATE IN THE NEUTRAL VACANCY OF SILICON

J.C. MALVIDO*, P.V. MADHAVAN** AND J.L. WHITTEN**
* 375-4H South End Avenue, New York, NY 10280
** Department of Chemistry, State University of New York at Stony Brook, Stony Brook, NY 11794

ABSTRACT

A method for constructing Jahn-Teller coupled vibronic states is developed by projecting distorted electronic eigenstates onto undistorted states and, with the aid of irreducible tensor methods, the force constants can be identified with reduced matrix elements of the normal mode expansion of the vibrational Hamiltonian. A scheme is then realized for estimating the force constants as well as the total vibronic energy from the calculated energies of the distorted electronic states. An application of the method to the ^{1}E ⊗ ε vibronic state of the neutral silicon vacancy is presented as an illustration.

Effective Vibronic Hamiltonian

Three assumptions form the basis of this work:

1) Validity of the adiabatic separability of electronic and nuclear degress of freedom.

2) Nuclear distortions away from the symmetric configuration are small compared to the internuclear distances.

3) The vibronic coupling involves a single normal mode and is governed by a spinless, non-relativistic Hamiltonian.

The first assumption is essential for the interpretation of the Jahn-Teller theorem[1], i.e., to the concept of an electronic state, degenerate by reason of symmetry, as a distinguishable vibronic component inducing a vibrational distortion which in turn reduces the local nuclear symmetry and lowers the total energy. The second point is generally valid in a solid. The third assumption permits a concise vibronic model that focuses on spatially resolved degeneracies.

We first isolate the nuclear kinetic energy from the 'static' Hamiltonian H_e in the vibronic Hamiltonian:

$$H_v(\vec{r}_e, \vec{q}_\lambda) = T_N + H_e(\vec{r}_e, \vec{q}_\lambda) \tag{1}$$

where $\vec{r}_e$ = totality of electronic co-ordinates; $\vec{q}_\lambda$ = degenerate normal mode vector $(\ldots, q_{\lambda\lambda_i}, \ldots)$ transforming as the λ-irreducible representation (irrep.). Define the electronic state for a distorted nuclear configuration,

$$H_e(\vec{q}_\lambda)|\sigma(\vec{q}_\lambda)\rangle = \mathcal{E}_\sigma(\vec{q}_\lambda)|\sigma(\vec{q}_\lambda)\rangle \tag{2}$$

The dependence on $\vec{r}_e$ is understood and will be dropped. Moreover, define the undistorted state $|\mu, T\gamma\rangle$

$$H_e(\vec{q}_\lambda = 0)|\mu, T\gamma\rangle = \mathcal{E}_{\mu T}|\mu, T\gamma\rangle \tag{3}$$

$(T\gamma)$ is the label for a multiplet term transforming as the γ-row of the T-irrep. within the μth electronic shell, i.e., the shell configuration which dominates the many-electron structure of the multiplet.
Finally, define the shell-Hamiltonian

$$H_{\mu\mu'}(\vec{q}_\lambda) = \sum_{T\gamma} \sum_{T'\gamma'} \mathcal{E}^{\mu\mu'}_{T\gamma;T'\gamma'}(\vec{q}_\lambda)\ |\mu, T\gamma\rangle \times \langle\mu', T'\gamma'| \tag{4a}$$

where

$$\mathcal{E}^{\mu\mu'}_{T\gamma;T'\gamma'}(\vec{q}_\lambda) = \langle\mu, T\gamma|H_e(\vec{q}_\lambda)|\mu', T'\gamma'\rangle \tag{4b}$$

It follows that $\quad H_e = \sum_{\mu\mu'} H_{\mu\mu'} \quad$ and hence,

$$\mathcal{E}_\sigma(\vec{q}_\lambda) = \sum_{\mu T\gamma} \sum_{\mu'T'\gamma'} \mathcal{E}^{\mu\mu'}_{T\gamma;T'\gamma'}(\vec{q}_\lambda)\langle\sigma(\vec{q}_\lambda)|\mu, T\gamma\rangle$$
$$\times \langle\mu', T'\gamma'|\sigma(\vec{q}_\lambda)\rangle \tag{5}$$

The function $\mathcal{E}^{\mu\mu'}_{T\gamma;T'\gamma'}$ $(\vec{q}_\lambda)$ represents a projected potential surface that describes the coupling between the $(\mu, T\gamma)$ and $(\mu', T'\gamma')$ multiplets by a distortion along $\vec{q}_\lambda$. As the distortion $|\vec{q}_\lambda| \to 0$, the state $|\sigma(\vec{q}_\lambda)\rangle$ adiabatically approaches a unique undistorted state $|\mu, T\gamma\rangle$. For small distortions it then suffices to approximate $|\sigma(\vec{q}_\lambda)\rangle$ as an admixture of symmetry-compatible multiplets within the μth shell. Consequently,

$$\mathcal{E}_\sigma(\vec{q}_\lambda) \sim \sum_{T\gamma} \sum_{T'\gamma'} \mathcal{E}^{\mu}_{T\gamma;T'\gamma'}(\vec{q}_\lambda)\langle\sigma(\vec{q}_\lambda)|\mu, T\gamma\rangle$$
$$\times \langle\mu', T'\gamma'|\sigma(\vec{q}_\lambda)\rangle \tag{6}$$

Clearly, for increased distortions, neighbouring shells would contribute and correction terms from Eq. 5 would be needed. $H_{\mu\mu'} \equiv H_\mu$ is referred to as the effective vibronic Hamiltonian.

Harmonic Expansion

We expand $H_e(\vec{q}_\lambda)$ in Eq. 4b to second order to obtain

$$\mathcal{E}^{\mu}_{T\gamma;T'\gamma'}(\vec{q}_\lambda) = \mathcal{E}_{\mu T}\,\delta_{TT'}\,\delta_{\gamma\gamma'} + \sum_\lambda q_{\lambda\lambda}\,f^{\lambda\lambda}_{T\gamma;T'\gamma'}(\mu)$$

$$+\ \tfrac{1}{2}\sum_{\lambda\lambda'} q_{\lambda\lambda}q_{\lambda\lambda'}\,S^{\lambda\lambda\lambda'}_{T\gamma;T\gamma'}(\mu) \tag{7}$$

where the force constants are given by

$$f^{\Lambda\lambda}_{T\gamma;T'\gamma'}(\mu) = \langle \mu, T\gamma \,|\, \partial He/\partial q_{\Lambda\lambda} \,|\, \mu', T'\gamma' \rangle \tag{8a}$$

$$S^{\Lambda\lambda\lambda'}_{T\gamma;T'\gamma'}(\mu) = \langle \mu, T\gamma \,|\, \partial^2 He/\partial q_{\Lambda\lambda}\partial q_{\Lambda\lambda'} \,|\, \mu', T'\gamma' \rangle \tag{8b}$$

The geometric content can be isolated from the dynamic one by the application of the Wigner-Echart theorem[2]

$$f^{\Lambda\lambda}_{T\gamma;T'\gamma'}(\mu) = \bar{f}^{\Lambda}_{T;T'}(\mu)\, V\begin{pmatrix} T & \Lambda & T' \\ \gamma & \lambda & \gamma' \end{pmatrix} \tag{9a}$$

$$S^{\Lambda\lambda\lambda'}_{T\gamma;T'\gamma'}(\mu) = \sum_{\Omega\omega} n(\Omega)\, \bar{S}^{\Omega}_{T;T'}(\mu)\, V\begin{pmatrix}\Lambda & \Lambda & \Omega \\ \lambda & \lambda' & \omega\end{pmatrix} V\begin{pmatrix}\Omega & T & T' \\ \omega & \gamma & \gamma'\end{pmatrix} \tag{9b}$$

where $\Omega \in \Lambda \otimes \Lambda$ and is such that a permutation of λ and λ' does not change the sign in V (:::), a Wigner V-coefficient for the point group[2]; $n(\Omega)$ denotes the dimension of the Ω-irrep. Equations 6 and 7 form the basic model equations for relating the force constants of the potential surface to the computed energies $\mathcal{E}_{\sigma}(\vec{q}_{\Lambda})$.

Application to the E $\otimes$ ε State of the Neutral Silicon Vacancy

Removal of a neutral atom from an otherwise perfect silicon crystal creates a vacancy defect characterized by unsaturated, dangling bonds from each of the four atoms neighbouring the vacancy. These bonds are tetrahedrally directed into the vacancy center. A theoretical treatment of the electronic states arising from this localized configuration that is based on a 16 atom cluster model of the crystal is given in Ref. 3. Symmetrization of the dangling bonds gives orbitals transforming as the A_1 and T_2 irreps. of the Tetrahedral group (T_d). The electronic shell-configuration $a_1^2 t_2^2$ gives rise to multiplet terms ^1E, $^3T_1$, $^1A_1$ and $^1T_2$. By introducing configuration interaction (CI) coupling to multiplet terms of the same symmetries derived from $a_1 t_2^3$ and t_2^4 shell configurations, the following spectral ordering is obtained for the lower vacancy states: $^1E < ^3T_1 < ^5A_2 < ^1T_2 < ^1A_1$. A degenerate ^1E ground state implies coupling to a degenerate vibrational mode capable of splitting the degeneracy. Such a mode, also of E symmetry, is shown in Fig. 1. A distortion along these modes reduces the local symmetry (to D_{2d} and D_2, respectively), and, while resolving the ^1E degeneracy, admixes the ^1E and $^1A_1$ multiplets. Since the $^1A_1$ state is energetically well removed from the ^1E state (4.4eV.), it is ignored as a first approximation.

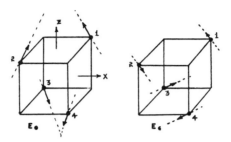

Fig. 1 Vibrational E Modes

The effective $E \otimes \epsilon$ vibronic Hamiltonian can be derived from Eqs. 4, 7 and 10 to give

$$H_{E \otimes \epsilon}(\vec{q}_e) = |E\theta\rangle\langle E\theta| \{ \mathcal{E}_0 - \tfrac{1}{2} \bar{f}^e_{E;E} q_{e\theta} + \tfrac{1}{4} \bar{S}^a_{E;E} (q^2_{e\theta} + q^2_{e\epsilon})$$

$$+ \tfrac{1}{4} \bar{S}^e_{E;E} (q^2_{e\theta} - q^2_{e\epsilon}) \}$$

$$+ |E\epsilon\rangle\langle E\epsilon| \{ \mathcal{E}_0 + \tfrac{1}{2} \bar{f}^e_{E;E} q_{e\theta} + \tfrac{1}{4} \bar{S}^a_{E;E} (q^2_{e\theta} + q^2_{e\epsilon})$$

$$- \tfrac{1}{4} \bar{S}^e_{E;E} (q^2_{e\theta} - q^2_{e\epsilon}) \}$$

$$+ (|E\theta\rangle\langle E\epsilon| + |E\epsilon\rangle\langle E\theta|) \{ \tfrac{1}{2} \bar{f}^e_{E;E} q_{e\epsilon} + \tfrac{1}{2} \bar{S}^e_{E;E} q_{e\theta} q_{e\epsilon} \} \quad (10)$$

where $|E\theta\rangle$ and $|E\epsilon\rangle$ are evaluated at $\vec{q}_e = 0$. Total electronic energies were computed for four distorted configurations: $q_{e\theta}$(au.) = ±.05, ±.10. Resulting energies and overlaps are tabulated in Table 1. These can be used to evaluate the matrix elements $\langle E\theta, \epsilon(\vec{q}_e) | H_{E \otimes \epsilon}(\vec{q}_e) | E\theta, \epsilon(\vec{q}_e)\rangle$ at the four values of $q_{e\theta}$. Including off-diagonal contributions, there are twelve equations and three parameters ($\mathcal{E}_0 = 0$ for scaling). The force constants were estimated using linear regression techniques and are tabulated below, in Table 1.

Table I. Relative Energies and Overlaps as a Function of Distortion. Asymmetry in Energies Between the Two Branches is Due to A_1 Mixing.

q_θ (au.)	-0.10	-0.05	0.00	0.05	0.10
$\mathcal{E}_{E\theta}(q_\theta)$ (eV.)	0.954	0.312	0.000	0.084	0.377
$\mathcal{E}_{E\epsilon}(q_\theta)$	0.584	0.139	0.000	0.330	0.880
$\langle E\theta(0)| E\theta(q_\theta)\rangle$	0.9944	0.9985	1.0000	0.9985	0.9934
$\langle E\epsilon(0)| E\epsilon(q_\theta)\rangle$	0.9995	0.9999	1.0000	0.9998	0.9990

$$\bar{f}^e_{E;E} = 0.155 \pm 0.027; \quad \bar{S}^a_{E;E} = 10.20 \pm 0.91; \quad \bar{S}^e_{E;E} = -0.112 \pm 0.60$$

Since $\bar{S}^e_{E:F}$ is insignificant, only $\bar{f}^e_{E:F}$ and $\bar{S}^a_{F:F}$ are needed to specify the Jahn-Teller potential surface, which, as a consequence, is degenerate at points of equal magnitude $|\vec{q}_e|$. The quadratic force constant is rather large; this is due to keeping the boundary atoms immobile as the vacancy atoms are distorted. Based on the above force constant estimates, we compute negligible Jahn-Teller stabilization energy and equilibrium displacement:

$$E_{JT} = 0.016eV.$$

$$q_0 = 0.008\mathring{A}.$$

REFERENCES

1. H. A. Jahn and E. Teller, Proc. Roy. Soc. A161, 220 (1937).
2. J. S. Griffith in "The Irreducible Tensor Method for Molecular Symmetry Groups" (Prentice-Hall, Englewood Cliffs, New Jersey, 1962).
3. J. C. Malvido and J. L. Whitten, Phys. Rev. B26, 4458 (1982).

CALCULATIONS OF RESISTIVITY AND SUPERCONDUCTING T_c IN TRANSITION METALS

P.B. ALLEN, T.P. BEAULAC , F.S. KHAN[+]
Dept. of Physics, SUNY, Stony Brook, NY 11794
     and
W.H. BUTLER, F.J. PINSKI[**], J.C. SWIHART[++]
Oak Ridge National Laboratory, Oak Ridge, TN 37830

ABSTRACT

     A survey is given of various electron-phonon effects which have been
calculated for the metals Nb, Mo, Ta, Pd, and Cu. These effects include
the mass enhancement λ, superconducting T_c, electrical and thermal resistivity, Hall coefficient, magnetoresistance, and the successfully tested
predictions of linewidths γ_Q of phonons. The calculations use local density
approximations (LDA) energy bands, experimental phonons, and the rigid
muffin tin (RMT) approximation. Mesh size noise is less than 1% and the
Bloch-Boltzmann integral equation has been solved to unprecedented accuracy.

1. INTRODUCTION

     Over the past 8 years, electron-phonon effects in transition metals
have been systematically calculated for the metals Nb,Mo,Ta,Pd, and Cu by
a Stony Brook-Oak Ridge consortium. A complete bibliography of this work is
given as table IV at the end of this report.
     The conductivity σ is the current per unit applied field E, or in terms
of the distribution $f(k)$ of occupied states $k \equiv (\underset{\sim}{k}n)$

$$\sigma = -(2e/\Omega) \sum_k v_{kx} \, df(k)/dE_x \qquad (1)$$

where the factor of 2 is for spin degeneracy. Lowest-order theory says
that $f(k)$ is the equilibrium Fermi distribution $f_0(k)$ displaced by an
amount $\delta_x = -eE_x \, \tau/\hbar$, yielding

$$\sigma^{(0)} = (2e^2\tau/\Omega) \sum_k v_{kx}^2 (-\partial f/\partial \varepsilon_k) = (n/m)_{eff} \, e^2\tau \qquad (2)$$

$$(n/m)_{eff} \equiv (2/\Omega) \sum_k v_{kx}^2 \, \delta(\varepsilon_k) \qquad (3)$$

where $(-\partial f/\partial \varepsilon_k)$ is approximated by $\delta(\varepsilon_k-\mu)$ and μ is set to zero. Thus σ
depends on two parameters, $(n/m)_{eff}$ which is easily calculated from band
theory, and $1/\tau$ which measures the scattering.
     In pure crystals, electron-phonon scattering dominates except at low T.
In the limit $T \gtrsim \theta_D$ there is a simple expression for the electron-phonon

[+]Permanent address: Dept. of Electrical Engineering, Ohio State Univ.,
   Columbus, Ohio 43210
[**]Permanent address: Dept. of Physics, Univ. of Cincinnati, Cincinnati,
   Ohio 45221
[++]Permanent address: Dept. of Physics, Indiana Univ., Bloomington,
   Indiana 47401.

scattering rate

$$\hbar/\tau = 2\pi\, \lambda_{tr}\, k_B T \qquad (4)$$

where λ_{tr} is an electron-phonon coupling constant very similar to the mass enhancement λ which determines the superconducting T_c:

$$\lambda_w = \frac{N(o) \displaystyle\sum_{kk'} w(k,k')|M(k,k')|^2\delta(\varepsilon_k)\delta(\varepsilon_{k'})/\hbar\omega_{k-k'}}{\displaystyle\sum_{kk'} w(k,k')\,\delta(\varepsilon_k)\delta(\varepsilon_{k'})} \qquad (5)$$

Here $N(o)$ is the (single spin) density of states per atom at the Fermi energy, $M(k,k')$ is the electron-phonon matrix element and $w(k,k')$ is a weight function equal to 1 for λ and $(v_{kx}-v_{k'x})^2$ for λ_{tr}. This equation follows rigorously when the Bloch-Boltzmann[1] equation for $f(k)$ is solved variationally[2] for σ in lowest order. These equations (2-5) reveal an intimate connection between T_c and σ. Both are described by integral equations (the Eliashberg[3] equations in the case of T_c) which have been rigorously justified to lowest order in the small parameter $N(o)\hbar\omega_D$ by many-body perturbation theory using procedures invented by Migdal[4]. Our aim has been to do numerical studies of σ and T_c simultaneously. By computing many different physical properties on the same basis we are able to test accuracy much more reliably than if we had focussed only on T_c.

We have also taken care to evaluate quantities like λ in such a way that numerical convergence errors (from finite mesh size) are less than 1%. This enables us to study the possibility[5] that there are systematic errors inherent in present day band theory which alter the value of quantities like λ. The sums in eq. (5) were performed using meshes of $\sim 5\times10^4$ k-points on the Fermi surface of each metal. An outline of the calculations is sketched in fig. 1. Energies ε_k and wavefunctions ψ_k were derived from KKR programs. Technically these were non-self-consistent Mattheiss-prescription energy bands[6], but for d-band elements these agree extremely well with experiment and very well with more sophisticated band theory. The phonon frequencies ω_Q came from Born-von Karman interpolations fitted to neutron scattering data.

The matrix element is formally

$$M(k,k') = \int dr dr'\; \psi_{k'}(r)^* \varepsilon^{-1}(r,r') \underset{\sim}{u}_Q\cdot\nabla V_0(r')\psi_k(r) \qquad (6)$$

but in practice the screening function $\varepsilon^{-1}(r,r')$ is too difficult to compute. Therefore we have used

$$M(k,k')_{RMT} = \int dr\; \psi_{k'}(r)^* \underset{\sim}{u}_Q\cdot\nabla V(r)\psi_k(r) \qquad (7)$$

where $V(r)$ is the total muffin tin potential around the atom at the origin, while $V_0(r)$ is the bare potential of this atom. This procedure, known as the rigid-muffin-tin (RMT) model[7], is expected to be very accurate in the core region, less accurate in the interstitial region, and totally fails to give the long range Friedel oscillations contained in (6). For d-band elements, the Fermi-surface wavefunctions $\psi_k(r)$ are fairly well localized near the core, so the model should be good. Copper, however, is a cross-over case, intermediate between d and s/p elements, and the procedure is questionable. To overcome this problem, we exploit the arbitrariness[8]

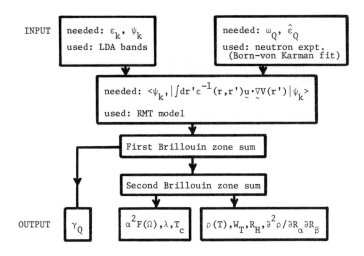

Fig. 1. Schematic outline of electron-phonon calculations

available in fixing the "muffin-tin zero." This arbitrariness is unimportant
in Nb, Mo, Ta, and Pd, but serious for Cu. We choose a fixed value for the
muffin-tin zero which gives a "forward scattering" matrix element M(k,k) in
accord with nearly free electron theory. This turns out to yield very
sensible answers for M(k,k'). One of the biggest virtues of the RMT model
is that the expression for M(k,k') turns out to be quite simple and depends
only on wave-function coefficients and scattering phase shifts $\delta_\ell(\epsilon_F)$[9] which
are fixed once the muffin-tin zero is chosen. There is no opportunity or
temptation for further adjustment.

2. PHONON LINEWIDTHS

An early triumph of this program was the successful prediction of
unanticipated structure in the phonon linewidth γ_{Qj} as a function of Q in
Nb and Pd (see table IV part B.) This linewidth describes the decay of a
phonon as it excites an electron from a state k just below ϵ_F to a state
k+Q just above ϵ_F. A complicated one-dimensional manifold of states k
contributes for fixed Q. As Q varies, γ_{Qj} can change quite rapidly through
a complicated interplay of Fermi surface geometry, wavefunction variation,
and phonon dispersion. Experimental observation of these effects is inhibited
by the low resolution and difficult background subtraction of inelastic
neutron scattering. Thus although experiment has qualitatively confimed our
calculations, detailed quantitative comparison is not possible in most cases.
Nevertheless, this success gave us considerable confidence in the correctness
of our algorithms and the validity of the model. Our predictions for Ta have
not yet been tested.

3. λ AND T_c

From γ_{Qj} it is straightforward to compute λ using the identity (table IV, A.1)

$$\lambda = (N\pi\hbar\, N(o))^{-1} \sum_{Qj} \gamma_{Qj}/\omega_{Qj}^2 \qquad (8)$$

Our calculations of λ and λ_{tr} are summarized in table 1. We have only small

Table I. Calculations of λ, λ_{tr}, and T_c

	present work		ref. 5	present	ref. 5	
	λ_{tr}	λ	λ	T_c^{calc}	T_c^{calc}	T_c^{expt}
Cu	0.116	0.111	-	-	-	-
Nb	1.07	1.12	1.3	12.1	17.4	9.2
Ta	0.57	0.88	0.9	7.0	8.0	4.5
Mo	-	0.40	0.4	0.8	0.8	0.91
Pd	0.46	0.41	0.5	0.3	1.4	<0.002

disagreements with the numerical values of Glötzel et al[5] also shown in table I. However, we have a major difference in interpretation. Glötzel et. al. state "...theory is incapable of producing reliable T_c's... The most probable reasons for this failure are [1] the rigid ion approximation ... or ...[2] conventional local density schemes..." Our view is that T_c's are adequately well accounted for, considering the sensitivity of T_c and the uncertainty in the Coulomb parameter μ^*. The computed T_c's of table I all assume $\mu^* = 0.13$, which is likely to be a significant underestimate in the case of Pd which has fairly long-lived spin fluctuations. Uncertainty in μ^* might account for the discrepancy between theory and experiment in Nb and Ta also. We find strong evidence that electron-phonon effects are well explained by [1] the RMT model, and [2] conventional local density schemes. We base this more on our transport calculations and γ_{Qj} predictions than on the sole criterion of T_c chosen in ref. 5.

4. ELECTRICAL RESISTIVITY

Few calculations have been made of $\rho(T)$ for transition metals which take the true band structure into account. Pioneering work was done by Yamashita and Asano,[10] who demonstrated feasibility. Their mesh of k and T points was coarse compared to ours, and permitted only semiquantitative comparison with experiment. Apart from this work and our own, we are aware of no quantitative work on the phonon-limited resistivity.

Table I shows an interesting regularity, namely that λ_{tr} and λ are usually quite similar. This was anticipated by Chakraborty et al[11] who proposed using eqs. (2-4) to estimate σ with λ_{tr} replaced by $\overline{\lambda}$ and derived from T_c experiments, and $(n/m)_{eff}$ derived from band theory. This procedure is now thoroughly vindicated, and it is a pity that band theorists seldom evaluate $(n/m)_{eff}$ which would be very useful in analyzing transport coefficients.

The calculation of $\rho(T)$ for $T < \theta_D$ as well as $T \gtrsim \theta_D$ requires a more elaborate scheme than given in eqs. (2-5). Rigorous procedures for solving the Bloch-Boltzmann equation are described in the papers listed in table IV parts A and E. Our results are summarized and compared with experiment[12] in fig. 2. Considering that no adjustable parameters have entered (apart from

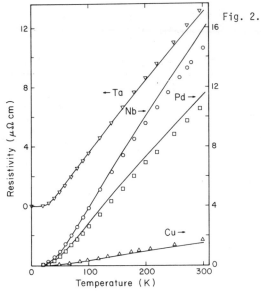

Fig. 2. Resistivity versus temperature. The solid lines for Nb,Pd, and Cu are calculations which include both ε-dependence and anisotropy; the curve for Ta is a lowest order variational calculation and thus somewhat too large for $T \lesssim 100K$. Data are from ref. 12.

the choice of muffin-tin zero in Cu) the spectacular agreement between theory and experiment can be taken as a refutation of the pessimism of ref. 5 and as confirmation of our view that band theory is fully capable of accounting for electron-phonon effects in transition elements.

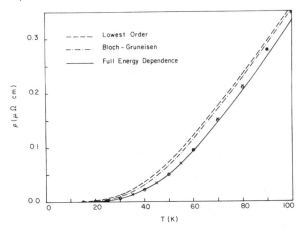

Fig. 3. Low T resistivity of Cu, showing the effect of ε-dependent corrections to the lowest-order solution. Also a Bloch-Gruneisen formula has been fitted to the theory in the regime $T \sim \theta_D$ and disagrees for $T < 100K$. Data are from refs. 13(0) and 14(x).

Figure 3 shows the low temperature part of $\rho(T)$ for Cu in more detail. This illustrates the role of corrections to the lowest order variational solution. In lowest order there is a relatively simple formula for $\rho(T)$ which consists of eqs. (2-4) with λ_{tr} replaced by $\lambda_{tr}(T)$ defined as

$$\lambda_{tr}(T) = 2 \int_0^\infty \frac{d\Omega}{\Omega} \, \alpha_{tr}^2 F(\Omega) \left(\frac{\hbar\Omega/2k_B T}{\sinh(\hbar\Omega/2k_B T)} \right)^2 \qquad (9)$$

where $\alpha_{tr}^2 F(\Omega)$ is a close analog of $\alpha^2 F(\Omega)$ used in T_c theory. The role of band theory is then to provide $(n/m)_{eff}$ and $\alpha_{tr}^2 F(\Omega)$. This approximation is remarkably accurate for $T \gtrsim \theta_D/2$, but at lower T interesting corrections enter which can be divided into two classes. The distribution function can be written as $f_0(k + \delta(\hat{k},\varepsilon))$ where δ is the displacement caused by the applied field. In lowest order, δ is a constant, but the exact solution exhibits both a dependence on the position on the Fermi surface, $\hat{k}$ (called the "anisotropy" effect) and a dependence on the elevation $\varepsilon-\mu$ above or below the Fermi energy (called the "energy dependence" effect.) It turns out to require no additional input from band theory to handle the second effect as accurately as desired (neglecting corrections of order $N(o)k_B T$.) However, the anisotropy effect requires additional tedious band theory input (e.g. a series of functions $\alpha_{JJ'}^2 F(\Omega)$.) Both are important as illustrated

Table II. Resistivity of Cu calculated at three temperatures at various levels of approximation.

	20K	40K	220K
Lowest order	0.00163	0.0360	1.067
energy dependence	0.00097	0.0210	1.061
anisotropy	0.00143	0.0352	1.063
complete solution	0.00083	0.0200	1.057

in fig.3 and table II. The total correction is a factor of 2 at 20K, but only 1% at 220K, for Cu. Our "complete" solution is actually a truncated series representation of the anisotropy effect which captures > 80% of the anisotropy correction, at a computational expense of a factor of $\sim$ 6 over the lowest order solution. It would not be easy to improve this calculation further.

5. DEVIATIONS FROM MATTHIESSEN'S RULE (DMR)

Matthiessen's rule states that $\rho_{tot}(T) = \rho_0 + \rho_{pure}(T)$ where ρ_0 is the residual resistivity which is sample dependent (usually dominated by impurities) and $\rho_{pure}(T)$, the experimental resistivity of the pure material, gives the temperature dependence. The lowest-order variational theory obeys Matthiessen's rule, but the higher order corrections do not, because the displacement $\delta(\hat{k},\varepsilon)$ is determined by impurity scattering if $\rho_{pure} < \rho_0$ and by phonon scattering if $\rho_{pure} > \rho_0$. These two scattering mechanisms have very different effects on δ (for example, when impurities dominate there is no ε-dependence) resulting in a well known source of DMR. Our calculations for Cu are shown in fig. 4. The simplest possible model for impurity scattering was used, namely a (k,k')-independent matrix element, which is a good starting point but cannot explain any differences between

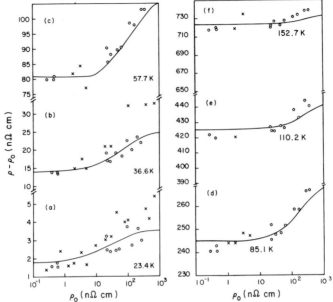

Fig. 4. $\rho(T)-\rho_0$ plotted versus ρ_0. Mattheissen's rule says the data should be on a horizontal line. In cases (a,b,c) the theory has not been adjusted. In cases (d,e,f) the theory has been shifted upwards by a constant correction of (4%,7%,8%). This shift compensates for a slight discrepancy between $\rho_{pure}(T)$ and experiment seen in fig. 2 at $T \gtrsim 80K$. Data are from refs. 15(0) and 16(x).

species of impurity.[19] Again, only trivial adjustments were made in the theory and the agreement with experiment is quite spectacular. This strongly suggests that the finer details of Bloch-Boltzmann theory are indeed meaningful, that band theory does justice to transport at a detailed level, and that our computational algorithms are adequate.

6. THERMAL CONDUCTIVITY

Although we have made calculations for Nb,Ta, and Pd as well, only results for Cu are discussed here. Figure 5 shows that once again close agreement with experiment is found. Corrections to the lowest order approximation are particularly important in $\kappa(T)$ at low T, because the ε-dependence of δ plays a special role in heat conduction. The Wiedemann-Franz law is accurately obeyed for $T \gtrsim \theta_D$ but not at low T.

7. MAGNETOTRANSPORT

The regime of weak applied magnetic fields is defined by $\omega_c \tau \ll 1$. Here the current can be written

$$j_\alpha = \sigma^{(o)}_{\alpha\beta}E_\beta + \sigma^{(1)}_{\alpha\beta\gamma}E_\beta H_\gamma + \sigma^{(2)}_{\alpha\beta\gamma\delta} E_\beta H_\gamma H_\delta + \ldots \tag{10}$$

In a cubic crystal the term linear in H , a third rank tensor, is determined by a single number, the Hall coefficient (R_H), while the quadratic part, a

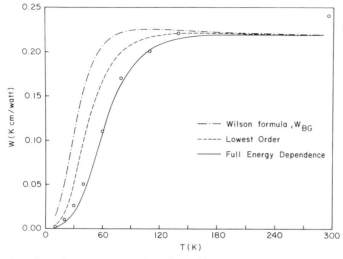

Fig. 5. Thermal resistance of Cu. Data are from ref. 13.

fourth rank tensor, contains three distinct numbers. We have calculated all four of these numbers as a function of temperature for Cu. The results for the Hall coefficient are shown in fig. 6. The common belief that

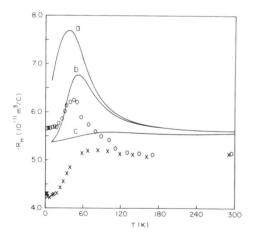

Fig. 6.

Hall coefficient in Cu. Curves (a,b,c) are calculated with resistance ratios (RRR) of (38500, 70,4). The data, from ref. 17, are for Ni impurities with RRR=110 (0) and Au impurities with RRR=27 (X).

$R_H = 1/nq$ and measures the carrier concentration and sign is actually correct only for elastic scattering around an ellipsoidal energy surface, a good approximation in many semiconductors, but meaningless for transition metals. The lowest order trial solution has a Fermi distribution which is shifted first by an amount $\delta = -eE\tau/h$ in the direction of the E field and then by an amount b $E \times H$ in the direction of the Lorentz force, where b must be determined from the Boltzmann equation. Unfortunately there is no variational principle for the Hall coefficient to guarantee that errors in R_H are higher order than errors in the distribution.

In relaxation time approximation the formula for the Hall coefficient is

$$R_H = (6e\hbar)^{-1} \sum_k v_k^3 \kappa_k \tau_k^2 (-\partial f/\partial \varepsilon_k) / [\sum_k v_k^2 \tau_k (-\partial f/\partial \varepsilon_k)]^2 \qquad (10)$$

where κ_k is the Gaussian curvature of the Fermi surface at point k. If $\tau_k = \tau$ is independent of k, it cancels from eq. (10) so that R_H is a purely geometrical object, independent of T, measuring a complicated average of curvature times velocity to the third power. This is understandable because in regions of large velocity and curvature, the Lorentz force can give a large redirection of the current. The lowest order trial solution reproduces this geometric answer. As can be seen in fig. 6, there is a large T-dependent correction for T < 100K, but at higher T, both theory and experiment give a T-independent value of R_H. At lower T the absence of a variational principle makes R_H difficult to compute, and agreement with experiment is only semi-quantitative. As in the DMR calculations, no effort was made to account for the anisotropy of the impurity scattering. The large difference between Ni and Au impurities shows the need for a more sophisticated treatment.

The lowest order T-Independent theory does quite well for many transition metals at high T. This is illustrated in table III. The palladium

Table III. Hall coefficients calculated in lowest order and expressed as $Z_{eff} = -(\Omega_{cell}/R_H e)$

	Z_{eff} theory	Z_{eff} expt
Cu	1.39	1.42
Nb	-1.50	-1.28
Pd_I	0.53	1.23
Pd_{II}	0.69	

results are off by a factor of 2 however, and show a significant dependence on the band structure (illustrated by calculations which used two different Mattheiss-prescription potentials, from relativistic (II) and non-relativistic (I) atomic charge calculations.) The factor of 2 discrepancy has been analyzed in paper G2 of table IV, and arises from the very large curvature anisotropy on the central portion of the Pd Fermi surface, which makes the result peculiarly sensitive to scattering anisotropy omitted in lowest order.

Finally the magnetoresistance calculations for Cu in paper G3 of table IV merit a comment. All three coefficients agree nicely with the available low T single crystal data. An as yet unchecked prediction was made, namely that the "Seitz coefficients," defined such that T-dependence cancels in lowest order, should actually exhibit a fairly strong T-dependence in the temperature interval $20K \lesssim T \lesssim 100K$. We hope this prediction will be checked.

8. CONCLUSION

When this work was started 8 years ago, it was not completely obvious that the Bloch-Boltzmann theory would work in detail for transition metals. The lesson to be drawn from our calculations is that accurate band theory combined with the Bloch-Boltzmann theory is spectacularly successful at accounting for a wide range of effects. Although there are important un-resolved questions[18] in the distinction between quasiparticles and density functional energy bands, there is no evidence that LDA bands fail to describe adequately the quasiparticles in transition elements. There are difficulties in explaining T_c to better than $\sim$ 50% reliability, but it is hard to believe that the sole source of this difficulty is uncertainty in electron-phonon theory.

Table IV. Publications of the Stony Brook-Oak Ridge Consortium on Electron-Phonon effects in Transition Metals

A. Formal Results

1. P.B. Allen, Phys. Rev. B 6, 2577 (1972). "Neutron Spectroscopy of Superconductors". (The relation between phonon linewidths and $\alpha^2F(\Omega)$).

2. P.B. Allen, Phys. Rev. B 13, 1416 (1976). "Fermi Surface Harmonics: A General Method for Non-Spherical Problems. Application to Boltzmann and Eliashberg Equations".

3. W.H. Butler, J.J. Olson, J.S. Faulkner, and B.L. Gyorffy, Phys. Rev. B 14, 3828 (1976). "Electron-Phonon Interaction in Cubic Systems: Application to Nb".

4. W.H. Butler and P.B. Allen, in Superconductivity in d- and f-Band Metals, D.H. Douglass, ed. (Plenum, New York, 1976). "Gap Anisotropy and T_c Enhancement: General Theory and Calculations for Nb, Using Fermi Surface Harmonics".

5. P.B. Allen, Phys. Rev. B 17, 3725 (1978). "New Method for Solving Boltzmann's Equation for Electrons in Metals".

6. F.J. Pinski, Phys. Rev. B 21, 4380 (1980). "Solutions to the Boltzmann Equation for Electrons in a Metal: Energy Dependence".

7. P.B. Allen, Z. Phys. B 47, 45 (1982). "Transition Temperature and Gap Anisotropy of Pure and Impure Superconductors".

8. W.H. Butler, Can. J. Phys. 60, 735 (1982). "Ideal Transport in Transition Metals: the Rigid Muffin Tin Approximation and Beyond".

B. Phonon Linewidths

1. W.H. Butler, H.G. Smith, and N. Wakabayashi, Phys. Rev. Lett. 39, 1004 (1977). Electron-Phonon Contribution to the Phonon Linewidth in Nb: Theory and Experiment".

2. W.H. Butler, F.J. Pinski, and P.B. Allen, Phys. Rev. B 19, 3708 (1979). "Phonon Linewidths and Electron-Phonon Interaction in Nb".

3. F.J. Pinski and W.H. Butler, Phys. Rev. B 19, 6012 (1979). "Calculated Electron-Phonon Contributions to Phonon Linewidths and to the Electronic Mass Enhancement in Pd.

4. R. Youngblood, Y. Noda, and G. Shirane, Phys. Rev. B 19, 6016 (1979). "Neutron Scattering Study of Phonon Linewidths in Pd". See also C4.

C. Superconducting T_c, Mass Enhancement

1. W.H. Butler, Phys. Rev. B 15, 5267 (1977). "Electron-Phonon Coupling in the Transition Metals: Electronic Aspects". (Calculations of η.)

2. F.J. Pinski, P.B. Allen, and W.H. Butler, J. de Physique Colloque C6 (Suppl. au v. 39), 472 (1978). "Calculated Electron-Phonon Coupling and

T_c of Transition Metals: Mo and Pd."

3. F.S. Khan and P.B. Allen, Phys. Stat. Sol. (b) 100, 227 (1980).
"Orthonormal Functions for Fermi Surfaces: A Test of Convergence for
Nb and Cu". (Anisotropy of λ_k.)

4. A. Aℓ-Lehaibi, J.C. Swihart, W.H. Butler and F. J. Pinski, "Electron-
Phonon Interaction Effects in Ta", to be published.

See also papers C2, C3, and E2.

D. p-Wave Pairing

1. F.J. Pinski, P.B. Allen, and W.H. Butler, in Superconductivity in d- and
f-Band Metals, H. Suhl and M.B. Maple, editors (Academic Press, New York,
1980), p. 215. "Exotic Pairing in Metals: Formalism and Application to
Nb and Pd".

See also paper F1.

E. Electron Lifetimes

1. F.S. Khan, P.B. Allen, and W.H. Butler, Physica 108B, 893 (1981).
"Phonon-Limited Lifetimes of Electrons in Pd".

2. F.S. Khan, P.B. Allen, W.H. Butler, and F.J. Pinski, Phys. Rev. B 26,
1538 (1982). "Electron-Phonon Effects in Cu:I Electron Scattering Rate
and Mass Enhancement".

F. Electrical and Thermal Conductivity

1. F.J. Pinski, P.B. Allen, and W.H. Butler, Phys. Rev. Lett. 41, 431 (1978).
"Electron-Phonon Contribution to Electrical Resistivity and Superconducting "p-Wave" Transition Temperature of Pd".

2. F.J. Pinski, P.B. Allen, and W.H. Butler, Phys. Rev. B 23, 5080 (1981).
"Calculated Electrical and Thermal Resistivity of Nb and Pd".

3. T.P. Beaulac, P.B. Allen, and F.J. Pinski, Phys. Rev. B 26, 1549 (1982).
"Electron-Phonon Effects in Cu:II: Electrical and Thermal Resistivities
and Hall Coefficient".

4. F.J. Pinski, W.H. Butler, and P.B. Allen, in Thermal Conductivity 16,
D.C. Larsen, editor (Plenum, New York, 1983), p. 155. "Ideal Thermal
Conductivity of Pd and Nb".

See also C4.

G. Hall Effect and Magnetoresistance

1. T.P. Beaulac, F.J. Pinski, and P.B. Allen, Phys. Rev. B 23, 3617 (1981).
"Hall Coefficient in Pure Metals: Lowest Order Calculation for Nb and Cu".

2. T.P. Beaulac and P.B. Allen, J. Phys. F 13, 383 (1983). "Calculation of
the Hall Coefficient in Pure Pd".

3. T.P. Beaulac and P.B. Allen, Phys. Rev. B 28, 5439 (1983). "Calculated
Magnetoresistivity of Cu".

H. Ultrasonic Attenuation

1. F.S. Khan, A. Auerbach, and P.B. Allen, Solid State Commun. 54, 135 (1985).
"Sound Attenuation in Metals: Reformulation of Pippard Theory to Include

Microscopic Band Structure and Scattering".

I. Reviews or Surveys

1. W.H. Butler, in Superconductivity in d- and f-Band Metals, edited by
   H. Suhl and M.B. Maple (Academic Press, New York, 1980) pp 443-453.
   "Progress in Calculations of the Superconducting Properties of Transition
   Metals".

2. P.B. Allen, in Dynamical Properties of Solids, edited by G.K. Horton and
   A.A. Maradudin (North Holland, Amsterdam, 1980) V.3 pp 95-196. "Phonons & T_c"

3. W.H. Butler, in Physics of Transition Metals 1980 (Inst. Phys. Conf.
   Proc. 55, 1981) pp 505-514. "The Rigid Muffin Tin Approximation for
   the Electron-Phonon Interaction in Transition Metals".

4. P.B. Allen and B. Mitrović, in Solid State Physics, edited by
   H. Ehrenreich, F. Seitz, and D. Turnbull (Academic Press, New York,
   1982) V. 37, pp 1-92. "Theory of Superconducting T_c."

ACKNOWLEDGEMENTS

   Work at Stony Brook was supported by NSF grant no. DMR84-20308. Research
at Oak Ridge was sponsored in part by the Division of Materials Sciences,
U.S.D.O.E. under contract no. DE-AC05-840R21400 with Martin Marietta
Energy Systems, Inc. Research at Indiana was supported by NSF grant no.
DMR81-17013.

REFERENCES

1. F. Bloch, Z. Phys. 52, 555 (1928).

2. J.M. Ziman, Electrons and Phonons (Oxford Univ. Press, Oxford 1980).

3. G.M. Eliashberg, Zh. Eksp. Teor. Fiz. 38, 966 (1960) [Soviet Physics -
   JETP 11, 696 (1960)]; D.J. Scalapino, J.R. Schrieffer, and J.W. Wilkins,
   Phys. Rev. 148, 263 (1966).

4. A.B. Migdal, Zh. Eksp. Teor. Fiz. 34, 1438 (1958)[Soviet Physics - JETP
   7, 996 (1958)].

5. D. Glötzel, D. Rainer, and H.R. Schober, Z. Phys. B35, 317 (1979).

6. L.F. Mattheiss, J.H. Wood, and A.C. Switendick, in Methods in Computational Physics, edited by B.J. Alder et al., V.8, Energy Bands in Solids
   (Academic Press, N.Y. 1968, p.63). For Ta, the bands are self-consistent.

7. S.K. Sinha, Phys. Rev. 169, 477 (1968).

8. M.J.G. Lee and V. Heine, Phys. Rev. B5, 3839 (1971).

9. G.D. Gaspari and B.L. Gyorffy, Phys. Rev. Lett. 28, 801 (1972).

10. J. Yamashita and S. Asano, Prog. Theor. Phys. 51, 317 (1974).

11. B. Chakraborty, W.E. Pickett, and P.B. Allen, Phys. Rev. B14, 3227 (1976).

12. J. Bass, in Landolt-Börnstein Tables, New Series, Gp III v 15a Metals:
    Electronic and Transport Phenomena (Springer-Berlin 1982.)

13. G.K. White and S.B. Woods, Phil. Trans. Roy. Soc. London A251, 273 (1959).

14. R.A. Matula, J. Phys. Chem. Ref. Data $\underline{8}$, 1147 (1979).

15. J.S. Dugdale and Z.S. Basinski, Phys. Rev. $\underline{157}$, 552 (1967).

16. B. Lengeler, W. Schilling, and H. Wenzl, J. Low Temp. Phys. $\underline{2}$, 59 (1970).

17. J.S. Dugdale and L.D. Firth, J. Phys. C$\underline{2}$, 1272 (1969).

18. W.E. Pickett, Comments on Sol. State Phys. $\underline{12}$, 1 (1985); ibid, in press.

19. A CPA-based calculation of resistivity in AgPd alloys which includes anisotropy has been reported by W.H. Butler and G.M. Stocks, Phys. Rev. B$\underline{29}$, 4217 (1984).

MICROSCOPIC PHENOMENA OF MACROSCOPIC CONSEQUENCES:
INTERFACES, GLASSES, AND SMALL AGGREGATES

UZI LANDMAN,* R. N. BARNETT,* C. L. CLEVELAND,* W. D. LUEDTKE* AND M. W. RIBARSKY*
DAFNA SCHARF,** AND JOSHUA JORTNER**
*School of Physics, Georgia Institute of Technology, Atlanta, Ga. 30332 USA
**School of Chemistry, Tel Aviv University, 69978 Tel Aviv, Israel

ABSTRACT

Computer simulations open new avenues in investigations of complex material systems and phenomena. The methodology, development and applications of molecular dynamics simulations are discussed and the wealth of microscopic information revealed via simulations of interphase-interfaces, liquid metals, metallic glasses and small aggregates are demonstrated.

I. PROLOGUE

Basic understanding of the structure and dynamics of materials and their properties often requires knowledge on a microscopic level, of the underlying energetics and interaction mechanisms, whose consequences we observe and measure. The degree of microscopic detail with which we probe physical phenomena is determined mainly by the resolution of our experimental tools, by the ability to found the theoretical analysis on microscopic principles and by the complexity, hence solubility, of the model. Critical testing and assessment of the validity of theoretical models frequently depend upon the accessibility and availability of experimental data with sufficient resolution (either as a source of values for some of the model parameters or for comparative purposes) and by the level of simplifications, dictated by considerations of feasibility and practicality, which render the model soluble.
Computer simulations, where the evolution of a physical system is simulated, with refined temporal and spatial resolution, via a direct numerical solution of the equations of motion, are in a sense computer experiments which open new avenues in investigations of the microscopic origins of material phenomena [1,2]. These methods alleviate certain of the major difficulties which hamper other theoretical approaches, particularly for complex systems such as those characterized by a large number of degrees of freedom, lack of symmetry, nonlinearities and complicated interactions. In addition to comparisons with experimental data, computer simulations can be used as a source of physical information, which is not accessible to laboratory experiments, and in some instances the computer experiment itself serves as a testing ground for theoretical concepts. The main issues involved in critical assessment of simulation studies are: (i) the faithfulness of the simulated model, focusing mainly on our knowledge of the interaction potentials, (ii) the spatial dimension, i.e., finite size of the computational cell and imposed periodic boundary conditions in the case of simulations of extended systems, and (iii) the finite time span of the simulation. The physical size and time extent employed in a simulation are constrained by the available computational resources, and their adequacy depends upon characteristics of the specific system and phenomenon being studied, such as the state of aggregation, range of interactions, characteristic relaxation times, the scale of intrinsic spatial and temporal fluctuations and ambient conditions.
Guided by the above considerations, we illustrate in the following the richness of microscopic information revealed via computer simulations for material systems and phenomena chosen to demonstrate the versatility and

wide-range applicability of these methods. In the second section, molecular dynamics (MD) simulations of metallic systems, with special consideration of the various microscopic contributions to the total energy, are discussed and results for an equilibrium liquid metal and for the formation and properties of a metallic glass are shown. The equilibrium structure and dynamics of crystal-melt interfaces for systems characterized by pair interactions and with the inclusion of three-body potentials, and model simulations of melting and rapid solidification processes, are presented in the third section. In the final section, classical and quantum simulations of small atomic clusters are discussed via studies of the evolution of electronically excited rare-gas clusters and electron interaction with charged alkali-halide clusters, respectively.

2. LIQUID METALS AND METALLIC GLASSES

While for rare-gas, molecular and ionic systems a satisfactory description of the energetics may be given in terms of pairwise (and higher order) interactions, it is well known that the cohesive energy of a metal contains, in addition, density-dependent contributions, which are structure independent and that the metallic effective pair potentials themselves depend on conduction-electron density [3]. Therefore, in order to perform realistic simulations of metallic systems and, in particular, investigations of processes which involve changes in temperature, pressure and/or volume, such as solidification and melting, we have developed [4] a new MD method which allows for volume and shape variations and incorporates explicitly the dependence on density of the "volume energy" and the effective pair potentials.

Consider a metallic alloy with N_σ particles of species σ and a total number of particles $N = \Sigma_\sigma N_\sigma$ contained in a computational cell (CC) of volume Ω. The periodic replications of the CC are labeled by $\underset{\sim}{l} = (l_1, l_2, l_3)^T$ with $l_\alpha = 0, \pm 1, \pm 2 \ldots$ ($\alpha = 1,2,3$), and the position of particle j in cell $\underset{\sim}{l}$ is given by $\underset{\sim}{r}_j(\underset{\sim}{l}) = \underset{\sim}{H} \cdot (\underset{\sim}{s}_j + \underset{\sim}{l})$, where $\underset{\sim}{s}_j = (s_{j1}, s_{j2}, s_{j3})^T$, $-\frac{1}{2} < s_{j\alpha} < \frac{1}{2}$ ($\alpha = 1,2,3$), and $\underset{\sim}{H}$ is a 3x3 matrix with $\det(\underset{\sim}{H}) = \Omega$. To allow for temporal and shape variations, the components of the matrix $\underset{\sim}{H}$ are taken as dynamical variables. The Ansatz Lagrangian is obtained by replacement of all the terms in the kinetic energy which involve $\dot{\underset{\sim}{H}}$ with $\frac{1}{2} W Tr(\dot{\underset{\sim}{H}}^T \dot{\underset{\sim}{H}})$, where W has the dimension of mass, yielding

$$L = \frac{1}{2} \sum_i m_i \dot{\underset{\sim}{s}}_i^T \cdot \underset{\sim}{G} \cdot \dot{\underset{\sim}{s}}_i + \frac{1}{2} W Tr(\dot{\underset{\sim}{H}}^T \dot{\underset{\sim}{H}}) - E_\Omega(r_s) - \frac{1}{2} \sum_{i,j,\underset{\sim}{l}} \phi_{ij}^{(2)}(r_s, r_{ij}(\underset{\sim}{l})) - P_{ext}\Omega,$$
(1a)

where $\underset{\sim}{G} = \underset{\sim}{H}^T \cdot \underset{\sim}{H}$, $r_{ij}(\underset{\sim}{l}) \equiv |\underset{\sim}{r}_l - \underset{\sim}{r}_j(\underset{\sim}{l})|$, r_s is the electron density parameter and P_{ext} is the external pressure. The structure independent contribution, E_Ω, which together with the effective pair-interactions yields the total cohesive energy, is given by

$$E_\Omega = \sum_\sigma N_\sigma [Z_\sigma E_{el}(r_s) + \phi_\sigma^{(1)}(r_s)] \quad ,$$
(1b)

where Z_σ is the valence number, E_{el} is the energy of the uniform electron gas, and $\phi_\sigma^{(1)}$ is a single-particle contribution. $\phi^{(1)}$, and the density-dependent effective pair potential $\phi^{(2)}$ are derived via pseudopotential theory [3,4] and their specific form depends upon the choice of ionic pseudopotneitals (model, local, nonlocal). The above Lagrangian extends that of Parrinello and Rahman [5] to include the volume energy $E_\Omega(r_s)$, and the volume dependence in the pair potentials. The restriction to volume variation only (Andersen's Lagrangian [6]) is obtained by setting $\underset{\sim}{H} = L\underset{\sim}{\hat{H}}$, where $\underset{\sim}{\hat{H}}$ is a constant matrix and L is a dynamical variable with the dimension of length. The equations of motions are derived from the Lagrangian in a straightforward manner [4].

With the definitions $V_{ij}(r_s,X_{ij}(\underset{\sim}{1})) = \phi_{ij}^{(2)}(r_s,r_{ij}(\underset{\sim}{1}))$, where $X_{ij}(\underset{\sim}{1}) = r_{ij}(\underset{\sim}{1})/r_s$, and with the prime denoting derivatives with respect to r_s, the internal (virial) pressure is given by

$$P = \left[- r_s \; [E_{\Omega}'(r_s) + \frac{1}{2} \sum_{i,j,\underset{\sim}{1}}' V_{ij}'(r_s,X_{ij}(\underset{\sim}{1}))] - \sum_i m_i \dot{\underset{\sim}{s}}_i{}^T \cdot \underset{\sim}{G} \cdot \dot{\underset{\sim}{s}}_i \right] (3\Omega)^{-1} , \qquad (2)$$

where the corresponding terms are the volume, pair-potential and kinetic contributions. An advantage of the present formulation is that the CC responds dynamically to pressure fluctuations. In equilibrium $\langle P \rangle = P_{ext}$ at the correct density (thus alleviating the need to impose an external "electronic" pressure to achieve the right density).

2.1 Liquid Metal

To illustrate and test the method, we carried out simulations of liquid magnesium at several temperatures and external pressures. Since liquids do not resist shear stresses, random fluctuations in the CC shape may eventually result in undesirable cell shapes which lead to interactions between particle images. To alleviate this problem and to minimize any deviation in spherical symmetry due to the periodic boundary conditions, we have used a constant-shape CC with $\underset{\sim}{H}$ chosen so that $\underset{\sim}{H} \cdot \underset{\sim}{1}$ describes an fcc lattice. The single-particle and pair-particle potentials, $\phi^{(1)}$ and $\phi^{(2)}$, were obtained from a simplified Heine-Abarenkov ionic pseudopotential [7] with exchange correlation included via the Singwi et al. [8] local field correction. In addition, the Hartree contribution to $\phi^{(1)}$ was scaled by a parameter z_H. The pseudopotential parameters and z_H were chosen to fit exactly the zero-temperature cohesive energy, bulk modulus and density and to give the correct hcp crystalline structure.

We have calculated, in addition to average thermodynamic quantities, their mean square deviations for which the following relations hold in the isoenthalpic-isobaric ensemble [9]

$$c_p = \left| \frac{2}{3} - \frac{N\langle\Delta T^2\rangle}{\langle T\rangle^2} \right|^{-1} , \qquad B_s = \frac{\langle\Omega\rangle \; \langle T\rangle}{\langle\Delta\Omega^2\rangle} , \qquad (3)$$

where c_p is the isobaric specific heat per atom, B_s is the adiabatic bulk modulus, T the kinetic temperature is defined as twice the instantaneous kinetic energy per degree of freedom, and the angular brackets indicate a time average.

Simulations were performed for N = 500 particles, an integration time step 1.46×10^{-15} s at $P_{ext} = 0$, and at three temperatures (T(melting) = 922 K). The results obtained are shown in Table I along with available experimental values. The internal energy, U, is the sum of the potential- and kinetic-energy terms in the Lagrangian, but with omission of the term involving $\underset{\sim}{\hat{H}}$. The experimental values for U are obtained from the cohesive energy at T = 0, the ionization energy [11], the heat of melting [10], and $c_p(T)$ for liquid and crystalline phases [12]. The thermal expansion coefficient $\alpha = r_s^{-1}\partial r_s/\partial T$, is obtained from a two-parameter fit to the simulation results, $r_s = a_0 \times (5.404 \times 10^{-2} - 6.94 \times 10^{-6} T[K])^{-1/3}$. The agreement between experiment and simulation for U, r_s, and α is quite good considering the uncertainty in the experimental values and the fact that a simple local model pseudopotential fit to T = 0 properties was used in the simulation. Reliable experimental values for c_p are available only for near melting, and c_p is estimated [12] to be constant over the temperature range of our simulations, while our results indicate that c_p decreases with increasing T. Finally, we have calculated the diffusion constants, D, via the velocity autocorrelation functions, and the electric resistivity, ρ, and thermopower, Q, using the Faber-Ziman theory [13], and the static structure factors obtained from the simulation (the corresponding pair-correlation function at T = 960 K is in good agreement with experiment [14]). Our results of a positive Q and negative $d\rho/dT$ are typical of divalent liquid metals. Experimental values for ρ and Q at melting are 27 $\mu\Omega$ cm and 1.5 μV/k, respectively, [15].

TABLE I. Simulation results and comparison with available experimental data (expt). Angular brackets indicate time averages. All simulation results correspond to $<P> = P_{ext} = 0$.

$<T>$	960 K	1070 K	1150 K
U/N (expt) (Ry)[a]	-1.762	-1.759	-1.757
$<U/N>$	-1.7496	-1.7464	-1.7443
r_s (expt) (a_0)[b]	2.741	2.759	2.772
$<r_s>$	2.764	2.779	2.790
α(expt) (10^{-5} K^{-1})[b]	5.63	5.73	5.83
α	4.88	4.97	5.03
c_p/k_B (expt)[c]	3.9	3.9	3.9
$[2/3 - (<T>^2/N<\Delta T^2>)^{-1}]^{-1}$	3.8	4.7	3.8
B_S (expt) (mRy/a_0)	1.75	1.67	1.60
$<\Omega>$ $k_B T>/<\Delta\Omega^2>$	1.76	1.68	1.70
D (10^{-5} cm^2/sec)	2.4	3.6	4.9
ρ ($\mu\Omega$ cm)	16.5	16.0	15.8
Q (μV/K)	1.45	2.47	2.82

[a]References 10, 11 and 12.

[b]Reference 11.

[c]Reference 12.

2.2 Metallic Glasses

Metallic glasses, i.e., the amorphous solid phases formed when segregation and crystallization are avoided by ultra-rapid cooling of the liquid alloys, are the subject of much recent research effort due to their unique physical properties, which are of joint scientific and technological interest [16]. Metallic glasses are distinct from many other glass forming materials due to the lack of bond-network entanglement commonly present in nonmetal based glasses. Their stability and ease of formation have been correlated with electronic effects [17] and with metal alloy chemistry and eutectic composition [18].

Disordered materials present an immense theoretical challenge due mainly to the lack of structural periodicity. The major questions confronting simulations [19] of metallic glasses are, (i) the interaction potentials, and (ii) the short-time scale of computer simulations in comparison to the laboratory ones. Short of performing simulations employing laboratory cooling rates, which is beyond the capability of current computers, an assessment of the correspondence between computer simulated and laboratory prepared glasses may be provided via simulations which employ a <u>realistic</u> description of the energetics for <u>specific</u> metallic systems for which experimental data is available. To this end we have employed [20] the MD method described in the previous section, modified to enable cooling at a specified rate and dynamical adjustment of the potentials in accord with the dynamically varying density.

A most sensitive probe of the nature of interatomic forces, and the structure (on various length scales) and dynamics of a glass-forming alloy, is provided by the static and dynamic structure factors obtained from neutron scattering experiments [16b]. To critically assess the structural and dynamical information obtained by our molecular dynamics simulation method, we have chosen to study the metallic glass $Ca_{67}Mg_{33}$, a simple metal glass for which neutron scattering data is available [21], and which has been investigated extensively using a random packing model [22].

The simulated system consisted of 500 particles, 333 Ca and 167 Mg atoms contained in a fixed-shape, variable volume, regular rhombic dodecahedral MD cell (see section 2.1). The interaction potentials were calculated using model ionic pseudopotentials (see section 2.1) for the range of densities covered by the simulation. For the pair interactions the principal effects of the increase in density in going from the liquid (r_S = 3.350 a_0) to the glass (r_S = 3.136 a_0) are a decrease in depth of the potential minima and a shift to shorter interatomic distance of the Friedel oscillations in the long-range part. In addition, the relative depths of the potentials are altered. The "electronic pressure", $r_S[\partial E_\Omega(r_S)/\partial r_S]/\Omega$, increases from 2.0x 10^{-4} Ry/a_0^3 in the liquid to 2.3x10^{-4} Ry/a_0^3 in the glass. Throughout we use energy (and temperature) in Rydbergs and length in Bohr radius, a_0. The time unit (tu) is 1.46x10^{-15} sec.

The simulation was performed in four stages (all at zero external pressure): (1) An equilibrium run for the hot liquid alloy at T = .01 Ry (1580^0 K, approximately twice the experimental eutectic temperature), for 3x10^4 tu; (2) a "cooling" run of 3x10^4 tu, $\dot{T}_0$ = 3x10^{-7} Ry/tu (3.2x10^{13} ^{0}K/sec), during which the glass transition occurred; (3) an "equilibrating" run of 14x10^4 tu in duration under temperature control at T_0 = .00191 Ry (300 ^{0}K), during which structural relaxation of the glass occurred; and (4) an equilibrium run for the glass at T = 0.00191 Ry (without temperature control) of 9x10^4 tu. The MD integration time step size for stages (1) and (2) was Δt = 1 tu, while in the glass it was Δt = 2.5 tu.

In Fig. 1(a) the mean squared displacement, $R^2(t)$, and the kinetic temperature, $T(t)$ as functions of time are shown, while Fig. 1(b) shows the (per particle) total internal energy and the potential energies for the two species as functions of temperature (Fig. 1(b) was obtained by performing Gaussian weighted local averages in time (Gaussian width of 1500 tu)). From the break in the slopes of energy versus T, we determine the glass transition temperature to be T_g = 0.0036 Ry (570 ^{0}K). A similar break in the slope of the volume, Ω, versus T occurs at the same temperature. This value of T_g is about 0.7 times the experimental eutectic temperature, which is a typical value for measured glass transition temperatures. Estimates of the specific heats of the supercooled alloy and glass can be obtained from the slopes of the total internal energy versus T in the two regions, yielding c_p/k_B = 4 and 3.3, respectively.

The structure and dynamics of the equilibrium glass were investigated via the radial pair-distribution function, $g(r)$, and static structure factors, $S(q)$, as well as the density of states and dynamic structure factors, $S(q,\omega)$. For the glass we observed correspondence between the locations of minima in the partial $g(r)$'s and maxima in the corresponding pair-potentials as discussed by Hafner [18]. Using the known neutron scattering lengths, we have calculated the neutron-weighted static and dynamic structure factors, which may be compared directly with the experimental results [21]. The pair-distribution function exhibits a split second peak typical of a glass and,

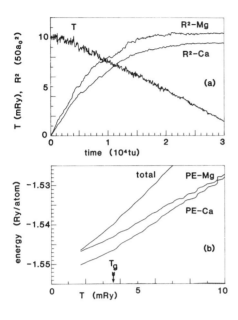

Fig. 1: System evolution during cooling: (a) Temperature, T, and mean squared displacement, R², versus time; (b) mean total energy and mean potential energies, PE, versus temperature.

in addition, there are indications of some amount of local order extending to at least the third neighbor shell. The positions of the first and second peaks of the static structure factor, $Q_1 = 1.12$ a_0^{-1} and $Q_2 = 1.92$ a_0^{-1}, are in good agreement with the reported experimental peak positions [21] ($Q_1 = 1.13$ a_0^{-1} and $Q_2 = 1.94$ a_0^{-1}). The density of the simulated alloy, $N/\Omega = .00387$ a_0^{-3} is also in good agreement with the experimental value [22] of .00398 a_0^{-3} (a steepest descent quench of our glass to 0 °K results in a density of .00397 a_0^{-3}). Comparison of $S(q,\omega)$ [20] with the experimental results [21] shows general agreement of both lineshapes and frequency range. Of particular interest are the increases in amplitude at low frequency for $q \approx Q_1$, $2Q_1$, which has been attributed to transverse excitations, suggesting the existence of quasi-zone boundaries in the glass [23]. From the positions of the peaks in $S(q,\omega)$ for the lower q's, we may read points on the dispersion curve for longitudinal phonons which are in agreement with the experimental results and with the theoretical dispersion curves obtained via the relaxed random packing model [22].

Having established via direct comparison with experimental data that our MD simulation, in which the full energetics of the metallic system is self-consistently incorporated provides a faithful description of the structure and dynamics of the metallic glass system, we turn now to a brief discussion of new observations relating to glass dynamics. Examination of the mean squared displacement, $R^2(t)$, at the "equilibrated glass", stage (4), shown for the individual species in Fig. 2 revealed time ranges when $R^2(t)$ increased significantly then returned to the original lower value. These changes are accompanied by variations in system properties, such as temperature, mean species potential energies (see Fig. 2), and average shear stress. By performing "steepest descent quenches" [24] for these different time spans, we determine that the system possesses at least four distinct, nearly degenerate, local potential minima ranging from -1.551966 to -1.551942 Ry per atom. Comparison of the atomic positions associated with these potential minima reveals that they result from structural rearrangements in which, typically, one atom displaces by 3 to 4 a_0 (about half the average nearest neighbor distance), the surrounding near-neighbor atoms move less, and the region of rearrangement extends over about three or four nearest neighbor shells of the central atom, or about 15 to 20% of the total volume of the MD CC. We emphasize that the transitions between these accessible local potential minima do not lead to an annealing or relaxation of the glass (over extended times), and that the underlying topology of the configurational energy is similar to that invoked in the "tunneling level" models of low-temperature glass behavior [25]

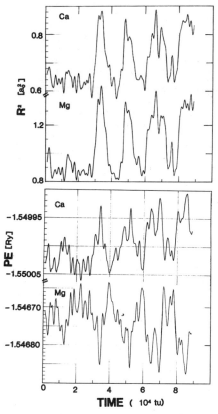

Fig. 2: Species mean square displacements, R^2, and potential energies, PE, per atom versus time for the related glass.

(though the nature of our transitions is different). Further studies of the influence of these configurational transitions on glass properties and the dynamics near the glass transition are in progress.

3. THE SOLID-LIQUID INTERFACE

The structure and dynamics of solid-vapor and solid-liquid interfaces have attracted a recent surge of interest because of improved experimental and theoretical techniques for probing phenomena such as surface melting, roughening and interface structure and morphology at equilibrium and during growth. The development of a microscopic theory of the solid-liquid interface is complicated since the structural and dynamical properties of the liquid are perturbed by the presence of the solid. Three main theoretical approaches which were developed and applied in investigations of solid-liquid interfaces are: (1) Model approaches [26], which follow and extend the original work of Bernal [27]. (2) Statistical mechanics theories employing liquid-state integral equation methodology in juxtaposition with perturbation theory [28]. (3) Monte Carlo and molecular dynamics simulation methods [1,2]. All three treatments result in a diffuse, layer structured in the liquid, equilibrium interface.

Molecular dynamics simulations employing pair-wise 6-12 Lennard-Jones interaction potentials [1,2] have shown that the extent of the transition region (typically 5 layers in width) and the manner in which the particle density and interlayer spacings vary across the interface depend on the crystalline face exposed. Additionally, other physical properties, such as the diffusion constant and resistance to shear (parallel to the crystal plane) vary gradually across the interface. In simulations of the nonequilibrium processes of liquid-phase epitaxy [29] and of rapid solidification following melting by an intense, short, heat pulse [2,30], it was found that the moving crystallization front is characterized by a similar layered interface which establishes on a fast time-scale. Direct evidence for stratified interface structures at the crystallization front was found recently via dynamic light scattering spectroscopy of ice [31] and salol [32], and a diffuse transition region was invoked in the analysis of laser annealing processes [33] and the growth of dendrites into undercooled liquids [34].

The network of low-angle grain boundaries that forms during zone-melting recrystallization of silicon has been interpreted in terms of the morphology of the solid-melt interface [35]. Ample evidence exists that surfaces of crystalline silicon become faceted upon melting, and the solid-melt interface of a growing Si crystal establishes itself on the (111) crystal planes [36]. While improved models have been proposed to understand the solid-liquid interface morphology at equilibrium, the simple two-layer model of Jackson [37] has been remarkably successful in formulating the criteria for facet formation. Although silicon satisfies the Jackson criterion, it is a borderline case, and one might therefore expect a transition near thermal equilibrium. Motivated by the above considerations, we have performed MD simulations and in situ observations of the melting and equilibrium structure of the Si (100)-melt interface, which provide evidence for a (111) faceting transformation [38].

Due to the directional bonding (covalent) characteristic to tetrahedral semiconductors a model of the potential for these materials must go beyond the often used pair-interactions, via the inclusion of nonadditive (angle-dependent) contributions (3-body and higher order). In our simulations, we have employed optimized two and three-body potentials, V_2 and V_3, respectively, which have yielded [39] a rather adequate description of the structural properties of crystalline and liquid silicon. The MD technique, which we used, allows for dynamical variations in particle density and structural changes via the Ansatz Lagrangian of Parrinello and Rahman [5], extended to include 3-body interactions, and planar two-dimensional (2D) periodic boundary conditions. In the following, length is expressed in units of σ = 2.0951 Å, energy and temperature in units of ε = 50 kcal/mole, and the time unit tu = 7.6634×10^{-14} sec. The integration time step is 1.5×10^{-2} tu.

We start with a silicon crystal consisting of N_L dynamic layers (with N_p particles per layer), exposing the (001) face (the z axis is taken parallel to the (001) direction, and the 2D cell is defined by the (110) and ($\bar{1}$10) directions). The bottom layer of the crystal (layer number 1) is positioned in contact with a static silicon substrate. Simulations for two systems were performed: (i) N_L = 28, N_p = 36 (for which results are shown), and (ii) N_L = 24, N_p = 144, yielding similar results (the N_L values were chosen so as to minimize the static substrate effects at the interface). Following prolonged equilibration, crystal-melt coexistence was established at an average kinetic temperature of 0.0662+0.0016 (compared to the experimental melting temperature of silicon at 0.0669 $\equiv$ 1683 °K).

The equilibrium crystal-melt interface exhibits a pronounced structure, demonstrated by the sample particle trajectories (recorded for 2000 Δt) shown in Fig. 3(a), viewed along the ($\bar{1}$10) direction (denoted by a circle on the bottom left), where the breakup into alternating (111) and ($\bar{1}\bar{1}$1) crystalline planes is indicated (here and in the following Z^* = 1 $\equiv$ 18.14σ, and X^* = Y^* = 1 $\equiv$ 10.9σ). In the vicinity of the solid (facet) planes the interface exhibits a certain degree of diffuseness. To complement the picture, we show in Fig. 3(b) particle trajectories in the region of the 7th layer (ℓ = 7), projected onto the 2D plane, exhibiting solid and (partial) liquid characteristics. In extended runs we observed that the morphology of the interface fluctuates and that the interface is rough at all times, exhibiting a tendency to faceted configurations such as that shown in Fig. 1.

Further insights are provided by the equilibrium particle density profile (Fig. 4(a)) and potential energies (Fig. 4(b-d)) which show both the crystal-melt and melt-vacuum interfaces. Focusing on the former interface, we observe the opposing trends in the behavior of the 2-body and 3-body potentials (V_2 and V_3, respectively), which when added yield the result shown in Fig. 4(b). In Fig. 4(e) contours of V_3 for $V_3 \leqslant 0.45$ are shown, complementing the picture of the interface. By restricting the value of V_3, only solid-like regions are captured (the appearance of the V_3 contour map does not change significantly with small changes in the cutoff value). In situ observations of the equilibrium solid-melt interface by means of a microscope equipped with a visible and near-infrared video camera revealed a faceted structure of the same crystallography as predicted by the MD simulations [38].

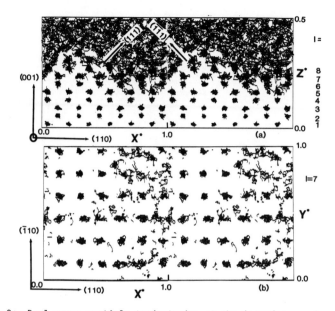

Fig. 3: Real space particle trajectories at the interface region recorded at equilibrium for 2000 Δt. (a) Viewed along the $(\bar{1}10)$ direction, and (b) trajectories for particles at the region of the 7th layer ($\ell = 7$) viewed from the (001) direction. $Z^* = 1 \equiv 18.14\sigma$ and $X^* = Y^* = 1 \equiv 10.9\sigma$. The 2D computational cell ($0 \leq X^*, Y^* \leq 1$) is replicated along the X^* (110) direction to aid visualization.

4. SMALL CLUSTERS

Small clusters, i.e., finite aggregates containing 3-500 particles, exhibit unique physical and chemical phenomena, which are of both fundamental and technological significance, and provide ways and means to explore the "transition" from molecular to condensed-matter systems [40,41]. Theoretical studies of clusters were hampered by the relatively large number of particles, which renders the adaptation of molecular science techniques rather cumbersome, while the lack of translational symmetry inhibits the employment of solid-state methodology.

In the following we demonstrate the application of a newly developed quantum path integral MD (QUPID) method [42] and classical MD simulations to problems of (i) electron localization [43], and (ii) dynamics and energy pathways in electronically excited clusters [41], respectively.

Localized electron states in clusters are of considerable interest with regard to the nonreactive and reactive mechanisms of electron attachment [41]. Furthermore, quantum phenomena are expected to be pronounced in these systems, since the electron wavelength is comparable to the cluster size. Therefore, we used the QUPID method, which rests on an isomorphism between the path integral evaluation of the quantum partition function, and the calculation of thermodynamic equilibrium properties as averages over the phase-space trajectories generated by a corresponding classical Hamiltonian system [42-44].

We have chosen to study electron localization in alkali-halide clusters since the interionic interactions in these clusters [45] and adequate models of the interaction between an electron and the ionic constituents are avail-

282

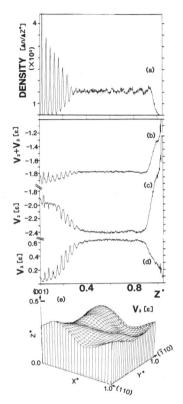

Fig. 4: Equilibrium particle density (a), $\overline{\Delta n/\Delta Z^*}$, where Δn is the number of particles with Z^* coordinates between Z^* and $Z^* + \Delta Z^*$, and potential energy profiles (b-d) versus distance along the (001) direction ($Z^* = 1 \equiv 18.14\sigma$). The total, and the two- and three-body potential energies are shown in b, c and d, respectively. A contour plot of V_3 for particles with $V_3 \leq 0.45\varepsilon$, exhibiting the solid-like region of the sample, recorded at the same time as the trajectories given in Fig. (3), is shown in (e).

able [44,46]. Simulations of electron attachment, at T = 300 °K, to positively charged clusters, $[Na_nCl_m]^{+(n-m)}$ with m = n-1 or n-2, have revealed: (1) Internal electron localization in moderately large clusters containing an anion vacancy accompanied by a small nuclear configurational rearrangement, thus establishing the dominance of short-range attractive interactions for this process (Fig. 5(a)). (2) Surface state localization in moderately large, vacancy free clusters. For $[Na_{14}Cl_{13}]^+$, localization around a surface Na^+ ion resulting in partial neutralization is shown in Fig. 5(b). (3) In small clusters novel effects of dissociative electron attachment and cluster isomerization induced by electron localization are predicted.

The microscopics of reactive (dissociative) and nonreactive vibrational relaxation (VR) following electronic excitation in clusters are important for the interpretation of experimental data [47] and as model systems for investigations of energy pathways in finite systems. Excitation of heavy rare-gas clusters (RGC) results in exciton formation, whose trapping involves the formation of a diatomic excimer at a highly vibrational state. The excitation of an equilibrated RGC is simulated [48] by an instantaneous switching-on of the intra-excimer potential [47], and of the interactions between the excimer and the ground state atoms characterized by an enhanced short-range repulsion. We find that the vibrational energy redistribution is governed by the mismatch-in characteristic frequencies between the local

Fig. 5: Ionic configurations and 2D projections of "electron distributions" for: (a) $[Na_{14}Cl_{12}]^{++}$ + e, yielding internal localization, and (b) $[Na_{14}Cl_{13}]^+$ + e, yielding surface localization. Small and large spheres correspond to Na^+ and Cl^- ions.

excimer mode and the remaining cluster modes, as well as by the local dilation of the structure around the excimer. This is manifested in two energy-
transfer time scales: (a) An ultrafast one due to repulsion ($\sim$ 200 fsec),
and (b) a "slow" one due to VR (10 - 10^2 psec). A gradual transition from
dissociative to nonreactive VR occurs with increasing cluster size. The
excimer "mode selective" excitation constitutes an example for the violation
of vibrational energy equipartitioning in large systems.

Acknowledgements: This work is supported by the U.S. DOE, Grant DE-FG05-86
ER45234 (UL), in part by Philips Laboratories, N.Y. (UL), and in part by the
United States Army through its European Research Office (JJ).

REFERENCES

[1] For a recent review see F. F. Abraham, J. Vac. Sci. Technol. B2, 534
(1984).
[2] U. Landman, R. N. Barnett, C. L. Cleveland and R. H. Rast, J. Vac. Sci.
Technol. A3, 1574 (1985).
[3] See for example, W. A. Harrison, Pseudopotentials in the Theory of
Metals, Benjamin, Reading, Mass. 1966).
[4] R. N. Barnett, C. L. Cleveland and U. Landman, Phys. Rev. Lett. 54,
1679 (1985).
[5] M. Parrinello and A. Rahman, Phys. Rev. Lett. 45, 1196 (1980) and J.
Appl. Phys. 52, 7182 (1981).
[6] H. C. Andersen, J. Chem. Phys. 72, 2384 (1980).
[7] R. N. Barnett, U. Landman and C. L. Cleveland, Phys. Rev. B28, 1685
(1983).
[8] K. S. Singwi, A. Sjolander, M. P. Tosi and R. H. Land, Phys. Rev. B1,
1044 (1970).
[9] J. M. Haile and H. W. Graben, Mol. Phys. 40, 1433 (1980).
[10] Karl A. Gshneider, Solid State Physics, (Academic, New York, 1964),
Vol. 16.
[11] Handbook of Chemistry and Physics, CRC Press, Cleveland, Ohio, 1975)
55th ed.
[12] P. Hultgren, R. L. Orr, P. D. Anderson and K. K. Kelley, Selected
Values of Thermodynamic Properties of Metals and Alloys, (Wiley, New
York, 1963).
[13] T. E. Faber, Theory of Liquid Metals, (Cambridge Univ. Press, Cambridge,
1972).
[14] Y. Waseda, The Structure of Non-Crystalline Materials, (McGraw-Hill,
New York, 1980).
[15] G. Bush and H. J. Guntherodt, in Ref. 10, Vol. 29.
[16] (a) Glassy Metals I, eds. H. I. Guntherodt and H. Beck (Springer, New
York, 1981); (b) Glassy Metals II, eds. H. Beck and H. I. Guntherodt,
(Springer, New York, 1983); (c) S. R. Nagel, Adv. Chem. Phys. 51, 227
(1982).
[17] S. R. Nagel and J. Tauc, Phys. Rev. Lett. 35, 380 (1975); see also ref.
16c.
[18] J. Hafner, Phys. Rev. B21, 406 (1981); see also ref. 16a.
[19] See for example, C. A. Angell, J. H. R. Clarke and L. V. Woodcock, Adv.
Chem. Phys. 48, 397 (1981); J. R. Fox and H. C. Andersen, J. Phys.
Chem. 88, 4019 (1984).
[20] R. N. Barnett, C. L. Cleveland and U. Landman, Phys. Rev. Lett. 55,
2035 (1985).
[21] J.-B. Suck, H. Rudin, H. J. Guntherodt and H. Beck, J. Phys. F 17,
1375 (1981); see also ref. 16(b), p. 217.
[22] J. Hafner, Phys. Rev. B27, 678 (1983); see also ref. 16.
[23] G. S. Grest, S. R. Nagel and A. Rahman, Phys. Rev. B29, 5968 (1984).

284

[24] T. A. Weber and F. H. Stillinger, Phys. Rev. B31, 1954 (1985).
[25] See review by J. L. Black in ref. 16(b), p. 167.
[26] A. Bonissent, *Interfacial Aspects of Phase Transformations,* ed. B
      Mutaftschiev (Reidel, Boston, 1982), p. 143.
[27] J. D. Bernal and S. King, *Physics of Simple Liquids,* (North-Holland,
      Amsterdam, 1968), p. 231.
[28] A. Bonissent and F. F. Abraham, J. Chem. Phys. 74, 1306 (1981).
[29] U. Landman, C. L. Cleveland and C. S. Brown, Phys. Rev. Lett. 45, 2032
      (1980); U. Landman et al. in *Nonlinear Phenomena of Phase Transitions
      and Instabilities,* ed. T. Riste (Plenum, N.Y., 1982), p. 379.
[30] C. L. Cleveland, U. Landman and R. N. Barnett, Phys. Rev. Lett. 49, 790
      (1982).
[31] P. Böni, J. H. Bilgram and W. Känzig, Phys. Rev. A28, 2953 (1983).
[32] U. Dürig, J. H. Bilgram and W. Känzig, Phys. Rev. A30, 946 (1984);
      O. N. Mesquita, D. G. Neal, M. Copic and H. Z. Cummins, Phys. Rev.
      B29, 2846 (1984).
[33] R. F. Wood, Phys. Rev. B25, 2786 (1982).
[34] A. D. Solomon, D. G. Wilson and V. Alexiades, Letters in Heat and Mass
      Transfer 9, 319 (1982).
[35] L. Pfeiffer, S. Paine, G. H. Gilmer, W. Van Saarloos and K. W. West,
      Phys. Rev. Lett. 54, 1944 (1985).
[36] G. K. Celler et al., *Energy Beam-Solid Interactions and Transient
      Thermal Processing,* eds. J. C. C. Fan and N. M. Johnson (North-
      Holland, N. Y., 1984), p. 409.
[37] D. P. Woodruff, *The Solid-Liquid Interface,* (Cambridge Univ. Press,
      1973), Chap. 3.
[38] U. Landman, W. D. Luedtke, R. N. Barnett, C. L. Cleveland, M. W.
      Ribarsky, E. Arnold, S. Ramesh, H. Baumgart, A. Martinez and B. Khan,
      Phys. Rev. Lett. 56, 155 (1986).
[39] F. Stillinger and T. Weber, Phys. Rev. B31, 5262 (1985).
[40] See papers in Ber. der Bunseges. Gesselsch. fur Phys. Chem. 88, (1984).
[41] J. Jortner, D. Scharf and U. Landman, *Energetics and Dynamics of
      Clusters,* (Proc. of the "Enrico Fermi" School, Varenna, 1985).
[42] D. Chandler and P. G. Wolynes, J. Chem. Phys. 79, 4078 (1981).
[43] U. Landman, D. Scharf and J. Jortner, Phys. Rev. Lett. 54, 1860 (1985).
[44] M. Parrinello and A. Rahman, J. Chem. Phys. 80, 860 (1984).
[45] See review by T. P. Martin, Phys. Rep. 95, 167 (1983).
[46] R. W. Shaw, Phys. Rev. 174, 769 (1968).
[47] K. Schwentner, E. E. Koch and J. Jortner, *Electronic Excitations in
      Condensed Rare Gases, Springer Tracts in Modern Physics* (Springer-
      Verlag, Heidelberg, 1985), Vol. 107.
[48] D. Scharf, J. Jortner and U. Landman (to be published).

285

Author Index

Alerhand, O.L.,37
Allan, D.C., 37
Allen, P.B., 13, 259
Anderson, M.P., 225
Angell, C.A., 85
Auerbach, D.J., 219

Bakker, A.F., 181, 219
Barnett, R.N., 273
Beaulac, T.P., 259
Biswas, R., 173
Broughton, J.Q., 13
Butler, W.H., 259

Car, R., 7
Carleton, H.R., 213
Carpenter, Jr., J.A., 191
Catlow, C.R.A., 55
Chaudhari, P., 95
Cheeseman, P.A., 85
Chen, T.C., 219
Ciccotti, G., 241
Cleveland, C.L., 273
Cohen, M.L., 107
Cormack, A.N., 55

de Hosson, J.Th.M., 137
de Lorenzi, G., 125
Dodson, B.W., 151

Eberhardt, J.J., 191

Foiles, S.M., 61
Freeman, A.J., 1
Freeman, C.M., 55
Frost, H.J., 233
Fu, C.L., 1

Grest, G.S., 225

Hafner, J., 73
Hamann, D.R., 173
Hao, Y.G., 167
Hay, P.J., 191
Haymet, A.D.J., 67
Hoover, W.G., 125

Jortner, J., 273

Kelly, P.J., 7
Khan, F.S., 259
Krakow, W., 43

Ladd, J.C., 125

Laird, B.B., 67
Lancon, F., 95
Landman, U., 273
Levi, A.A., 157
Lewis, G.V., 55
Lo, D.Y., 31
Luedtke, W.D., 273

Madhavan, P.V., 253
Malvido, J.C., 253
Marcus, P.M., 117
Martin, R.M., 21
Mele, E.J., 37
Mills, D.L., 27
Moran, B., 125
Moriarty, J.A., 125
Moruzzi, V.L., 117
Munshi, A.A., 219

Ogielski, A.T., 207
Oguchi, T., 1
Oshiyama, A., 7

Pandey, R., 247
Pantelides, S.T., 7
Paul, W.J., 219
Phifer, C.C., 85
Pinski, F.J., 259
Ponti Kis, V., 241

Ribarsky, M.W., 273
Rosato, V., 241
Roth, L.M., 167

Scharf, D., 273
Schwartz, K., 117
Shapiro, M.H., 31
Smith, D.A., 157
Srolovitz, D.J., 225
Swihart, J.C., 259

Taylor, P.A., 151
Thompson, C.V., 233
Tombrello, T.A., 31

Vail, J.M., 247
van de Walle, C.G., 21
Varadan, V.K., 101
Vitek, J.M., 247

Weber, T.A., 163
Wetzer, J.T., 157
Whitten, J.L., 253

Subject Index

acoustic phenomena, 101
algorithm oriented pro-
    cessor, 181
alkaline-earth oxide, 247
alloy, 61
aluminum, 137
amorphous structure, 95
angular distribution, 37
antiferromagnet, 207
artificial material, 1
atomic structure, 43
augmented plane wave, 1

barium titanate, 55
bcc/melt, 67
Benes permutation network,
    217
beta-tin structure, 173
Brillouin zone, 27

Cauchy relation, 137
ceramic, 55
chemical short range order,
    73
chromium, 1
cluster calculation, 167,
    247
coating, 191
cobalt, 117
color center, 247
composite material, 101
composition profile, 61
compressibility, 85
computation of structure,
    191
computer architecture,
    213, 217
conductivity, 13
constrained dynamics, 125
copper, 259
cost/performance ratio, 181
covalent bonding, 157
Cray-1, 213
crystal growth, 151

dedicated computer, 217
defect energy, 55
defect production, 241
density maximum, 85
density of states, 13
diamond, 157
diffusion, 7
    interface, 67
diffusivity, 85

digital frame store, 43
displacement shift com-
    plete, 137
DMA interface, 213
dopant, 7
dynamic charges, 31
dynamical scattering, 43

edge dislocation, 61
effective properties, 101
elasticdynamic phenomena,
    101
electrical resistivity, 259
electromagnetic phenomena,
    101
electron
    diffraction, 43
    energy loss, 27
    microscopy, 43
    -phonon effect, 259
electronic properties, 13
energy minimization, 163
equilibrium, 67

floating point unit, 217
flow, 125
fracture, 125

gadolinium zirconate, 55
gallium, 37
germanium, 157
glass transition, 85, 95
grain
    boundary, 191
    boundary motion, 233
    boundary structure, 137
    growth, 225
    size distribution, 225
    structure, 233

Hall coefficient, 259
hard sphere, 73
Hartree-Fock, 55, 247
heteroepitaxial growth, 151
heterojunction, 21
hierarchy of models, 191
high resolution, 43

image computation, 43
instruction set, 181
interaction energy, 151
interatomic
    forces, 125
    potential, 173

interface, 1, 21, 43, 273
ionic crystal, 247
iron, 1
irreducible tensor method,
    253
Jahn-Teller distortion, 253
Keating potential, 157, 163
kinetics, 225

Lagrangian, 273
Langevin molecular
    dynamics, 173
lattice dynamics, 27
lead, 259
Legendre pdynomial, 173
Lennard-Jones, 151, 241
linear regression
    techniques, 253
linear response theory, 73
liquid metal, 37, 273
local
    density functional
      approximation, 117
    densit functional
      theory, 1
    melting, 241
localized electronic
    state, 273
low energy channeling, 167

Magneli phases, 55
magnetic
    phases, 117
    properties, 1
magnetisation, 117
magnetoresistance, 259
materials by design, 191
melting, 95
memory organization, 181
mesoscopic scale, 125
metallic glasses, 273
microstructure, 233
misorientation, 137
model Hamiltonian, 31
molecular dynamics, 37,
    95, 125, 213, 217, 241
molybdenum, 259
Monte Carlo
    processor, 207
    simulation, 61
    technique, 225
morphology, 225
muffin-tin
    orbital, 1

potential, 117
multiple scattering, 101
multiplicity of boundary
    structure, 137
multiprocessing, 213, 217

negative pressure, 85
neutral vacancy, 253
nickel, 1, 27, 117
    -aluminum, 191
Ni-Cu alloy, 61
niobium, 259
nucleation, 233
number dependence, 125

one-component plasma, 73
order parameter, 241
oxides, 55

pair-wise forces, 137
parallel processing, 213,
    217
participation ratio, 13
path integral molecular
    dynamics, 273
perturbation theory, 31
phase
    diagram, 73, 117
    transformation, 125
phonon, 27, 31
    line width, 259
    π-bonded chain, 31
polycrystalline, 233
    solids, 225
prediction of properties,
    191
pseudopotential, 21, 73,
    107, 273

quasi crystal, 43

random field, 207
rigid-ion potential, 85
ring, 167
rutile, 55

scanning tunnelling
    microscopy, 167
Schottky disorder, 55
segregation, 61
self-interstitial, 7
semiconductor, 21, 107
silicon, 7, 13, 31, 157,
    163, 167, 173, 253

dioxide, 85
growth, 151
simple-hexagonal structure, 173
simulated annealing, 173
simulator, 213
small aggregate, 273
solid-liquid interface, 67
solid-solid phase trans-
   formation, 125
special purpose computer,
   181, 207
spin glasses, 207
spinodal, 85
spin-orbit, 21
spin-polarized band calculation, 117
sputtering, 37
static
   property, 67
   relaxation, 95
steepest-descent, 163
Stillinger-Weber poten-
   tial, 151, 163
structural unit model, 137
sulfur, 27
supercomputer, 181
superconductivity, 107
superlattice, 1
surface, 1, 27, 31
   conductivity, 31
   dimer, 163
   melting, 241
   reconstruction, 163
   reconstruction (7x7), 167

thermodynamic perturbation
   theory, 73
thin film, 233
tight-binding theory, 13, 31
tilt boundary, 157
T-matrix, 101
topological property, 233
total energy, 107
   calculation, 117
tribology, 191
twist boundary, 137
two-body potential, 37
two-dimensional, 95

vacancy, 7
valence band discontinuity, 21
vanadium, 1
vibration, 27

vibronic energy, 253

width-interface, 67
wurtzite structure, 173

yield ratio, 37
yield strength, 125